Pascale Charpin, Alexander Pott, Arne Winterhof (Eds.)
Finite Fields and Their Applications

Radon Series on Computational and Applied Mathematics

Volume 11

Preface

This book is based on the invited talks of the RICAM-Workshop on Finite Fields and Their Applications: Character Sums and Polynomials held at the Federal Institute for Adult Education (BIfEB) in Strobl, Austria, from September 2–7, 2012.

The topic of the book is the theory of *finite fields*. Finite fields play important roles in many application areas such as *coding theory, cryptography, Monte Carlo and quasi-Monte Carlo methods, pseudorandom number generation, quantum computing*, and *wireless communication*. In this book we will focus on *sequences, character sums*, and *polynomials* over finite fields in view of the above mentioned application areas.

The goal of this book is to give an overview of several recent research directions as well as to stimulate research in sequences and polynomials under the unified framework of character theory.

Chapters 1 and 2 deal with sequences mainly constructed via characters and analyzed using bounds on character sums. In Chapter 1 measures of pseudorandomness in view of applications to wireless communication are mentioned, whereas Chapter 2 contains a survey on measures of pseudorandomness from a more theoretical point of view where cryptography may be the most important application area.

Chapters 3, 5, and 6 deal with polynomials over finite fields. Chapter 3 gives an overview about results on polynomials with some properties described. Chapters 5 and 6 discuss polynomials which are suitable for cryptographic applications. Chapters 4 and 9 consider problems related to coding theory studied via finite geometry and additive combinatorics, respectively. Chapter 7 deals with quasirandom points in view of applications to numerical integration using quasi-Monte Carlo methods and simulation. Chapter 8 studies aspects of iterations of rational functions from which pseudorandom numbers for Monte Carlo methods can be derived. For Monte Carlo and quasi-Monte Carlo methods uniformly distributed sequences are needed. In many cases a measure for the uniform distribution, the discrepancy, can be estimated in terms of additive character sums.

All these chapters were reviewed and we wish to thank the anonymous referees for their precious help.

We also thank the other participants of the workshop listed below who contributed with excellent talks and made the workshop a great success: Jürgen Bierbrauer, Herivelto Borges, Nina Brandstätter, Claude Carlet, Francis Castro, Ayca Cesmelioglu, Stephen D. Cohen, Domingo Gomez-Perez, Cem Güneri, Jing He, Peter Hellekalek, Tor Helleseth, Roswitha Hofer, Leyla Isik, Jonathan Jedwab, Giorgos Kapetanakis, Daniel Katz, Alexander Kholosha, Peter Kritzer, Michel Lavrauw, Vsevolod Lev, Petr Lisonek, Florian Luca, Christian Mauduit, Wilfried Meidl, Sihem Mesnager, Sylvia Morris, Gary Mullen, Ferruh Özbudak, Buket Özkaya, Daniel Panario, Gottlieb Pirsic, Claudio Qureshi, Andras Sarközy, Kai-Uwe Schmidt, John Sheekey, Henning Stichtenoth,

Valentin Suder, David Thomson, Alev Topuzoglu, Simone Ugolini, Christiaan van de Woestijne, Joachim von zur Gathen, Qi Wang, and Qiang Wang.

More details on this workshop can be found on the webpage http://www.ricam.oeaw.ac.at/events/workshops/ffta2012/

We also thank the Radon Institute for Computational and Applied Mathematics (RICAM) of the Austrian Academy of Sciences for financial support, BIfEB for their hospitality, the publisher for the pleasant cooperation, our co-organizers Gary Mullen, Harald Niederreiter, and Daniel Panario, and last but not least Annette Weihs and Wolfgang Forsthuber for their great support during the preparation of the workshop.

Paris, Magdeburg, and Linz, December 2012

Pascale Charpin
Alexander Pott
Arne Winterhof

Contents

Guang Gong

Character Sums and Polyphase Sequence Families with Low Correlation, Discrete Fourier Transform (DFT), and Ambiguity

Abstract: We present a survey on the current status of the constructions of polyphase sequences with low correlation, discrete Fourier transform (DFT), and ambiguity in both time and phase domains, including some new insights and results. Firstly, we systematically introduce the concepts of phase-shift operators and ambiguity functions of sequences, and give a new construction of polyphase sequences from combinations of different indexing field elements and hybrid characters. We then present the constructions, some known and some new, of polyphase sequences with low degree polynomials, for their low correlation; DFT and ambiguity can be bounded by directly applying the Weil bounds. Thirdly, we introduce the Hadamard equivalence, restate the conjectured new ternary 2-level autocorrelation sequences, and present their Hadamard equivalence relations. Some open problems are presented.

Keywords: Polyphase Sequence, Character Sum, Finite Field, Time Shift, Phase Shift, Correlation, Discrete Fourier Transform, Ambiguity Function, 2-Level Autocorrelation

2010 Mathematics Subject Classifications: 94A12, 62H20, 11T24

Guang Gong: Department of Electrical and Computer Engineering, University of Waterloo, Waterloo, Ontario, Canada, e-mail: ggong@uwaterloo.ca

1 Introduction

Sequences with good correlation properties find many applications in wireless communications. In particular, low correlation is used for acquiring the correct timing information and distinguishing multiple users or channels, minimized discrete Fourier transform (DFT) spectra are for getting low peak-to-average power ratio (PAPR) for orthogonal frequency division multiplexing (OFDM) systems, and low ambiguity functions are for radar systems and signal processing schemes.

Correlation, DFT, and ambiguity of polyphase sequences are three properties used in practical systems for evaluating the performance of a communication system which employs polyphase sequences. We will give formal definitions for these three concepts in the later sections. However, mathematically, those three properties of polyphase sequences are determined by some exponential sums. Thus, sequences with low correlation, DFT, and ambiguity can be constructed directly using low degree polynomials where the Weil bounds are applicable. However, using additive character sums, there are many known sequences with low correlation or 2-level autocorrelation which correspond to high degree polynomials, whose correlation cannot be bounded by the Weil bound, neither their DFT nor their ambiguity.

In this survey, Section 2 is an introduction to basic definitions and concepts of sequences, and additive, multiplicative, and hybrid character sums. Section 3 introduces phase-shift operators, ambiguity functions, ambiguity signal sets, and the optimality of correlation, DFT, and ambiguity. Section 4 introduces polyphase sequences defined by the additive group $\mathbb{Z}_N$ and by additive and multiplicative characters under different indexing methods, and their corresponding exponential sums of the correlation, DFT, and ambiguity. Section 5 shows the constructions of three types of polyphase signal sets with low three metrics, namely, polyphase sequences defined by odd degree polynomials, polyphase sequences from power residue and Sidel'nikov sequences, and polyphase sequences from Weil representation and from general hybrid characters. In Section 6, we present four conjectures on ternary 2-level autocorrelation sequences, Hadamard equivalence of those conjectured sequences, and the exponential sums in terms of iterative decimation Hadamard transform. Section 7 addresses some open problems.

2 Basic Definitions and Concepts

2.1 Notations

The following notations will be used throughout this paper.

- $\mathbb{C}$ is the complex field, M a positive integer, $\mathbb{Z}_M = \{0, 1, \ldots, M-1\}$, and $\omega_M = e^{i\frac{2\pi}{M}}$ is a primitive complex M-th root of unity where $i = \sqrt{-1}$.
- p is a prime, n a positive integer, $q = p^n$, $\mathbb{F}_q$ the finite field with q elements, $\mathbb{F}_q^*$ the multiplicative group of $\mathbb{F}_q$, α a primitive element in $\mathbb{F}_q$, $\overline{\mathbb{F}}_q$ the algebraic closure of $\mathbb{F}_q$ and $\mathrm{Tr}_r^n(x) = x + x^{q_1} + \cdots + x^{q_1^{n/r-1}}$ $(q_1 = p^r)$ the trace function from $\mathbb{F}_{p^n}$ to $\mathbb{F}_{p^r}$ with $r \mid n$ where $\mathrm{Tr}_1^n(x)$ is denoted by $\mathrm{Tr}(x)$ for simplicity.
- For $x \in \mathbb{F}_q$, *the discrete logarithm* to the base α is defined by

$$\log_\alpha x = \begin{cases} t\,, & \text{if } x = \alpha^t\,,\ 0 \le t \le q-2\,, \\ 0\,, & \text{if } x = 0 \end{cases}$$

  or simply as $\log x$ if the context is clear.
- $\mathbf{a} = \{a(t)\}_{t\ge 0}$ where $a(t) \in \mathbb{Z}_M$, is called an M-ary sequence. If $a(t+N) = a(t)$, for all $t = 0, 1, \ldots$, then we say that N is a *period* of $\{a(t)\}$. The smallest integer with this property is called the *least period* of the sequence. Throughout this paper, when we say *the period* of the sequence we mean that it is the least period of the sequence for simplicity. If $\mathbf{a}$ has period N, then we use $(a(0), \ldots, a(N-1))$, a vector of dimension N, to represent the sequence.

- U is the set consisting of all complex sequences whose entries have magnitude 1, i.e. $\mathbf{a} = (a(0), a(1), \ldots) \in U$, $|a(t)| = 1$, for $t = 0, 1, \ldots$ Let $\mathbf{x} = (x_0, \ldots, x_{N-1})$ and $\mathbf{y} = (y_0, \ldots, y_{N-1})$ be two sequences in $\mathbb{C}^N$, the inner product of $\mathbf{x}$ and $\mathbf{y}$ is defined by $\langle \mathbf{x}, \mathbf{y} \rangle = \sum_{j=0}^{N-1} x_i y_j^*$ where y^* is the conjugate of the complex number y.

2.2 Polynomial Functions over $\mathbb{F}_q$

Any polynomial in $\mathbb{F}_q[x] = \{\sum_i c_i x^i, c_i \in \mathbb{F}_q\}$ is considered as a polynomial function mapping from $\mathbb{F}_q$ to $\mathbb{F}_q$. We may assume that the (algebraic) degrees of these polynomials are less than or equal to $q - 1$ because of $x^q = x, x \in \mathbb{F}_q$.

Property 2.1. *Let $f(x) = \sum_{i=0}^{q-1} c_i x^i$ and $g(x) = \sum_{i=0}^{q-1} d_i x^i$ be polynomials in $\mathbb{F}_q[x]$. Then $f(x) = g(x)$ for all $x \in \mathbb{F}_q$ if and only if $c_i = d_i$, for $i = 0, 1, \ldots, q-1$.*

For a function mapping from $\mathbb{F}_q$ to $\mathbb{F}_p$, where $q = p^n, n > 1$, we need some concepts on (cyclotomic) cosets. A coset containing r modulo $q - 1$ is defined as $C_r = \{r, rp, \ldots, rp^{n_r - 1}\} \subset \mathbb{Z}_{q-1}$ where n_r is the smallest positive integer such that $r \equiv rp^{n_r} \bmod (q-1)$. The smallest integer in C_r is called the *coset leader* of C_r. Note that $n_r \mid n$. Let $\Gamma(q)$ be the set consisting of all coset leaders modulo $(q - 1)$.

Let $\xi(x)$ be a mapping from $\mathbb{F}_q$ to $\mathbb{F}_p$, then $\xi(x)$ can be represented by

$$\xi(x) = \sum_{r \in \Gamma(q)} \mathrm{Tr}_1^{n_r}(\beta_r x^r) \tag{2.1}$$

where $\beta_r \in \mathbb{F}_{p^{n_r}}$ and $n_r = |C_r|$, the size of the coset containing r. This is called the *trace representation* of a function from $\mathbb{F}_q$ to $\mathbb{F}_p$. It can be computed in terms of the discrete Fourier transform (DFT) over $\mathbb{F}_q$ (see [14]). The trace representation of a function from $\mathbb{F}_q$ to $\mathbb{F}_p$ is unique.

Property 2.2. *The trace representation (2.1) of ξ satisfies $\xi(x) = 0$ for all $x \in \mathbb{F}_q$ if and only if $\beta_r = 0$ for all $r \in \Gamma(q)$.*

Note that for any function $\xi(x)$ from $\mathbb{F}_q$ to $\mathbb{F}_p$, we can find a polynomial $f(x)$ in $\mathbb{F}_q[x]$ with exponents being coset leaders modulo $(q - 1)$, i.e.

$$f(x) = \sum_{r \in \Gamma(q)} c_r x^r\,, \quad \text{such that } \xi(x) = \mathrm{Tr}\left(f(x)\right), \; x \in \mathbb{F}_q\,. \tag{2.2}$$

However, this representation is not unique except for the following case.

Property 2.3. *For $f(x) = \sum_{r \in \Gamma(q)} c_r x^r, c_r \in \mathbb{F}_q$, if for all $c_r \neq 0$, the coset leaders r have the full length n (i.e. $|C_r| = n$), then $f(x) = 0$ if and only if $c_r = 0$ for all $r \in \Gamma(q)$.*

In this paper, in order to easily incorporate the process for directly applying the Weil bound, we use the form given in (2.2) for a function from $\mathbb{F}_q$ to $\mathbb{F}_p$. For the theory of finite fields and the basics of sequences with good correlation properties, the reader is referred to [14, 34].

2.3 Characters of Finite Fields

Let G be a finite Abelian group with identity 1. A character χ of G is a homomorphism from G into U (recalled that U is the multiplicative group of complex numbers with magnitude 1), i.e. a mapping from G into U with $\chi(g_1g_2) = \chi(g_1)\chi(g_2)$ for all $g_1, g_2 \in G$.

Definition 2.4 (Additive character). For each $j = 0, 1, \ldots, p-1$, the function ψ_j, given by

$$\psi_j(x) = e^{2\pi i j \operatorname{Tr}(x)/p} = \omega_p^{j \operatorname{Tr}(x)}, \quad x \in \mathbb{F}_q$$

defines an additive character of $\mathbb{F}_q$ as a character of the additive group of $\mathbb{F}_q$. We also denote it as $\psi(x)$ when $j = 1$. Furthermore,

$$\psi_j(x+y) = \psi_j(x)\psi_j(y), \quad \forall x, y \in \mathbb{F}_q .$$

Definition 2.5 (Multiplicative character). Let $M \mid (q-1)$. For each $j = 0, 1, \ldots, M-1$, a multiplicative character χ_j of order $M/\gcd(j, M)$, as a character of the multiplicative group $\mathbb{F}_q^*$ of $\mathbb{F}_q$, is defined by

$$\chi_j(\alpha^k) = e^{2\pi i jk/M} = \omega_M^{jk}, \quad \alpha^k \in \mathbb{F}_q^*$$

or equivalently

$$\chi_j(x) = \omega_M^{(j \log_\alpha x) \bmod M}, \quad x \in \mathbb{F}_q^* .$$

We extend the definition of χ_j at zero by $\chi_j(0) = 1$ throughout this paper if not stated otherwise. We denote χ^0 as the trivial multiplicative character, i.e. $\chi^0(x) = 1$ for all $x \in \mathbb{F}_q^*$, and $\chi_1(x)$ as $\chi^1(x)$ when we emphasize the case that $M = q-1$ and $j = 1$. Furthermore,

$$\chi_j(x \cdot y) = \chi_j(x)\chi_j(y), \quad \text{for all } x, y \in \mathbb{F}_q^* .$$

2.4 The Weil Bounds on Character Sums

The following three lemmas are from [54, 55] and Corollary 2.8 is an improved a variation from [50].

Lemma 2.6. *Let ψ be a nontrivial additive character over $\mathbb{F}_q$ and $f(x) = c_r x^r + \cdots + c_1 x + c_0 \in \mathbb{F}_q[x]$ with $\deg(f) = r \geq 1$, $\gcd(r, q) = 1$ and $f \neq g^p - g + c$ for all $g(x) \in \mathbb{F}_q[x]$ and $c \in \mathbb{F}_q$, then*

$$\left| \sum_{x \in \mathbb{F}_q} \psi(f(x)) \right| \leq (r-1)\sqrt{q} .$$

Lemma 2.7. *Let χ be a multiplicative character of $\mathbb{F}_q$ of order $M > 1$ and $\chi(0) = 0$. For $g \in \mathbb{F}_q[x]$, $g(x) \neq c \cdot h^M(x)$ for some $h \in \mathbb{F}_q[x]$, let d be the number of distinct roots of g in the algebraic closure $\overline{\mathbb{F}}_q$ of $\mathbb{F}_q$, then*

$$\Big|\sum_{x\in\mathbb{F}_q} \chi(g(x))\Big| \leq \begin{cases} (d-1)\sqrt{q} \\ (d-2)\sqrt{q}+1 & \text{if } M|\deg(g)\,. \end{cases}$$

However, in sequence design, it is more convenient to define $\chi(0) = 1$, as shown in [59]. Thus, the above lemma can be rewritten as follows in order to easily determine the correlation related properties of sequences, which is the version that we use in this paper.

Corollary 2.8. *With the notation in Lemmas 2.6 and 2.7, if we define $\chi(0) = 1$ and let e be the number of distinct roots of $g(x)$ in $\mathbb{F}_q$, then*

$$\Big|\sum_{x\in\mathbb{F}_q} \chi(g(x))\Big| \leq \begin{cases} (d-1)\sqrt{q}+e \\ (d-2)\sqrt{q}+1+e & \text{if } M|\deg(g)\,. \end{cases}$$

From Corollary 2.8 and the hybrid sum in [55], the following result follows immediately.

Lemma 2.9. *Let ψ be a nontrivial additive character of $\mathbb{F}_q$ and χ a nontrivial multiplicative character of $\mathbb{F}_q$ of order $M > 1$ with $\chi(0) = 1$. Let $f(x) \in \mathbb{F}_q[x]$ be a polynomial of degree r with the condition in Lemma 2.6, and $g(x) \in \mathbb{F}_q[x]$ with $g(x) \neq c \cdot h^M(x)$ and d distinct roots in $\overline{\mathbb{F}}_q$ and e distinct roots in $\mathbb{F}_q$. Then*

$$\Big|\sum_{x\in\mathbb{F}_q} \chi(g(x))\psi(f(x))\Big| \leq (r+d-1)\sqrt{q}+e\,. \tag{2.3}$$

We call the sum of (2.3) *a hybrid character sum.*

3 Correlation, DFT, and Ambiguity Functions

3.1 Operators on Sequences

We first define four operators on U, namely, decimation D_s, the time-shift L_τ, (linear) M phase shift P_w, and discrete Fourier transform (DFT) F. Given $\mathbf{a} = \{a(t)\} \in U$ and a fixed positive integer $H > 1$, for s, τ, w arbitrary integers, and $t = 0, 1, \ldots$, we define

$$\begin{aligned} &\text{Decimation:} && D_s[\mathbf{a}](t) := a(st) \\ &\text{Time shift:} && L_\tau[\mathbf{a}](t) := a(t+\tau) \\ &\text{Phase shift:} && P_w[\mathbf{a}](t) := \omega_H^{wt} a(t) \end{aligned}$$

The N points $(\tilde{a}(0), \tilde{a}(1), \ldots, \tilde{a}(N-1))$ of the DFT of $(a(0), \ldots, a(N-1))$ are defined as follows:

$$\tilde{a}(k) = F[\mathbf{a}](k) := \sum_{t=0}^{N-1} a(t)\omega_N^{-tk} \tag{3.1}$$

where we use the notation $\tilde{a}$ for $F[\mathbf{a}]$ for simplicity. The inverse DFT (IDFT) is given by

$$a(t) = \frac{1}{N}\sum_{k=0}^{N-1} \tilde{a}(k)\omega_N^{tk} .$$

The time-shift and phase-shift operators capture the characteristics of signals when they are transmitted through physical channels. At the receiver's side, due to the possible Doppler's effect and multipath propagation of the channel, the received signal (or sequence) could be both time-shifted and phase-shifted. Also, one could receive multiple shifted signals. For the phase shift, the received signal could have nonlinear phase shifts. However, it is difficult to build a model which captures all these factors in real communication systems. Thus, we simplify the scenario to only considering the linear phase shift.

Furthermore, in sequence design, the choice of H for the phase-shift operator is determined by the algebraic structure over which the sequence is defined. In other words, we restrict the values of H to N, q, p, $q-1$ or $p-1$ depending on the sequences defined over $\mathbb{Z}_N$, $\mathbb{F}_p$ or $\mathbb{F}_q$, additively or multiplicatively. Those phenomena will be elaborated clearly in the next section. Note that this restriction is convenient for theoretical studies. However, it may be not the case in practice.

It is worth pointing out that sequence $\mathbf{a}$ may not be periodic and N may not be the period of $\mathbf{a}$ when the DFT is applied. The definition of DFT is very general and is applied to any finite segment of a sequence with an infinite length in U. This is the typical case in the application of orthogonal frequency division multiplexing (OFDM) communications. However, in this paper, we assume that $\mathbf{a}$ has the period N. For the theory of digital communications the reader is referred to [41, 42].

Definition 3.1. For two sequences $\mathbf{a}$ and $\mathbf{b}$ with period N, if $\mathbf{b} = D_s\mathbf{a}$ with $\gcd(s, N) = 1$ or $\mathbf{b} = L_\tau\mathbf{a}$ or $\mathbf{b} = P_w\mathbf{a}$, we say that $\mathbf{a}$ and $\mathbf{b}$ are *decimation equivalent* or *time-shift equivalent*, or *phase-shift equivalent*. We denote them as $\mathbf{b} \sim_T \mathbf{a}$, $T \in \{D_s, L_\tau, P_w\}$. Otherwise, they are *decimation distinct* or *time-shift distinct* or *phase-shift distinct*.

3.2 Correlation Functions

Let $S \subset U$ consist of the sequences with period N. For two sequences $\mathbf{a} = \{a(t)\}$ and $\mathbf{b} = \{b(t)\}$ in S, the *crosscorrelation* between $\mathbf{a}$ and $\mathbf{b}$ is defined by

$$C_{\mathbf{a},\mathbf{b}}(\tau) = \sum_{t=0}^{N-1} a(t)b(t+\tau)^* , \tag{3.2}$$

or equivalently,

$$C_{\mathbf{a},\mathbf{b}}(\tau) = \langle \mathbf{a}, L_\tau \mathbf{b} \rangle \,. \tag{3.3}$$

If $\mathbf{a} = \mathbf{b}$, then the crosscorrelation function becomes the *autocorrelation function*, denoted as $C_{\mathbf{a}}(\tau)$. The maximum correlation of S is defined by

$$C_{\max} := \max\{AC_{\max}, CC_{\max}\}$$

where the maximum autocorrelation is

$$AC_{\max} := \max\{|C_{\mathbf{a}}(\tau)| : \mathbf{a} \in S, 1 \le \tau \le N-1\}$$

and the maximum crosscorrelation is

$$CC_{\max} := \max\{|C_{\mathbf{a},\mathbf{b}}(\tau)| : \mathbf{a}, \mathbf{b} \in S, \mathbf{a} \ne \mathbf{b}, 0 \le \tau \le N-1\} \,.$$

We call $C_{\mathbf{a}}(\tau)$ for $\tau \not\equiv 0 \bmod N$ *out-of-phase autocorrelation* of $\mathbf{a}$.

Definition 3.2. Let $\mathbf{a} = \{a(t)\}$ where $a(t) \in \mathbb{Z}_{M'}$, then a *modulated polyphase sequence* of $\mathbf{a}$, denoted as $\omega^{\mathbf{a}}$, is defined as $\omega_M^{\mathbf{a}}$, i.e.

$$\omega^{\mathbf{a}} := \omega_M^{\mathbf{a}} = (\omega_M^{a(0)}, \omega_M^{a(1)}, \ldots, \omega_M^{a(N-1)})$$

where it is not necessary that M and M' be equal. If both $\mathbf{a}$ and $\mathbf{b}$ are M'-ary sequences, then the crosscorrelation of $\mathbf{a}$ and $\mathbf{b}$ is defined through their modulated sequences, given as

$$C_{\mathbf{a},\mathbf{b}}(\tau) = \sum_{t=0}^{N-1} \omega_M^{a(t)-b(t+\tau)} \,.$$

Furthermore, the phase-shift operator is applied to the modulated sequence of $\mathbf{a}$, i.e. $P_w(\mathbf{a}) = \{\omega_H^{wt}\omega_M^{a(t)}\}$.

Note that the decimation and time-shift operators do not change the image sets of the autocorrelation functions provided some conditions are satisfied. Specifically, we have the following results.

Property 3.3. *For* $\gcd(s, N) = 1$, $D_s(\mathbf{a})$, $L_\tau(\mathbf{a})$, *and* $\mathbf{a} + c$ *where* c *is constant have the same autocorrelation properties as* $\mathbf{a}$. *Furthermore, the autocorrelation function of* $\mathbf{b} := P_w(\mathbf{a})$ *is given by* $C_{\mathbf{b}}(\tau) = \omega_H^{-w\tau} C_{\mathbf{a}}(\tau)$.

We say that sequence $\mathbf{a}$ is a *perfect sequence* if $C_{\mathbf{a}}(\tau) = 0$ for $\tau \not\equiv 0 \bmod N$ and $\mathbf{a}$ is an *(ideal) two-level autocorrelation sequence* if $C_{\mathbf{a}}(\tau) = -1$ for $\tau \not\equiv 0 \bmod N$.

According to Property 3.3, if $\mathbf{a}$ is a perfect or 2-level autocorrelation sequence, then $D_s\mathbf{a}$, $L_\tau\mathbf{a}$, and $\mathbf{a} + c$ are perfect or 2-level autocorrelation sequences where $1 \le s < N$ with $\gcd(s, N) = 1$. Furthermore, if $\mathbf{a}$ is perfect then $P_w(\mathbf{a})$ is perfect.

Example 3.4. We consider one binary sequence and one ternary sequence.

(a) Let $M = M' = 2$, $H = 3$ and $\mathbf{a} = (1101100)$. We denote ω_3 by ω for simplicity. Then $D_3(\mathbf{a})$, $L_2(\mathbf{a})$, and $P_2(\mathbf{a})$ are given as follows:

$$\begin{aligned}
\mathbf{a} &= (1101100)\\
(-1)^{\mathbf{a}} &= (-1,-1,1,-1,-1,1,1)\\
D_3(\mathbf{a}) &= (1100011)\\
L_2(\mathbf{a}) &= (0110011)\\
P_2(\mathbf{a}) &= (-1,-\omega^2,\omega,-1,-\omega^2,\omega,1)\\
\{C_{\mathbf{a}}(\tau)\} &= (7,-1,-5,3,3,-5,-1)\\
\{C_{D_3(\mathbf{a})}(\tau)\} &= (7,3,-1,-5,-5,-1,3)\\
C_{L_2(\mathbf{a})}(\tau) &= C_{\mathbf{a}}(\tau)\\
\{C_{P_2(\mathbf{a})}(\tau)\} &= (7,-\omega,-5\omega^2,3,3\omega,-5\omega^2,-1)
\end{aligned}$$

(b) Let $M = M' = 3$, $H = 3$ and $\mathbf{a} = (1,0,1,1,2,0,2,2)$ where $a_i \in \mathbb{Z}_3$. Then $D_3(\mathbf{a})$, $L_2(\mathbf{a})$, and $P_2(\mathbf{a})$ are given as follows:

$$\begin{aligned}
\mathbf{a} &= (1,0,1,1,2,0,2,2)\\
\omega^{\mathbf{a}} &= (\omega,1,\omega,\omega,\omega^2,1,\omega^2,\omega^2)\\
D_3(\mathbf{a}) &= (1,1,2,0,2,2,1,0)\\
L_2(\mathbf{a}) &= (1,1,2,0,2,2,1,0)\\
P_2(\mathbf{a}) &= (\omega,\omega^2,\omega^2,\omega,\omega,\omega,\omega^2,\omega)
\end{aligned}$$

They all have the same autocorrelation as **a**:

$$C_{\mathbf{a}}(\tau) = \begin{cases} 8 & \tau \equiv 0 \bmod 8 \\ -1 & \tau \not\equiv 0 \bmod 8\,. \end{cases}$$

3.3 Ambiguity Functions

Definition 3.5. The *auto and cross ambiguity functions* of **a** and **b** of period N in U are defined as two-dimensional autocorrelation and crosscorrelation functions in both time and phase, given by

$$G_{\mathbf{a}}(\tau,w) = \langle \mathbf{a}, P_w L_\tau \mathbf{a}\rangle \quad \text{and} \quad G_{\mathbf{a},\mathbf{b}}(\tau,w) = \langle \mathbf{a}, P_w L_\tau \mathbf{b}\rangle\,, \qquad 0 \le \tau < N\,,\ 0 \le w < H\,.$$

Thus, the autocorrelation and crosscorrelation functions are equal to their respective auto and cross ambiguity functions for the case $w = 0$.

Definition 3.6. A set S is called an (N, r, σ) *(correlation) signal set* if each sequence in S has period N, there are r time-shift distinct sequences in S, and both the maximum magnitude of out-of-phase autocorrelation values and crosscorrelation values are upper bounded by σ.

Definition 3.7. A set S is called an (N, r, σ, ρ) *ambiguity signal set* if it is an (N, r, σ) correlation signal set, all r sequences are both time-shift distinct and phase-shift distinct, and both the maximum magnitude of out-of-phase auto ambiguity functions and cross ambiguity functions are upper bounded by ρ, i.e.

$$|G_{\mathbf{a}}(\tau, w)| \le \rho\,, \quad (\tau, w) \ne (0,0)\,,$$
$$|G_{\mathbf{a},\mathbf{b}}(\tau, w)| \le \rho\,, \quad \mathbf{a} \ne \mathbf{b} \in S\,.$$

For $(\tau, w) \ne (0,0)$, $G_{\mathbf{a}}(\tau, w)$ is referred to as the *out-of-phase auto ambiguity*. Similar to the correlation, we denote by $AG_{\max}$ the maximum magnitude of the out-of-phase auto ambiguity and by $CG_{\max}$ the maximum magnitude of the cross ambiguity functions of any two distinct sequences in S. If we define $G_{\max} = \max\{AG_{\max}, CG_{\max}\}$, then $G_{\max} \le \rho$.

Definition 3.8. If $\mathbf{u} = \{u(t)\} \in U$ and there exists an M-ary sequence $\mathbf{a} = \{a(t)\}$, $a(t) \in \mathbb{Z}_M$ such that $u(t) = \omega_M^{a(t)} v(t)$ where $\mathbf{v} \in U$, then we say that $\mathbf{u}$ has an *M-ary factor sequence.*

According to this definition, the phase-shifted sequence of $\mathbf{u}$ is a sequence in U which has an H-ary factor sequence $\{\omega_H^{wt}\}$. In other words, the phase-shifted sequence of $\mathbf{u}$ is the term-by-term product sequence of $\{\omega_H^{wt}\}$ and $\{u(t)\}$.

Example 3.9. We assume that $\mathbf{a} = (1,0,1,1,2)$, $M' = M = 3$, $H = 3$, and $\omega = \omega_3$. Then we list a few values of auto ambiguity function of $\mathbf{a}$ as follows:

$$\begin{aligned}
\mathbf{a} &= (1,0,1,1,2) & G_{\mathbf{a}}(\tau,w) &= \langle \mathbf{a}, P_w L_\tau \mathbf{a}\rangle\\
\omega^{\mathbf{a}} &= (\omega, 1, \omega, \omega, \omega^2) & G_{\mathbf{a}}(0,1) &= -\omega\\
P_1(\mathbf{a}) &= (\omega, \omega, 1, \omega, 1) & G_{\mathbf{a}}(0,2) &= -\omega^2\\
P_2(\mathbf{a}) &= (\omega, \omega^2, \omega^2, \omega, \omega) & G_{\mathbf{a}}(1,0) &= -1\\
L_1(\omega^{\mathbf{a}}) &= (1, \omega, \omega, \omega^2, \omega) & G_{\mathbf{a}}(1,1) &= 2\omega\\
P_1L_1(\omega^{\mathbf{a}}) &= (1, \omega^2, 1, \omega^2, \omega^2) & G_{\mathbf{a}}(1,2) &= -1\\
P_2L_1(\omega^{\mathbf{a}}) &= (1, 1, \omega^2, \omega^2, \omega) & &
\end{aligned}$$

Property 3.10. *Let S_1 be an (N, r, σ, ρ) ambiguity signal set and let $S_2 = \{P_w\mathbf{u} = \{\omega_H^{wt}u(t)\} \mid w \in \mathbb{Z}_H, \mathbf{u} \in S_1\}$. Then S_2 is an (N, Hr, ρ) correlation signal set.*

Note that the concept of ambiguity functions is strongly related to Costas arrays, introduced by Costas in [6], and extensively studied in the literature, see [9, 13, 15], just to list a few. Up to now, systematically, there have been only two constructions.

One is the Welch construction using $\mathbb{F}_p$ and the other is the Lempel–Golomb construction using $\mathbb{F}_{2^n}$, which correspond to power residue and Sidel'nikov sequences, respectively. In the remainder of the chapter, we restrict ourselves to a subset of U in which each sequence has at most two different M-ary factor sequences. From Property 3.10, given an ambiguity signal set, we can obtain a correlation signal set with the same correlation and the size increased to a multiple of the size of the given ambiguity signal set.

3.4 Convolution and Correlation

For $\mathbf{a}, \mathbf{b} \in U$ with period N, the correlation function between $\mathbf{a}$ and $\mathbf{b}$ is equal to convolution between $\mathbf{a}$ and $\mathbf{b}$, denoted as $\mathbf{a} * \mathbf{b}$, i.e.

$$(\mathbf{a} * \mathbf{b})(\tau) = C_{\mathbf{a},\mathbf{b}}(\tau) = \sum_{t=0}^{N-1} a(t)b(t+\tau)^* \,. \tag{3.4}$$

From the signals and systems in digital communication [41], we have the following relation with their respective DFTs.

Property 3.11. *Let* $\mathbf{a}$ *and* $\mathbf{b}$ *be two complex sequences in* U. *Then*

DFT of the convolution: $\widetilde{C}_{\mathbf{a},\mathbf{b}}(-k) = \widetilde{a}(k)\,\widetilde{b}(k)^*$

Parseval Identity: $\sum_{t=0}^{N-1} |a(t)|^2 = \frac{1}{N}\sum_{k=0}^{N-1} |\widetilde{a}(k)|^2 = N\,.$

3.5 Optimal Correlation, DFT, and Ambiguity

From the Welch bound [56], an (N, r, σ, ρ) ambiguity signal set S where $r > 1$ has the maximum correlation, maximum DFT spectra $F_{\max} = \max\{|\widetilde{a}(k)| : \forall \mathbf{a} \in S, \forall k\}$ (it can be considered as crosscorrelation of two sequences), and maximum ambiguity being at least the square root of N, i.e.

$$C_{\max}, F_{\max}, G_{\max} \geq \sqrt{N}\,, \quad \text{for large } N.$$

Thus the best we aim for is to find signal sets for which those values are upper bounded by $c\sqrt{N}$ for a small constant $c \geq 1$.

Thus, we have the following criteria for measuring correlation, DFTs, and ambiguity properties of ambiguity signal sets:

$$C_{\max} \leq c_1\sqrt{N} \tag{3.5}$$

$$F_{\max} \leq c_2\sqrt{N} \tag{3.6}$$

$$G_{\max} \leq c_3\sqrt{N} \tag{3.7}$$

where the c_is are constants satisfying $1 \le c_i < \log\log N$ which do not depend on N. Therefore, in order to avoid repeatedly saying that the maximum of correlation functions, DFTs, and ambiguity functions of the sequences in an ambiguity signal set satisfy (3.5)–(3.7), we simply refer to each inequality above as a *metric*. For example, a metric for good correlation of a signal set means its maximum correlation is upper bounded by (3.5).

4 Polyphase Sequences for Three Metrics

In this section, we introduce polyphase sequences constructed from the additive group $\mathbb{Z}_N$ or finite fields. We will separate the sequences obtained from $\mathbb{F}_p$ additive and multiplicative characters and $\mathbb{F}_q$ additive and multiplicative characters, since their constructions are slightly different and easily confused. For polyphase sequences over a Galois ring, the reader is referred to [25] and references listed there.

4.1 Sequences from the Additive Group of $\mathbb{Z}_N$ and the Additive Group of $\mathbb{Z}_p$

Let

$$f(x) = c_r x^r + \cdots + c_1 x + c_0, c_i \in \mathbb{Z}_N .$$

An N-ary sequence $\mathbf{a} = \{a(t)\}_{t\ge 0}$ with period N from the $\mathbb{Z}_N$ additive group and its modulated sequence are defined below.

$$\begin{aligned} a(t) &= f(t) \in \mathbb{Z}_N \quad &&\text{[an } N\text{-ary sequence]} \\ \omega^{a(t)} &= \omega_{kN}^{a(t)} &&\text{[a } (kN)\text{-phase modulated sequence]} \end{aligned} \tag{4.1}$$

where k is a positive integer. When $N = p$, a p-ary sequence $a(t)$ and its modulated sequence are defined as follows:

$$\begin{aligned} a(t) &= f(t) \in \mathbb{F}_p &&\text{[a } p\text{-ary sequence]} \\ \omega^{a(t)} &= \psi_1(a(t)) = \omega_p^{a(t)} &&\text{[a } p\text{-phase modulated sequence]} \end{aligned} \tag{4.2}$$

where $\psi_1(x) = \omega^x, x \in \mathbb{F}_p$, an additive character defined in Section 2. We say that the sequence $\mathbf{a}$ is an *additive sequence over* $\mathbb{Z}_N$ or an *additive sequence over* $\mathbb{F}_p$ if $N = p$. Note that the modulated sequence given in (4.2) is a special case of (4.1) when $N = p$ and $k = 1$.

4.1.1 Frank–Zadoff–Chu (FZC) Sequences

Proposition 4.1 (Frank, Zadoff, Chu [4, 10]). *For $c \in \mathbb{Z}_N$ with $\gcd(c, N) = 1$, we define $\mathbf{a}_c = \{a_c(t)\}$ and its modulated sequence, denoted as $\mathbf{b}_c = \{b_c(t)\}$ as follows:*

$$a_c(t) = f(t), t \in Z_N \quad \textit{and} \quad b_c(t) := \omega_{2N}^{a_c(t)}$$

where

$$f(x) = \begin{cases} cx^2\,, & N \text{ even} \\ cx(x+1)\,, & N \text{ odd.} \end{cases}$$

Then $\{a_c(t)\}$ *is a perfect sequence, called a Frank–Zadoff–Chu (FZC) sequence. If* $N = p$, *let* $f(x) = c_2x^2 + c_1x, c_i \in \mathbb{F}_p$, *where* $c_2 \neq 0$ *and* $p > 2$, *then the* p*-ary sequence* $\mathbf{a}$ *defined in* (4.2) *is a perfect sequence.*

A modulated FZC sequence $\{\omega^{a_c(t)}\}$ can be obtained in another way for N odd, which is presented by Sarwate [45].

Proposition 4.2 (Sarwate [45]). *Let* N *be odd.*

(i) *The modulated FZC sequence* $\mathbf{b}_c$ *can be represented by*

$$b_c(t) = (-1)^{ct}\omega_{2N}^{ct^2}\,. \tag{4.3}$$

(ii) *In the following, we assume* $S = \{\mathbf{b}_c \mid \gcd(c, N) = 1\}$.

(a) $\widetilde{b}_c(k) = b_1(c^{-1}k)^* \cdot \widetilde{b}_c(0), \forall k$.

(b) $|\widetilde{b}_c(k)| = \sqrt{N}, k = 0, 1, \ldots$

(c) *If* $c \neq d$ *and* $\gcd(c - d, N) = 1$, *then the crosscorrelation between* $\mathbf{b}_c$ *and* $\mathbf{b}_d$ *is given by* $|C_{\mathbf{b}_c,\mathbf{b}_d}(\tau)| = \sqrt{N}$.

(d) *Let* p *be the smallest prime divisor of* N *and* $S_1 = \{\mathbf{b}_{c^{-1}} \mid 1 \le c \le p-1, \gcd(c, N) = 1\}$. *Then* S_1 *is an* $(N, p-1, \sqrt{N})$ *correlation signal in which each sequence is perfect and* $F_{\max} = \sqrt{N}$.

Note that in [45] the condition in (c) is listed as $\gcd(d^{-1} - c^{-1}, N) = 1$, which is equivalent to the condition listed here.

Assertion (i) is straightforward. The proof for Assertion (ii) is to first prove (a) which can be done by directly expanding the DFT, then using the Parseval identity presented in Property 3.11 together with (a) to obtain (b). Taking the DFT of the correlation together with (b) to get (c), (d) follows directly from (c). The proof given by Sarwate was an earlier work to compute the correlation of sequences using a transform, i.e. using of DFT. In their work, Dillon and Dobbertin [8] also used this technique to prove the validity of the conjectured binary 2-level sequences.

It is worth pointing out that the results in Propositions 4.1 and 4.2 cannot be obtained by the Weil bound if N is not prime. Furthermore, the ambiguity of FZC sequences can reach N. However, up to now, this is the only class of perfect sequences with optimal correlation and DFT for an arbitrary odd integer N (see [45] for its optimality).

Remark 4.3. The result of (c) in Proposition 4.2(ii) shows that the crosscorrelation of two different FZC sequences with some condition is equal to $\sqrt{N}$. Sarwate proved this by showing that the DFT of the crosscorrelation of two FZC sequences is equal to an FZC sequence up to a scalar factor $\pm N$. If the definition of DFT in (3.1) had a factor

$1/\sqrt{N}$, then the DFT of the crosscorrelation of two FZC sequences were equal to an FZC sequence. This is the reason that FZC sequences are considered superior to other known sequences with good correlation. Note that the elements of an FZC sequence belong to $\mathbb{Z}_N$ and its modulated sequence is defined by the $(2N)$-th primitive complex root of unity.

Remark 4.4. From the identity (4.3) in Proposition 4.2(i), an FZC sequence for odd N can be considered as being defined by a kind of hybrid characters, which resembles a new type of sequences constructed from the Weil representation (see Section 5).

4.1.2 Another Class for $\mathbb{Z}_N$

Another class of quadratic phase sequences is defined by Alltop [1].

Proposition 4.5. *Let N be odd, p be the smallest prime factor of N, $f_c(x) = cx^2$ with $1 \leq c < p$, $a(t) = ct^2 \in \mathbb{Z}_N$, and the modulated sequence be defined by*

$$b_c(t) = \omega_N^{a(t)} = \omega_N^{ct^2} .$$

Let $S = \{\{b_c(t)\} \mid 1 \leq c < p\}$. Then $C_{\max} = F_{\max} = \sqrt{N}$.

Note that the ambiguity of this set can reach N whether or not N is a prime.

4.1.3 Sequences from $\mathbb{F}_p$ Additive Characters

For $\mathbb{F}_p$ additive sequences, the phase-shift operator is defined by additive characters. If the defining polynomials over $\mathbb{F}_p$ have degrees at most d, then all their correlations, DFTs, and phase-shifts are determined by a polynomial with a degree of at most d. Thus, the three metrics can be bounded directly by the Weil bound, Lemma 2.6, on the additive characters. This is straightforward, so we omit them here. However, there is one special case, which is given below.

Alltop sequences ([1, 37]). Let S consist of all $\mathbb{F}_p$ additive sequences given by $f_c(x) = cx^3 + x$, $c \in \mathbb{F}_p^*$ and $p > 3$. Then $C_{\max} = \sqrt{p}$ given in [1], which is better than the bound given by Lemma 2.6. From Lemma 2.6, we have $F_{\max}, G_{\max} \leq 2\sqrt{p}$.

4.2 Sequences from $\mathbb{F}_p$ Multiplicative Characters

Let $M \mid (p-1)$, $f(x) \in \mathbb{F}_p[x]$, and χ be a multiplicative character of order M with $\chi(0) = 1$. An M-ary sequence $\mathbf{u} = \{u(t)\}$ with period p from the multiplicative structure of $\mathbb{F}_p$ and its modulated sequence are defined as follows:

$$\begin{aligned} u(t) &= c \log_\alpha f(t) \bmod M\,, \quad c \neq 0 \in \mathbb{Z}_M \quad &&\text{[an } M\text{-ary sequence]} \\ \omega_M^{u(t)} &= \chi_c(f(t)) &&\text{[a modulated sequence]} \end{aligned} \tag{4.4}$$

(Recall that $\log_\alpha 0 = 0$, defined in the beginning of Section 2.) The sequence $\mathbf{u}$ is called an $\mathbb{F}_p$ *multiplicative sequence*. It has period p, the same as the additive case. If $f(x) = x$, then $\{u(t)\}$ is called a *power residue* sequence. If $M = 2$ and $p \equiv 3 \bmod 4$, then the complement of $u(t)$ is a *quadratic residue sequence (or Legendre sequence)* with 2-level autocorrelation.

Example 4.6. Let $p = 7$ and $\alpha = 3$ be a primitive element in $\mathbb{F}_7$. Then we have sequences from $\mathbb{F}_7$ additive and multiplicative structures listed below (note that $\log_\alpha 0 = 0$):

Additive:	$a(t) = t^2$	$\{a(t)\} = (0,1,4,2,2,4,1)$
Multiplicative: $M = 6$	$u(t) = \log_3 t$	$\{u(t)\} = (0,0,2,1,4,5,3)$
$M = 3$	$u_1(t) = \log_3 t \bmod 3$	$\{u_1(t)\} = (0,0,2,1,1,2,0)$
$M = 2$	$u_2(t) = \log_3 t \bmod 2$	$\{u_2(t)\} = (0,0,0,1,0,1,1)$ $1 + \mathbf{u}_2 = (1,1,1,0,1,0,0)$ a quadratic residue sequence of period 7

A phase shift of $\mathbf{u}$ is defined in terms of multiplicative characters of $\mathbb{F}_p$ with order M, given by

$$P_w[\mathbf{u}](t) = \omega_M^{w \log_\alpha t}\omega_M^{u(t)} = \chi_w(t)\chi(f(t)), \quad t, w \in \mathbb{F}_p . \tag{4.5}$$

We have the following relationship of correlation, DFT and ambiguity with character sums.

Proposition 4.7. *For $\mathbb{F}_p$ multiplicative sequences* $\mathbf{u}$, *defined in* (4.4), *and* $\mathbf{v}$, *given by* $v(t) = d\log_\alpha g(t) \bmod M, 1 \le d < M$ *where $g(x) \in \mathbb{F}_p[x]$, their correlation, DFT, and ambiguity function are given by*

$$C_{\mathbf{u},\mathbf{v}}(\tau) = \sum_{t\in\mathbb{F}_p} \chi_c(f(t))\chi_d(g(t+\tau))^* = \sum_{t\in\mathbb{F}_p} \chi(h_0(t)) \tag{4.6}$$

$$\text{where } h_0(x) = f(x)^c g(x+\tau)^{M-d}$$

$$\tilde{u}(k) = \sum_{t\in\mathbb{F}_p} \chi_c(f(t))\omega_p^{-tk} = \sum_{t\in\mathbb{F}_p} \chi_c(f(t))\psi(-tk) \tag{4.7}$$

$$G_{\mathbf{u},\mathbf{v}}(\tau, w) = \sum_{t\in\mathbb{F}_p} \chi_c(f(t))[\chi_d(g(t+\tau))\chi_w(t)]^* = \sum_{t\in\mathbb{F}_p} \chi(d_0(t)) \tag{4.8}$$

$$\text{where } d_0(x) = h_0(x)x^{M-w} .$$

From Proposition 4.7, we know that both correlation and ambiguity can be bounded by the Weil bound on multiplicative characters when both f and g have low degrees, and their DFTs can be bounded by the Weil bound on hybrid character sums as long as f has low degree. For a power residue sequence with $f(x) = x$, from (4.7), both $f(x)$ and $(-kx)$ have degree 1. According to Lemma 2.9, the DFT of a power residue sequence is bounded by $\sqrt{p}+1$.

In order to facilitate a quicker assessment for the condition in Corollary 2.8 and Lemma 2.9, we introduce the following concepts (see [22]), which will be also used in the $\mathbb{F}_q$ case. For $f \in \mathbb{F}_q[x]$ ($q = p$ or $q = p^n$), if $f(x) = c \cdot h^M(x)$, then we say that $f(x)$ is an *M-th power multiple* in $\mathbb{F}_q[x]$. Thus, the following assertions are equivalent.

(i) $f(x)$ is an M-th power multiple in $\mathbb{F}_q[x]$.

(ii) The factorization of $f(x)$ is $f(x) = c(x-y_1)^{e_1}\dots(x-y_s)^{e_s}$ where $y_i \in \overline{\mathbb{F}}_q, c \in \mathbb{F}_q$ where $M \mid e_i$ for all i.

Using this result, we can easily see that ambiguity of a power residue sequence cannot be bounded since $d_0(x)$ in (4.8) could be an M-th power multiple. In detail, $d_0(x) = x^c(x+\tau)^{M-d}x^{M-w}$ where $1 \le c, d, w < M$. By choosing $c = d + w \bmod M$ with $c \ne d \bmod M$ and $\tau = 0$ we have $d_0(x) = x^{2M}$, an M-th power multiple. So, $G_{\mathbf{u},\mathbf{v}}(0, w) = p$.

In the following, we summarize the correlation, DFT, and ambiguity of power residue sequences in a theorem where the last two results are from [53].

Theorem 4.8. *Let $M \mid (p-1)$, an M-ary power residue sequence $\mathbf{u}_c = \{u_c(t)\}$ whose elements are defined by $u_c(t) = c\log_\alpha t \bmod M, t \in \mathbb{F}_p, M \mid (p-1)$, and let $S = \{\mathbf{u}_c : 1 \le c < M\}$. Then $\mathbf{u}_c$ has the period p.*

(i) *For any sequence $\mathbf{u} \in S$, the autocorrelation function of $\mathbf{u}$ is bounded by (Sidel'nikov [48], Lempel et al. [33], Sarwate [44])*

$$|C_{\mathbf{u}}(\tau)| \le 3, \tau \not\equiv 0 \bmod p\,.$$

In particular, for $M = 2$,

$$C_{\mathbf{u}}(\tau) \in \begin{cases} \{-1\} & \tau \not\equiv 0 \bmod p \quad \text{if } p \equiv 3 \bmod 4 \\ \{1,-3\} & \tau \not\equiv 0 \bmod p \quad \text{if } p \equiv 1 \bmod 4\,. \end{cases}$$

(ii) *For any two sequences $\mathbf{u} \in S$ and $\mathbf{v} \in S$, their crosscorrelation function is bounded by $|C_{\mathbf{u},\mathbf{v}}(\tau)| \le \sqrt{p}+2$ (Kim et al. [29]).*

(iii) *The DFT is bounded by $|\tilde{u}(k)| \le \sqrt{p}+1$.*

(iv) *The cross ambiguity function has a peak value p.*

4.3 Sequences from $\mathbb{F}_q$ Additive Characters

We now introduce sequences defined by $\mathrm{Tr}(f(x))$, a function from $\mathbb{F}_q$ to $\mathbb{F}_p$ in (2.2), which are the most popular sequences in both theory and practice. We assume that $f(0) = 0$ in this case. Let $\mathbf{a} = \{a(t)\}$ whose elements are defined by

$$a(t) = \mathrm{Tr}(f(\alpha^t))\,, \quad t = 0, 1, \dots \tag{4.9}$$

Then **a** is a sequence over $\mathbb{F}_p$ with period $N \mid (q-1)$ where α is a primitive element in $\mathbb{F}_q$, and we also say that $\mathbf{a} = \{a(t)\}$ is defined by $f(x)$. The equation (2.1) is also called a *trace representation* of the sequence $\{a(t)\}$. If $f(x) = x^d$ with $(d, q-1) = 1$, then **a** is an *m-sequence* over $\mathbb{F}_p$ with period $p^n - 1$, i.e.

$$m\text{-sequence} \longleftrightarrow \mathrm{Tr}(x^d), \quad \gcd(d, q-1) = 1 .$$

Any m-sequence has a 2-level autocorrelation function. If $\mathbf{b} = \{b(t)\}$ where $b(t) = a(dt)$, then **b** is a d-decimation of **a**. Thus, any sequence of period $N \mid (q-1)$ can be obtained by summing up different decimations on a shifted m-sequence with trace representation $\mathrm{Tr}(x)$.

The study of correlation of the sequences defined in (4.9) has been around for more than five decades. Since the polynomials considered here usually have high degrees, they cannot be bounded by the Weil bounds. Each case with low correlation was found by a special method for manipulating the exponential sums (see [25]). In general, DFT and ambiguity for those sequences with low correlation are unknown. One exceptional example is the Kasami small set of the sequences of period $2^n - 1$ (n even), which are defined by polynomials $\mathrm{Tr}(x + cx^d), c \in \mathbb{F}_q$ where $d = 2^{n/2} + 1$ have a very high degree, but the DFT is bounded, see [32].

We may write $f(x) = \sum_{r \in \Gamma(q)} c_r x^r, c_r \in \mathbb{F}_{p^{n_r}}$ (recall that $\Gamma(q)$ consists of all the coset leaders modulo q in Section 2.2). Let

$$S = \{\{a(t)\} : c_r \in \mathbb{F}_{p^{n_r}}\}$$

where $a(t)$ are defined by (4.9). The set S corresponds to a cyclic code. Determining the maximum correlation is equivalent to finding the minimum distance of this code. This connection also indicates the hardness for determining the three metrics of this set if $f(x)$ is too general. From now, for the sequences defined by (4.9), we restrict ourselves to the case that the sequences have period $N = q - 1$.

The sequences defined here are p-ary sequences, so the phase shift is defined through an additive character $\psi(x) = \omega_p^{\mathrm{Tr}(x)}$ (see Section 2), given by

$$\begin{aligned} P_w[\mathbf{a}](t) &= \psi(\mathrm{Tr}(\alpha^{t+w}))\omega_p^{a(t)} = \psi(\alpha^{t+w} + a(t)) \\ &= \psi(\alpha^{t+w} + f(\alpha^t)), t, w = 0, 1, \ldots \end{aligned} \tag{4.10}$$

where the last identity comes from (4.9). The following proposition presents correlation, DFT, and ambiguity of an $\mathbb{F}_q$ additive sequence using character sums.

Proposition 4.9. *With $f(x)$ and* **a** *above, let $b(t) = \mathrm{Tr}(g(\alpha^t))$ where $g(x) \in \mathbb{F}_q[x]$ with $g(0) = 0$ and the exponents of x in g belong to $\Gamma(q)$. Recall that χ^1 is a multi-*

plicative character of order $q-1$.

$$C_{\mathbf{a},\mathbf{b}}(\tau)=\sum_{x\in\mathbb{F}_q}\psi(f(x))\psi(g(\alpha^\tau x))^*-1=\sum_{x\in\mathbb{F}_q}\psi(h_1(x))-1 \tag{4.11}$$

$$\text{where } h_1(x)=f(x)-g(\alpha^\tau x),\tau\in\mathbb{Z}_{q-1}\,.$$

$$\tilde{a}(k)=\sum_{t=0}^{q-2}\omega_p^{\mathrm{Tr}(f(\alpha^t))}\omega_{q-1}^{-tk}=\sum_{x\in\mathbb{F}_q}\psi(f(x))\chi^1(x^{-k})-1,k\in\mathbb{Z}_{q-1}\,. \tag{4.12}$$

$$G_{\mathbf{a},\mathbf{b}}(\tau,w)=\sum_{t=0}^{q-2}\psi(f(\alpha^t))[\psi(g(\alpha^{t+\tau}))\psi(\alpha^{t+k})]^*=\sum_{x\in\mathbb{F}_q}\psi(d_1(x))-1 \tag{4.13}$$

$$\text{where } d_1(x)=h_1(x)-\alpha^w x,\ \tau,w\in\mathbb{Z}_{q-1}\,.$$

From (4.12), when $f(x)=x$ we immediately see that the DFT is less than or equal to $\sqrt{q}$ by the Weil hybrid sum in Lemma 3. In other words, the DFT of an m-sequence defined by $\mathrm{Tr}(x)$ is upper bounded by $\sqrt{q}$. This is reported in several papers, for example [40]. From Proposition 4.9, three metrics can be obtained using the Weil bounds only when the polynomials have low degrees.

The sequence **a** has 2-level autocorrelation if and only if the exponential sum in (4.11) is equal to zero. For the binary case, i.e. $p=2$, all known 2-level autocorrelation sequences with period 2^n-1 are presented in [14]. For nonbinary cases, i.e. $p>2$, not so many constructions are known. We will introduce them in Section 6 and present some conjectures on zero exponential sums.

4.4 Sequences from $\mathbb{F}_q$ Multiplicative Characters

Let $f(x)\in\mathbb{F}_q[x]$. For $M\mid(q-1)$, and $1\le c<M$, recall that the definition of a multiplicative character $\chi_c(\alpha^x)=\omega_M^{cx}$ from Definition 2.5. An M-ary $\mathbb{F}_q$ multiplicative sequence $\mathbf{u}_c=\{u_c(t)\}$ and its modulated sequence are defined as

$$u_c(t)=c\log_\alpha f(\alpha^t) \bmod M \quad\text{and}\quad \omega_M^{u_c(t)}=\chi_c(f(\alpha^t))\,,\quad t=0,1,\ldots$$

Let $S=\{\mathbf{u}_c:1\le c<M\}$. Then each sequence in S has period $q-1$ and $|S|=M$. If $f(x)=x+1$, we have

$$u_c(t)=c\log_\alpha(\alpha^t+1) \bmod M \quad\text{and}\quad \omega_M^{u_c(t)}=\chi_c(\alpha^t+1)\,,\quad t=0,1,\ldots \tag{4.14}$$

$\{u(t)\}$ is called a *Sidel'nikov sequence* and $\{\omega_M^{u_c(t)}\}$ is its modulated sequence.

Example 4.10. Let $p=2$, $n=4$ and $\mathbb{F}_{2^4}$ be defined by a primitive polynomial $t(x)=x^4+x+1$, and let α be a primitive element in $\mathbb{F}_{2^4}$ with $t(\alpha)=0$. Below we list a binary m-sequence defined by $\mathrm{Tr}(x)$ and three Sidel'nikov sequences for $M=15,5,3$.

Additive	$\mathbf{a} = (000100110101111)$, $a(t) = \mathrm{Tr}(\alpha^t)$	$f_1(x) = x$
Multiplicative	$u(t) = \log(\alpha^t + 1) \bmod M$	$f_2(x) = x + 1$
$M = 15$	$\mathbf{u} = (0,4,8,14,1,10,13,9,2,7,5,12,11,6,3)$	
$M = 5$	$\mathbf{u}_1 = (0,4,3,4,1,0,3,4,2,2,0,2,1,1,3)$	
$M = 3$	$\mathbf{u}_2 = (0,1,2,2,1,1,1,0,2,1,2,2,2,0,0)$	

The phase shift of an $\mathbb{F}_q$ multiplicative sequence is defined by multiplicative characters of order M, i.e.

$$P_w[\mathbf{u}](t) = \omega_M^{wt}\omega_M^{u(t)} = \chi_w(\alpha^t)\chi(f(\alpha^t)), \quad 0 \le t < q-1, \quad 1 \le w < M.$$

We now present the correlation, DFT, and ambiguity of an $\mathbb{F}_q$ multiplicative sequence in terms of character sums.

Proposition 4.11. *With $f(x)$ as above, let $g(x) \in \mathbb{F}_q[x]$ and $v(t) = d\log_\alpha g(\alpha^t) \bmod M$. Then*

$$C_{\mathbf{u},\mathbf{v}}(\tau) = \sum_{x\in\mathbb{F}_q} \chi_c(f(x))\chi_d(g(\alpha^\tau x))^* - 1 = \sum_{x\in\mathbb{F}_q} \chi(h_2(x)) - 1 \tag{4.15}$$

$$\text{where } h_2(x) = f(x)^c g(\alpha^\tau x)^{M-d}.$$

$$\tilde{u}(k) = \sum_{t=0}^{q-2} \chi_c(f(\alpha^t))\omega_{q-1}^{-tk} = \sum_{x\in\mathbb{F}_q} \chi_c(f(x))\chi^1(x^{-k}) - 1. \tag{4.16}$$

$$G_{\mathbf{u},\mathbf{v}}(\tau, w) = \sum_{t=0}^{q-2} \chi_c(f(\alpha^t))[\chi_d(g(\alpha^{t+\tau}))\chi_w(\alpha^t)]^* = \sum_{x\in\mathbb{F}_q} \chi(d_2(x)) - 1 \tag{4.17}$$

$$\text{where } d_2(x) = h_2(x)x^{M-w}.$$

From Proposition 4.11, all three metrics can be obtained by the Weil bound on multiplicative characters by Corollary 2.8 when both f and g have low degrees. Thus, for $f(x) = x+1$, i.e. the Sidel'nikov case, from (4.16), the polynomial $f(x)$ has -1 as a root and the polynomial x has zero as a root, but both in $\mathbb{F}_p$. Applying Lemma 2, the magnitude of the DFT is upper bounded by $\sqrt{q} + 1$. Similarly, from (4.17), $d_2(x)$ has three distinct roots, where two are in $\mathbb{F}_p$ and there is no M-th power multiple in $\mathbb{F}_q[x]$. Again using Corollary 2.8, the magnitude of the auto and cross ambiguity functions are upper bounded by $2\sqrt{q} + 3$. We summarize these results in the following theorem where the last part is taken from [53].

Theorem 4.12. *With the notation defined above, each Sidel'nikov sequence has period $q - 1$ and S is an $(q - 1, M, \sqrt{q} + 3, 2\sqrt{q} + 2)$ ambiguity set, whose properties are listed below in detail.*

(i) *The autocorrelation function is bounded by (Sidel'nikov [48], Lempel et al. [33], Sarwate [44])*

$$|C_{\mathbf{u}}(\tau)| \le 4, \tau \not\equiv 0 \bmod (q-1).$$

Table 4.1: Four types of sequences and characters

	$\mathbb{F}_p$	$\mathbb{F}_q$
Additive: p-ary sequence		
Sequence	$a(t) = f(t)$	$a(t) = \mathrm{Tr}(f(\alpha^t))$
Correlation	$\psi(f_0(x))$	$\psi(h_1(x))$
DFT	$\psi(f_1(x))$	$\psi(f(x))\chi^1(x^{-k})$
Ambiguity	$\psi(f_2(x))$	$\psi(h_1(x) + \alpha^w x)$
	f_i's have the same degree	$h_1(x) = f(x) - g(\alpha^\tau x)$
Multiplicative: M-ary sequence		
Sequence	$u(t) = \log_\alpha f(t) \bmod M$	$u(t) = \log_\alpha f(\alpha^t) \bmod M$
Correlation	$\chi(h_0(x))$	$\chi(h_2(x))$
DFT	$\chi_c(f(x))\psi(-kx)$	$\chi_c(f(x))\chi^1(x^{-k})$
Ambiguity	$\chi(h_0(x)x^{M-w})$	$\chi(h_2(x)x^{M-w})$
	$h_0(x) = f(x)^c g(x+\tau)^{M-d}$	$h_2(x) = f(x)^c g(\alpha^\tau x)^{M-d}$

In particular, for $M = 2$,

$$C_{\mathbf{u}}(\tau) \in \begin{cases} \{-2, 2\} & \tau \not\equiv 0 \mod (q-1) \quad \text{if } q \equiv 3 \bmod 4 \\ \{0, -4\} & \tau \not\equiv 0 \mod (q-1) \quad \text{if } q \equiv 1 \bmod 4 . \end{cases}$$

(ii) *The crosscorrelation is bounded by* $\sqrt{q} + 3$ *(Kim et al.* [28]*).*
(iii) *The DFT is bounded by* $|\tilde{u}(k)| \le \sqrt{q} + 1$ *(Yu et al.* [57]*).*
(iv) *Ambiguity are bounded by* $|G_{\mathbf{u},\mathbf{v}}(\tau, w)| \le 2\sqrt{q} + 2$ *for any* (τ, w) *when* $\mathbf{u} \ne \mathbf{v}$ *and* $(\tau, w) \ne (0, 0)$ *when* $\mathbf{u} = \mathbf{v}$.

We summarize these four types of sequences and their corresponding characters in Table 4.1 where ψ is a nontrivial additive character, χ is a multiplicative character of $\mathbb{F}_p$ or $\mathbb{F}_q$ of order M, and χ^1 is a multiplicative character of $\mathbb{F}_q$ of order $q-1$.

Note that $\mathbb{F}_p$ or $\mathbb{F}_q$ additive sequences have p phases in their modulated sequences. On the other hand, $\mathbb{F}_p$ or $\mathbb{F}_q$ multiplicative sequences have M phases in their modulated sequences. For $\mathbb{F}_p$ additive and $\mathbb{F}_q$ multiplicative sequences, correlation, DFT, and ambiguity only involve $\mathbb{F}_p$ additive characters and $\mathbb{F}_q$ multiplicative characters, respectively. However, for $\mathbb{F}_p$ multiplicative and $\mathbb{F}_q$ additive sequences, the DFTs are hybrid character sums.

Remark 4.13. The auto and cross ambiguity functions of an $\mathbb{F}_q$ additive sequence are equal to the Hadamard transform of their respective autocorrelation and crosscorrelation functions. We will come back to the Hadamard transform in Section 6.

4.5 Sequences Defined by Indexing Field Elements Alternatively

For $\mathbb{F}_p$ sequences, regardless of whether they are additive or multiplicative, the elements are indexed by $0, 1, \ldots, p-1$ in the additive group of $\mathbb{F}_p$. On the other hand, for $\mathbb{F}_q$ sequences, both additive and multiplicative cases, the elements are indexed by $1, \alpha, \ldots, \alpha^{q-1}$ in the multiplicative group of $\mathbb{F}_q$. Thus, for the $\mathbb{F}_p$ case, we have two different indexing methods to define the sequences: the field elements are indexed by $0, 1, \ldots, p-1$, as shown in Sections 4.2 and 4.3, or indexed by $1, \alpha, \ldots, \alpha^{p-2}$, where α is a primitive element of $\mathbb{F}_p$, as shown in Sections 4.4 and 4.5 when $q = p$. However, for the $\mathbb{F}_q$ case with $q > p$, we also have an alternative definition for indexing $\mathbb{F}_q$ elements. In order to do so, we need a one-to-one correspondence between a p-adic integer $t = 0, 1, \ldots, q-1$ $(q = p^n)$, and the elements in $\mathbb{F}_q$ as shown below.

$$t = \sum_{i=0}^{n-1} t_i p^i, t_i \in \mathbb{F}_p \longleftrightarrow \alpha_t = \sum_{i=0}^{n-1} t_i \alpha^i \longleftrightarrow \mathbf{t} = (t_{n-1}, \ldots, t_0) .$$

These four alternative definitions of sequences are given in Table 4.2.

For example, Golay sequences are ordered by the additive group of $\mathbb{F}_{2^n}$ [7, 39].

Example 4.14. Let $n = 3$ and $f(x_0, x_1, x_2) = x_0x_1 + x_1x_2$. Then a binary Golay sequence, denoted by $\mathbf{b} = \{b(t)\}$ is defined as follows:

$$\begin{aligned} b(t) &= f(t) \text{ where } t = t_2 2^2 + t_1 2 + t_0, \mathbf{t} = (t_2, t_1, t_0), t = 0, 1, \ldots, 7, \text{ and} \\ f &= (f(0,0,0), f(1,0,0), f(0,1,0), f(1,1,0), f(0,0,1), f(1,0,1), \\ &\quad f(0,1,1), f(1,1,1)) \\ &= (0,0,0,1,0,0,1,0). \end{aligned}$$

In other words, the sequence $\mathbf{b}$ consists of the values of f in the truth table in Table 4.3.

Let F_{2^3} be the finite field defined by a primitive polynomial $g(x) = x^3 + x + 1$ over $\mathbb{F}_2$ and α be a primitive element of F_{2^3} satisfying $g(\alpha) = 0$. Using the finite field discrete Fourier transform (DFT) (see [14]), we can obtain the trace presentation of $\mathbf{b}$,

Table 4.2: Four types of sequences by indexing the elements of $\mathbb{F}_p$ and $\mathbb{F}_q$ alternatively

	$\mathbb{F}_p$	$\mathbb{F}_q$
Additive	$b(t) = f(\alpha^t)$	$b(t) = \mathrm{Tr}(f(\alpha_t))$
Multiplicative	$v(t) = \log_\alpha f(\alpha^t) \bmod M$ $t = 0, \ldots, p-2$	$v(t) = \log_\alpha f(\alpha_t) \bmod M$ $t = 0, 1, \ldots, q-1$
Period	$p-1$	q

Table 4.3: Truth table of the Golay sequence **b**

Index t	α_t	x_2	x_1	x_0	f
0	$\alpha_0 = 0$	0	0	0	0
1	$\alpha_1 = 1 = \alpha^0$	0	0	1	0
2	$\alpha_2 = \alpha$	0	1	0	0
3	$\alpha_3 = \alpha^3$	0	1	1	1
4	$\alpha_4 = \alpha^2$	1	0	0	0
5	$\alpha_5 = \alpha^6$	1	0	1	0
6	$\alpha_6 = \alpha^4$	1	1	0	1
7	$\alpha_7 = \alpha^5$	1	1	1	0

which is given by

$$B(x) = \mathrm{Tr}(h(x)) \quad \text{where} \quad h(x) = \alpha^6 x + \alpha^3 x^3 \tag{4.18}$$

and $\mathrm{Tr}(x) = x + x^2 + x^4$ is the trace function of F_{2^3}. Then $b(t) = \mathrm{Tr}(h(\alpha_t)), t = 0, 1, \ldots, 7$ where the conversions between t and α_t are given in Table 4.3.

If we evaluate $B(x) = \mathrm{Tr}(h(x))$ over $\{\alpha^t : t = 0, 1, \ldots, 6\}$, i.e. let $a(t) = B(\alpha^t)$, then we obtain $\{a(t)\} = (0001100)$. In other words, for a given function $B(x)$ in (4.18), using the additive character, we have two sequences:

Indexed in the additive group of $\mathbb{F}_{2^3}$	Indexed in the multiplicative group of $\mathbb{F}_{2^3}$
$b(t) = B(t)$	$a(t) = B(\alpha^t)$
$\{(-1)^{b(t)}\} = (1, 1, 1, -1, 1, 1, -1, 1)$	$\{(-1)^{a(t)}\} = (1, 1, 1, -1 - 1, 1, 1)$
Period: 8	Period: 7

Autocorrelation

$\{C_{\mathbf{b}}(\tau)\} = (8, 0, 0, 4, 0, 4, 0, 0)$	$\{C_{\mathbf{a}}(\tau)\} = (7, 3, -1, -1, -1, -1, 3)$

In general, the sequences indexed in the additive group of $\mathbb{F}_{2^n}$ have period 2^n or a factor of 2^n. However, it is not easy to find the polynomials or Boolean functions such that the sequences defined through the additive group of $\mathbb{F}_q$ for having low correlation, DFT, and ambiguity, since no known bounds can be applied to those cases. Even for the Golay sequences, the autocorrelation of an individual sequence is unknown, although the sum of the autocorrelation functions of a Golay pair sequences is equal to zero. Some preliminary treatments about correlation of the sequences defined by $\mathbb{F}_q$ additive group order can be found in [14]. It could be interesting to see whether some new sequences with good correlation, DFT, and ambiguity could arise from those classes.

5 Sequences with Low Degree Polynomials

From Propositions 4.7, 4.9, and 4.11 or the summaries in Table 4.1 in the previous section, we know that three metrics can be bounded by the Weil bounds on additive character sums, multiplicative character sums, and hybrid sums when the defining polynomials have low degrees. In the last section, we also introduced sequences with low degree polynomials from additive characters over $\mathbb{F}_p$, and $f(x) = x$ in $\mathbb{F}_p$ or $f(x) = x + 1$ in $\mathbb{F}_q$ through multiplicative characters. In this section, we continue our journey along this line with emphasis on the constructions of ambiguity signal sets with low degree polynomials.

5.1 Methods for Generating Signal Sets from a Single Sequence

For $\mathbb{F}_q$ additive sequences, a general method is to apply the decimation-and-add operation, as explained in Section 2.3, which has been attracting researchers since the end of 1950s. For those constructions, most of them are the sum of two m-sequences. Golomb constructed binary m-sequences using linear feedback shift registers in the middle of the 1950s [12] and Zieler extended them to $\mathbb{F}_q$ [62] with period $q - 1$. The multiterm sequences from the decimation-and-add operations on an m-sequence have been marked by various footprints from many researchers, for example, [56] (1974), [23] (1976), [31] (1991), to list just a few.

Investigation of $\mathbb{F}_q$ multiplicative sequences including the $\mathbb{F}_p$ has occurred more recently. These can be classified as follows.

(i) Amplitude scaling operation: $y_c(t) = cu(t)$, where c is a constant (see [29] for the case that $u(t)$ is a power residue sequence and [28] for a Sidel'nikov sequence $u(t)$).

(ii) Shift-and-add with/without inverse: $y_{(c,d,\tau)}(t) = cu(t) + du(\delta t + \tau)$ where $\delta = \pm 1$, c, d and τ are constant (first observed in [61] and later proved in [43] in 2006 for $\delta = 1$).

(iii) Interleaved method: A Sidel'nikov sequence of period $q^2 - 1$, writing it as an $q - 1$ by $q + 1$ array, then taking some columns to form a signal set. (The result in [59] shows that this can be done when $u(t)$ is a Sidel'nikov sequence.)

In the remainder of this section, we first introduce the signal sets with low odd degree polynomials, then present sequences from power or Sidel'nikov sequences using the above operations. Thirdly, we present the sequences from the Weil representation and their extensions, and finally, we give a new construction of sequences from combinations of different indexing field elements and hybrid character sums.

5.2 Sequences with Low Odd Degree Polynomials

In this subsection, we consider the following polynomial functions of odd degrees over $\mathbb{F}_q$, which have been considered extensively in the literature, for example, in [38, 40, 46], for constructing signal sets using $\mathbb{F}_q$ additive characters. Let

$$d < \begin{cases} \sqrt{q}\,, & \text{if } p = 2 \\ \sqrt{q}/2\,, & \text{if } p > 2 \end{cases} \tag{5.1}$$

$$T_0 = \left\{ \sum_{i=2}^{d} b_i x^{2i-1} + x + 1 \mid b_i \in \mathbb{F}_q \text{ with (5.1)} \right\} \tag{5.2}$$

$$T_1 = \left\{ \sum_{i=2}^{d} b_i x^{2i-1} + x \mid b_i \in \mathbb{F}_q \text{ with (5.1)} \right\} \tag{5.3}$$

$$T_2 = \left\{ \sum_{i=3}^{d} b_i x^{2i-1} + x^3 \mid b_i \in \mathbb{F}_q \text{ with (5.1)} \right\}. \tag{5.4}$$

The polynomials in T_1 have recently been considered in [11] for deterministic extractors for affine random sources. By directly applying the Weil bounds, the authors show that the character sum $\sum_{x \in \mathbb{F}_q} \eta(f(x))$ is bounded when η is an additive character, where $p = 2$, or η is a multiplicative character of order 2 for $p > 2$. In the following, we investigate the size and three metrics of the signal set consisting of the sequences defined by polynomials in T_2 using $\mathbb{F}_q$ additive characters and T_0 for multiplicative characters.

5.2.1 $\mathbb{F}_q$ Additive Sequences with Low Odd Degree Polynomials

For the additive case, the exponents of the polynomials should belong to different cosets in order to generate time-shift distinct and phase-shift distinct sequences. According to Property 2.3 in Section 4, we need the following result. For easy reference, we also included a proof there.

Proposition 5.1. *The odd integers $2i - 1$'s with the condition in (5.2) belong to different cosets modulo $(q-1)$ for $p = 2$ and for $p > 2$, provided that $2i - 1 \not\equiv 0 \bmod p$. Furthermore, each coset of $2i - 1$ has the full size n.*

Proof. Case 1: $p = 2$. For $s = 2i - 1$, $i > 1$, since $d < \sqrt{q} = 2^{n/2}$, the binary representation of s is

$$(1, s_1, \ldots, s_{\lfloor n/2 \rfloor - 1}, \underbrace{0, \ldots, 0}_{m}), m = n - \lfloor n/2 \rfloor, s_i \in \{0, 1\}\,. \tag{5.5}$$

Thus, for any two different vectors in the form of (5.5), one cannot be obtained from the other by the shifting operator. Thus they are not in the same coset modulo $2^n - 1$.

Furthermore, the coset containing s has the full size n, since no binary vector given in (5.5) has a period less than n.
Case 2: $p > 2$. Since $s \neq p^i$, the p-ary representation of s is given as

$$(s_0, \ldots, s_{\lfloor n/2 \rfloor - 1}, \underbrace{0, \ldots, 0}_{m}), s_i \in \mathbb{F}_p, s_0 \neq 0 .$$

A similar argument can be made for this case, thus the assertion is true. □

Let

$$a(t) = \mathrm{Tr}(f(\alpha^t)), f \in T_2 , \quad \text{and} \quad S_2 = \{\{\omega_M^{a(t)}\} : f(x) \in T_2\} . \tag{5.6}$$

Using the Weil bounds in Lemmas 1 and 3, and Proposition 4.9, the following results on binary sequences follow immediately.

Theorem 5.2. *For $p = 2$ and $3 \nmid q - 1$, S_2 defined by* (5.6) *is an ambiguity signal set where the size of S_2 and three metrics are given by*

$$|S_2| = q^{d-2}, C_{\max}, G_{\max} \leq (2d-2)\sqrt{q} + 1 , \quad \text{and} \quad F_{\max} \leq (2d-1)\sqrt{q} .$$

Remark 5.3. For $p > 2$, $f(x)$ could contain monomial terms with even exponents. In other words, we may assume that there exists some $1 < i_0 \leq d$ such that $\gcd(i_0, q - 1) = 1$. Then we set

$$S_2' = \left\{\{a(t)\} : f(x) = \sum_{\substack{i=2 \\ i \neq i_0}}^{d} c_i x^i + x^{i_0} , \quad c_i \in \mathbb{F}_q\right\}$$

where $a(t) = \mathrm{Tr}(f(\alpha^t))$. Then the number of the phase-shift distinct sequences defined by S_2' and their three metrics are given by

$$|S_2'| = q^{d-2-\lfloor \frac{d}{p} \rfloor} , \quad C_{\max}, G_{\max} \leq (d-1)\sqrt{q} + 1 , \quad \text{and} \quad F_{\max} \leq d\sqrt{q} .$$

Note that i_0 is not unique, which is used to prove the phase-shift distinctness of the sequences in the set.

Remark 5.4. Note that the correlation bound obtained directly by applying the Weil bound is not as good as those obtained by a special method (see [25]). The following two cases show how far the bounds of $C_{\max}$ given by Theorem 5.2 are from the bound proved using a special technique.

(i) $f(x) = cx^3 + x, c \in \mathbb{F}_{2^n}$, and S is the set consisting of the sequences defined by all those polynomials. Then S is a Gold pair signal set (see [25]) where $C_{\max} \leq \sqrt{2q}$. However, $C_{\max}$ in Lemma 1 or Theorem 5.2 is bounded by $C_{\max} \leq 2\sqrt{q}$. However, the sequences in S are not phase-shift distinct. Thus, the ambiguity can reach $2^n - 1$.

(ii) $f(x) = c_2x^5 + c_1x^3 + x, c_i \in \mathbb{F}_{2^n}$. Then S is a triple error correction code where $C_{\max} \le \sqrt{8q}$. From Theorem 5.2, it is bounded by $C_{\max} \le 4\sqrt{q} + 1$. Note that not all sequences in S are phase-shift distinct. However, all sequences in S_2 for $d = 3$ are phase-shift distinct. Thus, the ambiguity function of the sequences in S can reach $2^n - 1$ except for $c_1 = 1$. In this case, $S = S_2$ for $d = 3$.

Note that ambiguity of the above Case 1 is not bounded and Case 2 is not bounded if $c_1 \neq 1$.

5.2.2 $\mathbb{F}_q$ Multiplicative Sequences with Low Odd Degree Polynomials

Next we look at the multiplicative case. Let

$$u(t) = c \log_\alpha f(\alpha^t) \bmod M\,, \quad f \in T_0\,, \quad \text{and}$$

$$S_0 = \{\{\omega_M^{u(t)}\} \mid \forall f \in T_0\,,\ 1 \le c < M\}\,.$$

We need to show that the associated polynomials are not M-th power multiples in $\mathbb{F}_q[x]$ where $M \mid (q-1)$. In the following, we only present the proof for the case of $M = 2$ and $d = 3$, i.e. the binary case. The proof for this case and the results on general M and d are reported in [52]. Note that the size of S_0 is equal to the number of phase-shift distinct sequences. In this case, as long as not all $d - 1$ coefficients in $f(x) \in T_0$, defined in (5.2), are equal zero, their corresponding sequences are phase-shift distinct. Thus $|S_0| = (q-1)q^{d-2}$. The following results are due to [52].

Theorem 5.5. *For any p, assume that $M = 2$ and $d < \sqrt{q}/2$. Then S_0 is an ambiguity signal set where the size and three metrics are given by*

$$\begin{aligned}
|S_0| &= (q-1)q^{d-2}\\
C_{\max} &\le (4d-4)\sqrt{q} + 4d\\
F_{\max} &\le (2d-1)\sqrt{q} + 2d - 1\\
G_{\max} &\le (4d-2)\sqrt{q} + 4d - 2\,.
\end{aligned}$$

Proof. According to Proposition 4.11 or Table 4.1, for $C_{\max}$, we only need to show *Claim 1:* $g(x) = f_1(x)^c f_2(\alpha^\tau x)^{M-d} = f_1(x)f_2(\alpha^\tau x), f_i \in T_0$ is not a square multiple in $\mathbb{F}_q[x]$.

For $G_{\max}$, from Proposition 4.9 or Table 4.1, we only need to prove *Claim 2:* $g(x)x^{M-w} = g(x)x^w$ is not a square multiple in $\mathbb{F}_q[x]$ where $g(x)$ is defined in Claim 1.

However, Claim 1 implies Claim 2. Thus, we only need to show that Claim 1 is true. We now show the case of $d = 3$. We assume that both f_1 and f_2 have degree 5. The other cases can be proved in a similar, but much simpler, way. Considering the coefficient of x and constant term, $f_2(\alpha^\tau x) = c \cdot f_1(x)$ will induce $\tau = 0$ and $f_1 = f_2$. Thus $f_1(x)$ and $f_2(\alpha^\tau x)$ have at least one different root in $\overline{F}_q$. If $f_1(x)f_2(\alpha^\tau x)$

is a square multiple, then we have the following three cases where the computations are in $\overline{\mathbb{F}}_q$. Note that $M \mid, q-1$ and now $M = 2$, so $p \neq 2$ in the following derivations.
Case 1: $f_1(x)$ and $f_2(\alpha^\tau x)$ share five roots. This case leads to $f_2(\alpha^\tau x) = c \cdot f_1(x)$, which is a contradiction.
Case 2: $f_1(x)$ and $f_2(\alpha^\tau x)$ share three roots, say, a, b and c. In this case, each of these two polynomials will have the form $h(x) = e(x-a)(x-b)(x-c)(x-d)^2$ since $f_1(x)f_2(\alpha^\tau x)$ is a square multiple. We now consider the coefficient of x^4 of $h(x)$, which is zero. Then we get $d = -2^{-1}(a+b+c)$, which implies that $f_1(x)$ and $f_2(\alpha^\tau x)$ share five roots. This is a contradiction.
Case 3: $f_1(x)$ and $f_2(\alpha^\tau x)$ share one root a. In this case, each of these two polynomials will have the form $t(x) = e(x-a)(x-b)^2(x-c)^2$. Since the coefficients of x^4 and x^2 of $t(x)$ are zero, we get

$$\begin{aligned} a + 2b + 2c &= 0\,, \\ 2(b^2c + c^2b) + 4abc + b^2a + c^2a &= 0\,. \end{aligned} \quad \Longrightarrow \quad \begin{aligned} b + c &= -2^{-1}a\,, \\ bc &= -(2^{-1}a)^2\,. \end{aligned}$$

Thus b and c are the roots of equation $x^2 + 2^{-1}ax - (2^{-1}a)^2 = 0$ over $\overline{F}_q$, which implies that $f_1(x)$ and $f_2(\alpha^\tau x)$ share five roots. A contradiction occurs.

Every case induces a contradiction here, so $f_1(x)f_2(\alpha^\tau x)$ is not a square multiple in $\mathbb{F}_q[x]$. Hence, the assertions are true. □

The results in Theorems 5.2 and 5.5 are true for $d < \sqrt{q}/2$. When $d < \log\log_2 q$ and $d < \log_2 n$ when $p = 2$, they satisfy the bounds given in (3.5)–(3.7).

5.3 Sequences from Power Residue and Sidel'nikov Sequences

5.3.1 Interleaved Structure of Sidel'nikov Sequences

We first present a result on Sidel'nikov sequences. Yu and Gong [60] studied the interleaved structure of Sidel'nikov sequences (for interleaved sequences, see [16]). They consider the case of M-ary Sidel'nikov sequences of period $q^2 - 1$ for $M \mid (q-1)$. By investigating the $(q-1) \times (q+1)$ array structure of the Sidel'nikov sequences, they proved that half of the column sequences correspond to the polynomials

$$f_j(x) = (\alpha^j x - 1)(\alpha^{qj}x - 1) = \alpha^{(q+1)j}x^2 - \mathrm{Tr}(\alpha^j) \cdot x + 1 \tag{5.7}$$

where $\mathrm{Tr}(x) = x + x^q$, $1 \le j \le \frac{q}{2}$ and $f_{q+1-j} = f_j$. Then f_j is irreducible over $\mathbb{F}_q$.

Example 5.6. Let $q = p = 7$, $M = 6$, $q^2 = 7^2$, and a finite field $\mathbb{F}_{7^2}$ be defined by a primitive polynomial $t(x) = x^2 + x + 3$. Then a 6-ary Sidel'nikov sequence $u(t)$ of period $q^2 - 1 = 48$ can be presented by a $(q-1) \times (q+1) = 6 \times 8$ array as follows:

$$(v_0, v_1, \ldots, v_7) = \begin{bmatrix} 4 & 1 & 5 & 0 & 5 & 1 & 5 & 1 \\ 2 & 4 & 4 & 2 & 2 & 2 & 5 & 4 \\ 2 & 4 & 3 & 3 & 1 & 0 & 4 & 4 \\ 0 & 5 & 0 & 3 & 5 & 2 & 3 & 5 \\ 4 & 1 & 3 & 1 & 2 & 3 & 0 & 1 \\ 0 & 0 & 5 & 2 & 1 & 3 & 3 & 0 \end{bmatrix}$$

where $v_j(t) = \log_\alpha f_j(t) \bmod M, 0 \le t \le q-1$ and $f_{8-j}(x) = f_j(x), j = 1, 2, 3$.

5.3.2 Sequences from Linear and/or Quadratic/Inverse Polynomials

Similar to (5.7), we formally define quadratic polynomials over $\mathbb{F}_p$ as follows, although these do not correspond to any interleaved structures of power residue sequences.

$$g_j(x) = (x - j\alpha)(x - j\alpha^p) = x^2 - j \cdot \mathrm{Tr}(\alpha) + j^2\alpha^{p+1}, \quad 1 \le j \le \frac{p-1}{2} \tag{5.8}$$

where α is a primitive element of $\mathbb{F}_p$. Then $g_j(x)$ is irreducible and $\deg(g_j(x)) = 2$. We now assume that $\mathbf{u}$ is defined by

$$u(t) = \begin{cases} \log_\alpha t \bmod M, M \mid (p-1) & \text{for } \mathbb{F}_p \\ \log_\alpha(\alpha^t + 1) \bmod M, M \mid (q-1) & \text{for } \mathbb{F}_q \end{cases} \tag{5.9}$$

and

$$v_j(t) = \begin{cases} \log_\alpha g_j(t) \bmod M, M \mid (p-1) & \text{for } \mathbb{F}_p \\ \log_\alpha f_j(\alpha^t) \bmod M, M \mid (q-1) & \text{for } \mathbb{F}_q \end{cases} \tag{5.10}$$

where the quadratic polynomials f_j and g_j are defined by (5.7) and (5.8), respectively. We can now write a unified set for both $\mathbb{F}_p$ and $\mathbb{F}_q$. For $r \in \{p, q\}$, we define the signal sets in Table 5.1, where $\delta_r = p$ when $r = p$ and $\delta_r = q - 1$ when $r = q$.

Note that the sequence $\{c_0 u(t) + c_1 u(t+\tau)\}$ in $\mathcal{A}_{2,r}$ corresponds to polynomial $f(x) = x^{c_0}(x+\tau)^{c_1}$ and the sequences in $\mathcal{Z}_{2,r}$ correspond to polynomials $(x+1)^{c_0}(\beta x^{-1} + 1)^{c_1}$. Thus their respective correlation functions correspond to a polynomial with four distinct roots and a polynomial with five distinct roots. Note that $\mathcal{A}_{1,p}$ is only a correlation signal set, but not an ambiguity signal set (see Theorem 4.8 in Section 4). From Theorems 4.8 and 4.12 in Section 4, we have the results listed in Table 5.2.

Together with Corollary 2.8 in Section 2, for the signal sets in Table 5.1, one only needs to show that each of the corresponding polynomials for correlation, DFT,

Table 5.1: Signal sets from power residue and Sidel'nikov sequences

Scalar	$\mathcal{A}_{1,r} = \{c\mathbf{u} \mid 1 \le c < M\}$, $\mathcal{Z}_{1,r} = \{c\{u(-t)\} \mid 1 \le c < M\}$
Shift-and-Add	$\mathcal{A}_{2,r} = \{\{c_0u(t) + c_1u(t+\tau)\} \mid 1 \le \tau \le \lfloor \frac{\delta_r}{2} \rfloor, c_0, c_1 \neq 0\}$ $c_0 < c_1$ if $\tau = \lfloor \frac{\delta_r}{2} \rfloor$ then $c_0 = 1$ for $r = p$, which are applied for all the cases below $\mathcal{A}_{2,0,r} = \{\{c_0u(t) + c_1u(t+\tau)\} \in \mathcal{A}_{2,r} \mid c_0 + c_1 \equiv 0\}$ $c_0 < c_1$ if $\tau = \lfloor \frac{\delta_r}{2} \rfloor$
Shift-and-Add Inverse	$\mathcal{Z}_{2,r} = \{\{c_0u(t) + c_1u(-t+\tau)\} \mid 1 \le \tau \le \lfloor \frac{\delta_r}{2} \rfloor,\ c_0, c_1 \neq 0\}$ $c_0 \neq c_1$ if $\tau = \lfloor \frac{\delta_r}{2} \rfloor$
Quadratic and Scalar	$\mathcal{B}_{2,r} = \{\{cv_j(t)\} \mid 1 \le c \le M-1, 1 \le j \le \frac{p-1}{2}$ or $\frac{q}{2}\}$, and
Quadratic	$\mathcal{B}_{2,0,r} = \{\{\frac{M}{2}v_j(t)\} \mid 1 \le j \le \frac{p-1}{2}$ or $\frac{q}{2}\}$ for M even

Table 5.2: Three metrics for $A_{1,r}$

$\lvert\mathcal{A}_{1,r}\rvert = M-1$	$C_{\max} \le \sqrt{r}+2+h_r$, where $h_r = \begin{cases} 0 & \text{for } r = p \\ 1 & \text{for } r = q \end{cases}$
$F_{\max} \le \sqrt{r}+1$ and $G_{\max} \le 2\sqrt{q}+2$	

and ambiguity is not an M-th power multiple in $\mathbb{F}_r[x], r \in \{p, q\}$. This can be easily done. Thus they are ambiguity signal sets where the three metrics are listed in Table 5.3. These ambiguity signal sets have been studied by a number of researchers. In particular, it is shown that $\mathcal{A}_{2,r}$ and $C_{\max}$ in [30] for $r = q$ and in [21] for $r = p$;

Table 5.3: Ambiguity signal sets with quadratic/inverse polynomials and their three metrics

Sets	$C_{\max}$	$F_{\max}$	$G_{\max}$	Sizes
$\mathcal{A}_{2,p}(P)$	$3\sqrt{p}+4$	$2\sqrt{p}+2$	$3\sqrt{p}+4$	$(M-1)\frac{p-1}{2}$
$\mathcal{A}_{1,r} \cup \mathcal{A}_{2,r}(S)$	$3\sqrt{q}+5$	$2\sqrt{q}+2$	$4\sqrt{q}+4$	$(M-1)+(M-1)^2\frac{q-2}{2}$
$\mathcal{A}_{1,r} \cup \mathcal{A}_{2,r} \cup$ $\mathcal{Z}_{1,r}(S) \cup \mathcal{Z}_{2,r}(S)$	$4\sqrt{q}+4$	$2\sqrt{q}+2$	$4\sqrt{q}+4$	$2(M-1)+$ $2(M-1)^2\frac{q-2}{2}$
$\mathcal{A}_{2,p} \cup \mathcal{B}_{2,p}(P)$	$3\sqrt{p}+4$	$2\sqrt{p}+2$	$4\sqrt{p}+2$	$(M-1)(p-1)$
$\mathcal{A}_{1,r} \cup \mathcal{A}_{2,r} \cup \mathcal{B}_{2,r}(S)$	$3\sqrt{q}+5$	$2\sqrt{q}+2$	$4\sqrt{q}+4$	$(M-1)\frac{q+2}{2}+$ $(M-1)^2\frac{q-2}{2}$
$\mathcal{A}_{2,0,p} \cup \mathcal{B}_{2,0,p}(P)$	$2\sqrt{p}+5$	$2\sqrt{p}+2$	$4\sqrt{p}+2$	$p-1$
$\mathcal{A}_{2,0,r} \cup \mathcal{B}_{2,0,r}(S)$	$2\sqrt{q}+6$	$2\sqrt{q}+2$	$4\sqrt{q}+4$	$M\frac{q-2}{2}+1$

$\mathcal{A}_{2,r} \cup \mathcal{Z}_{2,r}$ and $C_{\max}$ in [5] for $r = q$; $\mathcal{A}_{2,r} \cup \mathcal{B}_{2,r}$ and $C_{\max}$ in [60] for $r = q$, and $\mathcal{A}_{2,c,r}$ and $C_{\max}$ in [59] for both $r = p$ and $r = q$. The rest of the resultsare due to [53].

In Table 5.3, we use the notation P and S to indicate the sequences constructed from the power residue sequences and Sidel'nikov sequences, respectively, because the Sidel'nikov sequences can have both cases for $\mathbb{F}_q$ for $q = p^n, n > 1$ and $q = p$. The sizes of the signal sets constructed from Sidel'nikov sequences are only listed for the case $q = 2^n$ for simplicity. The sizes will be slightly different for $q = p^n$ where p is odd (see [53]). Note that the values given in Table 5.3 for $C_{\max}$, $F_{\max}$, and $G_{\max}$ are precise values and not bounds.

5.4 Sequences from Hybrid Characters

5.4.1 Sequences Using Weil Representation and Their Generalizations

By using the Weil representation, a signal set was constructed by Gurevich, Hadani, and Sochen [20], and three metrics of this signal set were proved by algebraic geometry. This is the first work to consider all the three metrics together for a signal set, which leads to the following new discoveries. A simple elementary construction for these sequences was found by Wang and Gong [51]. Let $g(x) = x$, $f(x) = bx^2 + x, b \in \mathbb{F}_p$, for $1 \le c < p - 1$, define

$$s_{c,b}(t) = \chi_c(g(t))\psi(f(t)) = \chi_c(t)\psi(bt^2 + t)\,.$$

Let

$$\Omega_{2,p} = \{\mathbf{s}_{c,b} \mid 1 \le c < M, b \in \mathbb{F}_p\}\,.$$

Wang and Gong [51] showed that $\Omega_{2,p}$ is the sequences constructed using the Weil representation by Gurevich, Hadina, and Sochen [20]. Shortly after that, Schmidt [47] gave a direct proof of three metrics of those sequences and the following generalized construction. Let $g(x) = x$, $f(x) = \sum_{j=2}^{d} b_j x^j + x$, where $d < p$, $1 \le c \le p - 2$ and $\mathbf{b} = (b_d, \ldots, b_2, 1)$ with $b_j \in \mathbb{F}_p$ and $b_1 = 1$, define

$$s_{c,\mathbf{b}}(t) = \chi_c(g(t))\psi(f(t)) = \chi_c(t)\psi\Big(\sum_{j=1}^{d} b_j t^j\Big)\,, \quad \mathbf{s}_{c,\mathbf{b}} = \{s_{c,\mathbf{b}}(t)\}$$

where χ is a multiplicative character of $\mathbb{F}_p$ of order $p - 1$. We may further extend the above sequences to a multiplicative character with order $M \mid (p - 1)$ as follows:

$$s_{c,\mathbf{b}}(t) = \omega_M^{c \log_\alpha t}\psi(f(t))\,. \tag{5.11}$$

Let

$$\Omega_{d,p} = \{s_{c,\mathbf{b}} | 1 \le c < M, b_i \in \mathbb{F}_p\}\,.$$

The phase-shift operation for hybrid sequences, defined by (5.11), is through additive characters in $\mathbb{F}_p$. By directly applying the Weil bound in Lemma 3, we obtain that $\Omega_{d,p}$ is an ambiguity signal set with the following parameters.

Table 5.4: Three metrics of hybrid sequences over $\mathbb{F}_p$

$C_{\max}$	$F_{\max}$	$G_{\max}$	$\lvert\Omega_{d,p}\rvert$
$3\sqrt{p}+2$	$2\sqrt{p}+1$	$3\sqrt{p}+2$	$(M-1)p,\ d=2$
$(d+1)\sqrt{p}+2$	$d\sqrt{p}+1$	$(d+1)\sqrt{p}+2$	$(M-1)p^{d-1},\ d>2$

Proposition 5.7. *With the above notation, if the phase-shift operation for hybrid sequences is defined by additive characters in $\mathbb{F}_p$, then we have the results listed in Table* 5.4.

Remark 5.8. For $d = 2$ and $M = p - 1$, the resulting correlation in (5.4), which appeared in [47], actually improved the bounds originally proved by using the Weil representation in [20, 51].

5.4.2 Generalization to $\mathbb{F}_q$ Hybrid Sequences

A generalization of the construction of $\Omega_{d,p}$ to $\mathbb{F}_q$ is given in [53] and as a special case, the product of a binary m-sequence and a ternary Sidel'nikov sequence is investigated in [27]. Let $M \mid (q-1)$, $g(x) = x + 1$, $u(t)$ be an M-ary Sidel'nikov sequence and $f(x) \in \mathbb{F}_q[x]$ be defined as

$$f(x) = bx^2 + x\,, \quad b \in \mathbb{F}_q \text{ where } p > 2 \text{ or} \tag{5.12}$$

$$f(x) = \sum_{i=2}^{d} b_i x^{2i-1} + x \in T_1\,, \quad \text{where } T_1 \text{ is defined in (5.3).} \tag{5.13}$$

Let

$$s_{c,\mathbf{b}}(t) = \chi(\alpha^t + 1)\psi_1(f(\alpha^t)) = \omega_M^{c\log_\alpha(\alpha^t+1)}\omega_p^{\mathrm{Tr}(f(\alpha^t))}$$

where χ is a multiplicative character of order $M \mid q-1$. We write $\mathbf{b} = (b_1, \ldots, b_d)$, where $b_1 = 1$. Let

$$\Omega_{2,q} = \{\mathbf{s}_{c,b} \mid 1 \le c < M, b \in \mathbb{F}_q\}\,, \quad f(x) \text{ in (5.12)}\,.$$

$$\Omega_{d,q,o} = \{\mathbf{s}_{c,\mathbf{b}} \mid 1 \le c < M, b_j \in \mathbb{F}_q\}\,, \quad f(x) \text{ in (5.13)}\,.$$

Note that the coset containing 2 modulo $q-1$ has the full size n.

Proposition 5.9. *With the above notation and the phase-shift operator is defined through additive characters of $\mathbb{F}_q$. Then $\Omega_{d,q}$ is an ambiguity signal set where the size and three metrics are given in Table* 5.5.

Example 5.10. The signal set $\Omega_{2,q}$ is an analog of the signal set from the Weil representation where $g(x) = x + 1$ and $f(x) = bx^2 + x$, $b \in \mathbb{F}_q$. For $\Omega_{2,3^2}$ ($q = 3^2$), let $\mathbb{F}_9$ be defined by a primitive polynomial $t(x) = x^2 + 2x + 2$ over $\mathbb{F}_3$ and $t(\alpha) = 0$,

Table 5.5: Three metrics of hybrid sequences over $\mathbb{F}_q$

Sets	$C_{\max}$	$F_{\max}$	$G_{\max}$	**Sizes**
$\Omega_{2,q}$	$3\sqrt{q}+3$	$2\sqrt{q}+1$	$3\sqrt{q}+3$	$(M-1)q$
$\Omega_{d,q,o}$	$2d\sqrt{q}+3$	$(2d-1)\sqrt{q}+1$	$2d\sqrt{q}+3$	$(M-1)q^{d-1}$

then α is a primitive element of $\mathbb{F}_9$. In this case, $\mathrm{Tr}(x)$ defines the m-sequence $\{0,1,1,2,0,2,2,1\}$ and $\mathrm{Tr}(bx^2)$ gives the shifts of 2-decimation of the m-sequence for different $b_2 \in \mathbb{F}_{3^2}$, i.e. $\{0,1,0,2,0,1,0,2\}$ and $\{1,2,2,1,1,2,2,1\}$.

A sequence in $\Omega_{2,9}$ is a term-by-term product of three sequences from each of the columns in Table 5.6 and the m-sequence. Note that this example is used only to illustrate the construction of the sequences through hybrid characters, since the bounds of the three metrics in this case are not meaningful.

Remark 5.11. Note that the phase-shift operation for the case for $\mathbb{F}_q$ is different from the case of $\mathbb{F}_p$. In Proposition 5.9, the phase shift of the $\mathbb{F}_q$ hybrid sequences is defined through additive characters of $\mathbb{F}_q$. However, it can also be defined through multiplicative characters of $\mathbb{F}_q$, since the field elements are indexed by the order of multiplicative group of $\mathbb{F}_q$. If so, the bound on ambiguity functions in Proposition 5.9 will be changed to $4\sqrt{q}+2$ and $(2d+1)\sqrt{q}+2$, respectively.

The known ambiguity signal sets with the sizes in the order of q^2 are $\mathcal{A}_{2,0,q} \cup \mathcal{B}_{2,0,q}$, $\Omega_{2,p}$ and $\Omega_{2,q}$ given by Tables 5.3, (5.4), and (5.5), which are collectively grouped in Table 5.7. We denote by P the number of phases in a sequence in the table.

Table 5.6: Component sequences for hybrid sequences from $\mathbb{F}_9$ for $M = 8$

$u(t) = \omega_8^{c\log_\alpha(\alpha^t+1)}$	$\psi(b_2t^2) = \omega_3^{\mathrm{Tr}(b_2t^2)}$
$(-1,\omega_8^2,\omega_8^7,\omega_8^6,1,\omega_8^3,\omega_8^5,\omega_8)$	$(1,\omega_3,1,\omega_3^2,1,\omega_3,1,\omega_3^2)$
$(1,-1,\omega_8^6,-1,1,\omega_8^6,\omega_8^2,\omega_8^2)$	$(\omega_3,1,\omega_3^2,1,\omega_3,1,\omega_3^2,1)$
$(-1,\omega_8^6,\omega_8^5,\omega_8^2,1,\omega_8,\omega_8^7,\omega_8^3)$	$(1,\omega_3^2,1,\omega_3,1,\omega_3^2,1,\omega_3)$
$(1,1,-1,1,1,-1,-1,-1)$	$(\omega_3^2,1,\omega_3,1,\omega_3^2,1,\omega_3,1)$
$(-1,\omega_8^2,\omega_8^3,\omega_8^6,1,\omega_8^7,\omega_8,\omega_8^5)$	$(\omega_3,\omega_3^2,\omega_3^2,\omega_3,\omega_3,\omega_3^2,\omega_3^2,\omega_3)$
$(1,-1,\omega_8^2,-1,1,\omega_8^2,\omega_8^6,\omega_8^6)$	$(\omega_3^2,\omega_3^2,\omega_3,\omega_3,\omega_3^2,\omega_3^2,\omega_3,\omega_3)$
$(-1,\omega_8^6,\omega_8,\omega_8^2,1,\omega_8^5,\omega_8^3,\omega_8^7)$	$(\omega_3^2,\omega_3,\omega_3,\omega_3^2,\omega_3^2,\omega_3,\omega_3,\omega_3^2)$
	$(\omega_3,\omega_3,\omega_3^2,\omega_3^2,\omega_3,\omega_3,\omega_3^2,\omega_3^2)$
	$(1,1,1,1,1,1,1,1)$

Table 5.7: Known ambiguity signal sets with the sizes in the order of q^2

Sets	$(C_{\max}, F_{\max}, G_{\max})$	Sizes	$(P, \min P)$
$\mathcal{A}_{2,0,q} \cup \mathcal{B}_{2,0,q}$	$(2\sqrt{q}+6, 2\sqrt{q}+2, 4\sqrt{q}+4)$	$M\frac{q-2}{2}+1$	$(M, 2)$
$\Omega_{2,p}$	$(3\sqrt{p}+2, 2\sqrt{p}+1, 3\sqrt{p}+2)$	$(M-1)p$	$(Mp, 2p)$
$\Omega_{2,q}$	$(3\sqrt{q}+3, 2\sqrt{q}+1, 3\sqrt{q}+3)$	$(M-1)q$	$(Mp, 6)$ when $p=2$

Thus $\mathbb{F}_q$ multiplicative sequences with quadratic polynomials have the best parameters in terms of correlation and DFT, and the other two are superior in terms of ambiguity.

5.5 A New Construction

As we introduced in Section 4.5, there are alternative four classes of sequences in terms of the different methods for indexing the elements of $\mathbb{F}_p$ and $\mathbb{F}_q$. Thus a sequence defined by hybrid characters could have a number of different combinations.

Construction 5.12. *Let*

$$s(t) = \eta(x(t))\,\sigma(y(t)),$$
$$\text{where } \eta, \sigma \in \{\psi_i, \chi_j \mid 0 < i < p, 0 < j < q-1\}, \text{ and } x(t), y(t) \in \mathcal{P},$$
$$\text{where } \mathcal{P} = \{a(t) \text{ and } u(t), \text{ defined in Table 1, and } b(t) \text{ and } v(t), \text{ defined in Table 4.2}\}.$$

The ambiguity signal sets presented in Section 5.4 for both $\mathbb{F}_p$ and $\mathbb{F}_q$ use $x(t) = a(t)$, $y(t) = u(t)$, $\eta = \psi$ and $\sigma = \chi$. Thus, it is interesting to see whether there exist some combinations which yield ambiguity signal sets with low three metrics. An example of this new construction is the sequences considered in [49], which are obtained by flipping a few bits of the following sequence

$$s(t) = \chi_i(u(t))\chi_j(v(t)) = (-1)^{s_1(t)}(-1)^{s_2(t)} \quad \text{where}$$
$$s_1(t) = \log_\alpha t \mod 2$$
$$s_2(t) = \log_\alpha^{(\alpha^t+1)} \mod 2, \; t = 0, 1, \ldots$$

where $u(t), v(t) \in \mathcal{P}$ in Construction 5.12, and both χ_i and χ_j are multiplicative quadratic characters of $\mathbb{F}_p$. Note that $\{s_1(t)\}$ has period p and $\{s_2(t)\}$ has period $p-1$ since it uses the multiplicative group order of $\mathbb{F}_p$. Thus $\{s(t)\}$ has period $(p-1)p$.

6 Two-Level Autocorrelation Sequences and Double Exponential Sums

In this section, we look at the sequences with ideal 2-level autocorrelation. Up to now in all the known multiplicative cases, there are only polynomials with degree 1, which produce sequences with optimal autocorrelation, i.e. power residue sequences for $\mathbb{F}_p$ multiplicative and Sidel'nikov sequences for $\mathbb{F}_q$ multiplicative. For $\mathbb{Z}_N$ additive, FZC sequences are perfect sequences with defining polynomials of degree 2. However, for $\mathbb{F}_q$ additive, there are several classes of sequences with 2-level autocorrelation with high degree polynomials. The work on binary 2-level autocorrelation sequences, i.e. $q = 2^n$, has been collected in [14] and no new sequences have come out since then. For the ternary case, i.e. $q = 3^n$, there are several conjectured sequences, whose validity has been established recently by Arasu, Dillion, and Player [3], but the proofs have not yet appeared. For $p > 3$, there are only two known classes of primary constructions (we will define this concept below), one is the class of m-sequences and the other is the Helleseth–Gong (HG) class.

In this section, we first introduce the concepts on prime 2-level autocorrelation sequences and Hadamard equivalence, then we show the conjectured ternary sequences and their alternative exponential sums in terms of the 2nd order decimation-Hadamard transform. We use the trace representation of $\mathbb{F}_q$ additive sequences in this section.

6.1 Prime Two-Level Autocorrelation Sequences

Let $f(x)$ be the trace representation of a p-ary sequence $\mathbf{a} = \{a(t)\}$, i.e. $a(t) = f(\alpha^t), t = 0, 1, \ldots$, with 2-level autocorrelation, i.e. $f(x)$ is a function from $\mathbb{F}_q$ to $\mathbb{F}_p$. Let m be a proper factor of n and $h(x)$ be a function from $\mathbb{F}_q$ to $\mathbb{F}_{p^m}$. We say that $h(x)$ is $\mathbb{F}_{p^m}$ *linear* if for $y \in \mathbb{F}_{p^m}$, and $x \in \mathbb{F}_q$, we have $h(xy) = y^d h(x)$ for some d with $1 \le d < p^m - 1$. If we can write

$$f(x) = g(x) \circ h(x)$$

where $g(x)$ is a polynomial nonlinear function from $\mathbb{F}_{p^m}$ to $\mathbb{F}_p$ (i.e. $g(x) \ne \mathrm{Tr}_1^m(ax)$, $\forall a \in \mathbb{F}_p$) such that the sequence defined by $g(x)$ has 2-level autocorrelation, and $h(x)$ is an $\mathbb{F}_{p^m}$ linear function from $\mathbb{F}_q$ to $\mathbb{F}_{p^m}$, then we say that $\mathbf{a}$ is a *composited 2-level autocorrelation sequence*. Otherwise, it is said to be a *prime 2-level autocorrelation sequence*.

Currently, there are only two known classes of F_{p^m} linear functions; one is given by $\mathrm{Tr}_m^n(x^d)$ with $1 \le d < q-1$ and $\gcd(d, q-1) = 1$, and the other is HG functions, which will be defined shortly. For example, GMW sequences or generalized GMW sequences are the composited 2-level autocorrelation sequences (see [14]). So, we only need to classify all prime 2-level autocorrelation sequences.

6.2 Hadamard Transform, Second-Order Decimation-Hadamard Transform, and Hadamard Equivalence

The Hadamard transforms of $f(x)$ and **a** are defined by

$$\widehat{f}(\lambda) = \sum_{x\in\mathbb{F}_q} \psi(\mathrm{Tr}(\lambda x))\psi(f(x))^* = \sum_{x\in\mathbb{F}_q} \omega_p^{\mathrm{Tr}(\lambda x)-f(x)}, \quad \lambda \in \mathbb{F}_q$$
$$\widehat{\mathbf{a}}(\tau) = \widehat{f}(\alpha^\tau) - 1, \quad \tau = 0, 1, \ldots \tag{6.1}$$

Note that the Hadamard transform of **a** is equal to the crosscorrelation of **a** and an m-sequence defined by $\mathrm{Tr}(x)$. From (6.1), the Hadamard transforms of $f(x)$ and **a** are determined by each other. From now on, we focus on the Hadamard transform of functions. The inverse formula for $f(x)$ is given by

$$\psi(f(\lambda)) = \frac{1}{q}\sum_{x\in\mathbb{F}_q} \psi(\mathrm{Tr}(\lambda x))\widehat{f}(x)^*, \quad \lambda \in \mathbb{F}_q .$$

For $(v, t) \in \mathbb{Z}_{q-1}^2$ and $\lambda \in \mathbb{F}_q$, the first-order and second-order decimation-Hadamard transforms (DHT) are defined as follows:

$$\text{1st order DFT:} \quad \widehat{f}(v)(\lambda) = \sum_{x\in\mathbb{F}_q} \psi(\mathrm{Tr}(\lambda x))\psi(f(x^v))^* = \sum_{x\in\mathbb{F}_q} \omega_p^{\mathrm{Tr}(\lambda x)-f(x^v)} .$$

$$\text{2nd order DFT:} \quad \widehat{f}(v,t)(\lambda) = \sum_{y\in\mathbb{F}_q} \psi(\mathrm{Tr}(\lambda y))\widehat{f}(v)(y^t)^* = \sum_{x,y\in\mathbb{F}_q} \omega_p^{\mathrm{Tr}(\lambda y)-\mathrm{Tr}(y^t x)+f(x^v)} .$$

In general, for any integer pair (v, t), $x \in \mathbb{F}_q$, $\widehat{f}(v,t)(x)$ may be just a complex number. However, if it satisfies the following condition:

$$\widehat{f}(v,t)(x) \in \{q\omega_p^i \mid i = 0, \ldots, p-1\}, \quad \forall x \in \mathbb{F}_q ,$$

then we can construct a function, say $g(x)$, from $\mathbb{F}_q$ to $\mathbb{F}_p$, whose elements are given by

$$\psi(g(x)) = \frac{1}{q}\widehat{f}(v,t)(x), \quad x \in \mathbb{F}_q . \tag{6.2}$$

In this case, we say that (v, t) is *realizable*, and $g(x)$ is a *realization* of $f(x)$ or **a**. If $g(x)$ is realized by $f(x)$ with (v, t), from (6.2), we have

$$\widehat{g}(\lambda) = \widehat{f}(v)(\lambda^t), \quad \lambda \in \mathbb{F}_q . \tag{6.3}$$

Definition 6.1. If f and g satisfy (6.3), then we say that f and g are *Hadamard equivalent*, written as $f \sim_H g$.

In order to determine the autocorrelation of the sequence defined by $g(x)$, we need the Parseval formula on the Hadadmard transform.

Property 6.2 (Parseval Formula).

$$\sum_{x \in \mathbb{F}_q} \psi(f(\lambda x))\psi(f(x))^* = \sum_{x \in \mathbb{F}_1} \hat{f}(\lambda x)\hat{f}(x)^*, \lambda \in \mathbb{F}_q .$$

From the Parseval formula, the autocorrelation of **a**, defined by $f(x)$, is equal to the autocorrelation of its Hadamard transform sequence. If one of them has 2-level autocorrelation, so does the other. Formally, we have

Property 6.3. *Let* **b** *be a sequence defined by* $g(x)$, i.e. $b(t) = g(\alpha^t)$, *where* $g(x)$ *is a realization of* **a**. *Then* **a** *has 2-level autocorrelation if and only if* **b** *has 2-level autocorrelation.*

Dillon and Dobbertin [8] used the Parseval formula to prove binary 2-level sequences. The concepts of the second order DHT were introduced in [17] by Gong and Golomb for the case that both v and t are coprime with $q - 1$ and these are extended to any integers in [58].

6.3 Conjectures on Ternary 2-Level Autocorrelation Sequences

Let $p = 3$, $n = 2m + 1$ and $d = 2 \cdot 3^m + 1$. Lin's Conjecture [35] is stated as follows.

Conjecture 6.4 (Lin [35]). *Let* $f(x) = \mathrm{Tr}(x + x^d)$ *and* $a(t) = f(\alpha^t)$. *Then* $\mathbf{a} = \{a(t)\}$ *has 2-level autocorrelation. In other words,*

$$\sum_{x \in \mathbb{F}_{3^n}} \omega_3^{\mathrm{Tr}(x+x^d) - \mathrm{Tr}(\alpha^\tau x + \alpha^{\tau d} x^d)} = 0\,, \quad \textit{for all} \quad \tau = 1, \ldots, q-2\,.$$

Conjecture 6.5 (Gong et al. [18]). *Let* $f(x) = \mathrm{Tr}(x)$ *and* (v, t) *be defined as follows:*

$$v = \begin{cases} 2(3^{m+1} - 1) & \textit{for } m \textit{ even} \\ -2(3^{m+1} - 3) & \textit{for } m \textit{ odd} \end{cases} \qquad \textit{and } t = \frac{3^n + 1}{4}\,.$$

Then (v, t) *is a realizable pair and* $\hat{f}(v, t)(\lambda)$ *realizes the conjectured two-term sequences in Conjecture* 6.4.

Note that in this case, $\gcd(v, q - 1) \neq 1$. So the realized sequence should be computed through the multiplexing method as presented in [58].

The following two conjectures involve the HG sequences.

Theorem 6.6 (Helleseth et al. [24]). *Let $n = (2m+1)k$ and s be an integer with $1 \le s \le 2m$ and $\gcd(s, 2m+1) = 1$. Define $b_0 = 1$, $b_{is} = (-1)^i$ and $b_i = b_{2m+1-i}$ for $i = 1, 2, \ldots, m$, where all indices of b_i are taken modulo $2m+1$. Let $u_0 = b_0/2 = (p+1)/2$ and $u_i = b_{2i}$ for $i = 1, 2, \ldots, m$. Define*

$$e(x) = \sum_{i=0}^{m} u_{m-i} x^{(q_1^{2i}+1)/2}, \quad q_1 = p^k .$$

Then the sequence $\{s(t)\}$ over $\mathbb{F}_p$ whose elements are defined by $s(t) = \mathrm{Tr}(e(\alpha^t))$ has an ideal 2-level autocorrelation for any p.

This is referred to as *HG sequences*, which was discovered by Helleseth and Gong [24]. Note in [24] there are two classes of 2-level autocorrelation sequences, however, they are decimation equivalent. When $p = 3$, $k = 1$, and $s = 2$, $e(x)$ becomes

$$e(x) = \mathrm{Tr}\left(\sum_{i=0}^{m} u_{m-i} x^{(3^{2i}+1)/2}\right).$$

Let

$$\delta = \begin{cases} 1 & m \text{ odd} \\ 2 & m \text{ even} \end{cases} \quad \text{and} \quad \epsilon = \begin{cases} 1 & m \text{ odd} \\ 0 & m \text{ even.} \end{cases} \tag{6.4}$$

The following conjecture illustrates that Lin's conjectured 2-term sequences are the realizations of HG sequences.

Conjecture 6.7 (Ludkovski and Gong [36]). *Let $u_0 = \frac{3^n-1}{2}$, $u_1 = \frac{3^m-1}{2}$, $f(x) = x + x^d$, defined in Conjecture 6.4, and $g(x) = \delta e(x)$. Then*

$$\widehat{f}(\lambda) = \widehat{g}(\lambda^{t^{-1}}), \quad \forall \lambda \in \mathbb{F}_{3^n} \tag{6.5}$$

where $t = \epsilon u_0 + u_1$. Thus, Lin sequences and HG sequences are Hadamard equivalent. The exponential sum equality of (6.5) is written as

$$\sum_{x \in \mathbb{F}_q} \omega^{\mathrm{Tr}(\lambda x - x - x^d)} = \sum_{x \in \mathbb{F}_q} \omega^{\mathrm{Tr}(\lambda^{t^{-1}} x) - g(x)}, \quad \forall \lambda \in \mathbb{F}_q .$$

The following conjecture indicates that the conjectured ternary 2-level autocorrelation sequences by Ludkovski and Gong in [36] can also be realized by HG sequences.

Conjecture 6.8. *Keep $g(x) = \delta e(x)$. Let*

$$v = \frac{3^n+1}{4} \quad \text{and} \quad t_j = \frac{3^{n-1}-1}{2} - 3^j ,$$

$$j = n-2, n-3, \ldots, \frac{n+1}{2}, \epsilon\frac{n+1}{2}, \frac{n-1}{2}$$

where ϵ is defined in (6.4) and let

$$\omega^{T_j(x)} = \frac{1}{q} \widehat{g}(v, t_j)(x), \quad x \in \mathbb{F}_q .$$

Table 6.1: Hadamard equivalence relations given by Conjectures 6.5–6.8

m-sequences	$\sim_H$	Lin conjectured 2-term sequences (Conjecture 6.5)
Lin conjectured 2-term sequences	$\sim_H$	HG-sequences (Conjecture 6.7)
HG-sequences	$\sim_H$	Ludkovski–Gong conjectured sequences (Conjecture 6.8)

Then $T_j(x)$ defines a 2-level autocorrelation sequence. Equivalently,

$$\sum_{x,y\in\mathbb{F}_q} \omega^{\mathrm{Tr}(\lambda y - y^{t_j}x)+g(x^v)} \in \{q, q\omega, q\omega^2\}\,, \quad \forall\lambda\in\mathbb{F}_q\,.$$

T_j contains exactly three classes B, C and D conjectured in [36].

The validity of all the conjectures has been verified for $n = 5, 7, 9, 11$, and 13 and probabilistically verified for $n = 15$. From Conjectures 6.5–6.8, we have the Hadamard equivalence relations listed in Table 6.1.

7 Some Open Problems

In this section, we summarize some unsolved problems on polyphase sequences with good correlation, DFT, and ambiguity properties, which were introduced in the previous sections.

7.1 Current Status of the Conjectures on Ternary 2-Level Autocorrelation

In [2] Arasu announced the validity of the 2-level autocorrelation property of Lin conjectured sequences, and Ludkovski and Gong conjectured sequences in an unpublished paper [3]. However, the Hadamard equivalences given by Table 6.1 still remain open. It is not clear whether those questions could be solved by the light of their approaches.

If Conjectures 6.5–6.8 are true, then there are only two classes of known ternary 2-level autocorrelation sequences. One is the class denoted by $\mathcal{T}_1$, consisting of all known ternary prime 2-level autocorrelation sequences and one type of HG sequences. Note that the ternary 2-level autocorrelation sequences in [26] is a special case of HG sequences. The other class, denoted by $\mathcal{T}_2$, is the remaining HG sequences. Conjectures 6.5–6.8 state that all the sequences in $\mathcal{T}_1$ are Hadamard equivalent, which still remain unsolved.

7.2 Possibility of Multiplicative Sequences with Low Autocorrelation

It is natural to ask whether there exist $\mathbb{F}_p$ or $\mathbb{F}_q$ multiplicative sequences with low autocorrelation, but the defining polynomials have high degrees. For example, $\mathbb{F}_p$ multiplicative sequences have the same autocorrelation as the power residue sequences and $\mathbb{F}_q$ multiplicative sequences have the same autocorrelation as the Sidel'nikov sequences. Those sequences, if there exist any, should be revealed using some special techniques. This could constitute an extremely challenge task.

7.3 Problems in Four Alternative Classes of Sequences and the General Hybrid Construction

As we showed in Section 4.5, corresponding to the four classes of sequences through $\mathbb{F}_p$ and $\mathbb{F}_q$ algebraic structures, there are the other four classes of the sequences using alternative indexing field elements of $\mathbb{F}_p$ and $\mathbb{F}_q$. Among those sequences, only Golay sequences [7], indexing the elements of $\mathbb{F}_{2^n}$ by an additive group, have been extensively investigated in the literature. Recently, it was reported in [19] that Golay sequences have a large zero autocorrelation zone. However, the other out-of-phase autocorrelation values have very large peaks. We ask whether there exist some classes of those sequences with low autocorrelation or with low crosscorrelation. The other open question is to find some class of the sequences with three metrics, constructed from combinations of different indexing field elements and hybrid characters in Construction 5.12.

8 Conclusions

We have introduced four types of polyphase sequences defined by $\mathbb{F}_p$ additive and multiplicative characters, and $\mathbb{F}_q$ additive and multiplicative characters, respectively. We have considered polyphase sequences with three metrics, namely, correlation, DFT, and ambiguity together. We have showed sequences with three metrics which are obtained from odd degree polynomials, power residue sequences, Sidel'nikov sequences, Weil representation sequences, and sequences from combinations of different indexing field elements and hybrid characters. We restated conjectured ternary sequences with 2-level autocorrelation and their Hadamard equivalence relation. Some open problems have been addressed.

References

[1] W. O. Alltop. Complex sequences with low periodic correlations. *IEEE Transactions on Information Theory* **26(3)** (1980), 350–354.

[2] K. T. Arasu. Sequences and arrays with desirable correlation properties. http://arhiva.math.uniri.hr/NATO-ASI/abstracts/arasu.pdf, 2011.

[3] K. T. Arasu, J. F. Dillon, and K. J. Player. Character sum factorizations yield perfect sequences, Preprint, 2010.

[4] D. C. Chu. Polyphase codes with good periodic correlation properties. *IEEE Transactions on Information Theory* **18(4)** (1972), 531–532.

[5] J. S. Chung, J. S. No, and H. Chung. A construction of a new family of M-ary sequences with low correlation from Sidelnikov sequences. *IEEE Transactions on Information Theory* **57(4)** (2011), 2301–2305.

[6] J. P. Costas. A study of a class of detection waveforms having nearly ideal range of doppler ambiguity properties. *Proceedings of the IEEE* **72(8)** (1984), 996–1009.

[7] J. A. Davis and J. Jedwab. Peak-to-mean power control in OFDM, Golay complementary sequences, and Reed-Muller codes. *IEEE Transactions on Information Theory* **45(7)** (1999), 2397–2417.

[8] J. F. Dillon and H. Dobbertin. New cyclic difference sets with Singer parameters. *Finite Fields and Their Applications* **10(3)** (2004), 342–389.

[9] T. Etzion. Combinatorial designs derived from Costas arrays. *Discrete Mathematics* **93(2–3)** (1991), 143–154.

[10] R. Frank, S. Zadoff, and R. Heimiller. Phase shift pulse codes with good periodic correlation properties. *IRE Transactions on Information Theory* **8(6)** (1962), 381–382.

[11] A. Gabizon and R. Raz. Deterministic extractors for affine sources over large fields, in: *Proceedings of the 46th Annual IEEE Symposium on Foundations of Computer Science*, FOCS '05, pp. 407–416. IEEE Computer Society, 2005.

[12] S. W. Golomb. Sequences with randomness properties. The Glenn L. Martin Company, Baltimore, MD, 1955.

[13] S. W. Golomb. Algebraic constructions for Costas arrays. *Journal of Combinatorics Theory (A)* **37(1)** (1984), 13–21.

[14] S. W. Golomb and G. Gong. *Signal Design for Good Correlation for Wireless Communication, Cryptography, and Radar*. Cambridge University Press, 2005.

[15] S. W. Golomb and G. Gong. The status of Costas arrays. *IEEE Transactions on Information Theory* **53(11)** (2007), 4260–4265.

[16] G. Gong. Theory and applications of q-ary interleaved sequences. *IEEE Transactions on Information Theory* **41(2)** (1995), 400–411.

[17] G. Gong and S. W. Golomb. The decimation-Hadamard transform of two-level autocorrelation sequences. *IEEE Transactions on Information Theory* **48(4)** (2002), 853–865.

[18] G. Gong, T. Helleseth, H. G. Hu, F. Huo, and Y. Yang. On conjectured ternary 2-level autocorrelation sequences. Progress Report, August 2011.

[19] G. Gong, F. Huo, and Y. Yang. Large zero autocorrelation zone of Golay sequences and 4^q-QAM Golay complementary sequences. Technical Report CACR 2011-16, University of Waterloo, 2011.

[20] S. Gurevich, R. Hadani, and N. Sochen. The finite harmonic oscillator and its applications to sequences, communication, and radar. *IEEE Transactions on Information Theory* **54(9)** (2008), 4239–4253.

[21] Y. K. Han and K. Yang. New M-ary sequence families with low correlation and large size. *IEEE Transactions on Information Theory* **55(4)** (2009), 1815–1823.

[22] G. H. Hardy and E. M. Wright. *An Introduction to the Theory of Numbers*, 5th ed. Oxford University Press, New York, 1980.

[23] T. Helleseth. Some results about the cross-correlation functions between two maximal length linear sequences. *Discrete Mathematics* **16(3)** (1976), 209–232.

[24] T. Helleseth and G. Gong. New nonbinary sequences with ideal two-level autocorrelation. *IEEE Transactions on Information Theory* **48(11)** (2002), 2868–2872.
[25] T. Helleseth and P. V. Kumar. Sequences with low correlation, in: *Handbook of coding theory*, volume 2, pp. 1765–1853. Elsevier, Amsterdam, The Netherlands, 1998.
[26] T. Helleseth, P. V. Kumar, and H. Martinsen. A new family of ternary sequences with ideal two-level autocorrelation function. *Designs, Codes and Cryptography* **23** (2001), 157–166.
[27] F. Hou. Sequences design for OFDM and CDMA systems. Master's Thesis, University of Waterloo, 2011.
[28] Y. J. Kim and H. Y. Song. Cross correlation of Sidelnikov sequences and their constant multiples. *IEEE Transactions on Information Theory* **53(3)** (2007), 1220–1224.
[29] Y. J. Kim, H. Y. Song, G. Gong, and H. Chung. Crosscorrelation of q-ary power residue sequences of period p, in: *IEEE International Symposium on Information Theory 2006*, pp. 311–315. IEEE, 2006.
[30] Y. S. Kim, J. S. Chung, J. S. No, and H. Chung. New families of M-ary sequences with low correlation constructed from Sidelnikov sequences. *IEEE Transactions on Information Theory* **54(8)** (2008), 3768–3774.
[31] P. V. Kumar and O. Moreno. Prime-phase sequences with periodic correlation properties better than binary sequences. *IEEE Transactions on Information Theory* **37(3)** (1991), 603–616.
[32] J. Lahtonen. On the odd and aperiodic correlation properties of the Kasami sequences. *IEEE Transactions on Information Theory* **41(5)** (1995), 1506–1508.
[33] A. Lempel, M. Cohn, and W. Eastman. A class of balanced binary sequences with optimal autocorrelation properties. *IEEE Transactions on Information Theory* **23(1)** (1977), 38–42.
[34] R. Lidl and H. Niederreiter. *Finite Fields*, 2nd ed. Cambridge University Press, 1997.
[35] A. H. Lin. *From cyclic Hadamard difference sets to perfectly balanced sequences*. Ph.D. thesis, University of Southern California, 1998.
[36] M. Ludkovski and G. Gong. New families of ideal 2-level autocorrelation ternary sequences from second order DHT, in: *Proceedings of the second International Workshop in Coding and Cryptography*, pp. 345–354. INRIA, 2001.
[37] H. D. Lüke. Large family of cubic phase sequences with low correlation. *Electronics Letters* **31(3)** (1995), 163.
[38] O. Moreno and P. V. Kumar. Minimum distance bounds for cyclic codes and Deligne's theorem. *IEEE Transactions on Information Theory* **39(5)** (1993), 1524–1534.
[39] K. G. Paterson. Generalized Reed-Muller codes and power control for OFDM modulation. *EEE Transactions on Information Theory* **46(1)** (2000), 104–120.
[40] K. G. Paterson and V. Tarokh. On the existence and construction of good codes with low peak-to-average power ratios. *IEEE Transactions on Information Theory* **46(6)** (2000), 1974–1987.
[41] J. G. Proakis. *Digital Communications*. McGraw-Hill, 2006.
[42] M. B. Pursley. *Introduction to Digital Communications*. Prentice Hall, 2005.
[43] J. J. Rushanan. Weil sequences: A family of binary sequences with good correlation properties, in: *Proceedings of IEEE International Symposium on Information Theory (ISIT)*, pp. 1648–1652. IEEE, 2006.
[44] D. Sarwate. Comments on "A class of balanced binary sequences with optimal autocorrelation properties" by Lempel et al. *IEEE Transactions on Information Theory* **24(1)** (1978), 128–129.
[45] D. Sarwate. Bounds on crosscorrelation and autocorrelation of sequences. *IEEE Transactions on Information Theory* **25(6)** (1979), 720–724.
[46] D. V. Sarwate and M. B. Pursley. Crosscorrelation properties of pseudorandom and related sequences. *Proceedings of the IEEE* **68(5)** (1980), 593–619.
[47] K. U. Schmidt. Sequence families with low correlation derived from multiplicative and additive characters. *IEEE Transactions on Information Theory* **57(4)** (2011), 2291–2294.

[48] V. M. Sidel'nikov. Some k-valued pseudo-random sequences and nearly equidistant codes. *Problemy Peredachi informatii (Problems on Information Transmission)* **5(1)** (1969), 16–22.

[49] M. Su and A. Winterhof. Autocorrelation of Legendre - Sidelnikov sequences. *IEEE Transactions on Information Theory* **56(4)** (2010), 1714–1718.

[50] D. Wan. Generators and irreducible polynomials over finite fields. *Mathematics of Computation* **66(219)** (1997), 1195–1212.

[51] Z. Wang and G. Gong. New sequences design from Weil representation with low two-dimensional correlation in both time and phase shifts. *IEEE Transactions on Information Theory* **57(7)** (2011), 4600–4611.

[52] Z. L. Wang and G. Gong. Cross correlation of binary sequence from multiplicative characters of polynomials. Preprint, March 2012.

[53] Z. L. Wang, G. Gong, and N. Y. Yu. Polyphase sequence families with low correlation from the bounds of character sums. CACR Technical Report, University of Waterloo, 2012.

[54] A. Weil. On some exponential sums. *Proceedings of the National Academy of Sciences of the United States of America* **34(5)** (1948), 204–207.

[55] A. Weil. *Basic Number Theory*, 3rd ed. Springer-Verlag, New York, 1974.

[56] L. Welch. Lower bounds on the maximum cross correlation of signals. *IEEE Transactions on Information Theory* **20(3)** (1974), 397–399.

[57] N. Y. Yu. New near-optimal codebooks associated with binary Sidel'nikov sequences, in: *The Proceedings of IEEE International Symposium on Information Theory Proceedings (ISIT)*, 2012.

[58] N. Y. Yu and G. Gong. Multiplexing realizations of the decimation-Hadamard transform of two-level autocorrelation sequences, in: *Coding and Cryptology*, volume 5557, pp. 248–258. Springer-Verlag, 2009.

[59] N. Y. Yu and G. Gong. Multiplicative characters, the Weil bound, and polyphase sequence families with low correlation. *IEEE Transactions on Information Theory* **56(12)** (2010), 6376–6387.

[60] N. Y. Yu and G. Gong. New construction of M-ary sequence families with low correlation from the structure of Sidelnikov sequences. *IEEE Transactions on Information Theory* **56(8)** (2010), 4061–4070.

[61] G. H. Zhang and Q. Zhou. Pseudonoise codes constructed by Legendre sequence. *Electronic Letters* **38(8)** (April 2002), 376–377.

[62] N. Zierler. Linear recurring sequences. *Journal of the Society for Industrial and Applied Mathematics* **7(1)** (1959), 31–48.

Katalin Gyarmati

Measures of Pseudorandomness

Abstract: In the second half of the 1990s Christian Mauduit and András Sárközy [86] introduced a new quantitative theory of pseudorandomness of binary sequences. Since then numerous papers have been written on this subject and the original theory has been generalized in several directions. Here I give a survey of some of the most important results involving the new quantitative pseudorandom measures of finite binary sequences. This area has strong connections to finite fields, in particular, some of the best known constructions are defined using characters of finite fields and their pseudorandom measures are estimated via character sums.

Keywords: Pseudorandomness, Well Distribution, Correlation, Normality

2010 Mathematics Subject Classifications: 11K45

Katalin Gyarmati: Department of Algebra and Number Theory, Eötvös Loránd University, Budapest, Hungary, e-mail: gykati@cs.elte.hu

1 Introduction

In the twentieth and twenty-first centuries various pseudorandom objects have been studied in cryptography and number theory since these objects are widely used in modern cryptography, in applications of the Monte Carlo method and in wireless communication (see [39]). Different approaches and definitions of pseudorandomness can be found in several papers and books. Menezes, Oorschot and Vanstone [95] have written an excellent monograph about these approaches. The most frequently used interpretation of pseudorandomness is based on complexity theory; Goldwasser [38] has written a survey paper about this approach. However, recently the complexity theory approach has been widely criticized. One problem is that in this approach usually *infinite* sequences are tested while in the applications only *finite* sequences are used. Another problem is that most results are based on certain unproved hypotheses (such as the difficulty of factorization of integers). Finite pseudorandom $[0,1)$ sequences have been studied by Niederreiter and others (see, for example, [103–106]). Niederreiter [107] also studied random number generation and quasi-Monte Carlo methods and their connections.

Research partially supported by ERC/AdG.228005, Hungarian National Foundation for Scientific Research, Grants No. K72731 and K100291 and the János Bolyai Research Fellowship.

In the second half of the 1990s, Christian Mauduit and András Sárközy [86] introduced a new constructive approach, in which the pseudorandomness of finite binary sequences is well characterized, and they also constructed binary sequences (and later other pseudorandom objects) with strong pseudorandom properties. In order to characterize the pseudorandomness of binary sequences Mauduit and Sárközy introduced new quantitative pseudorandom measures. Although earlier certain statistical tests (see, for example, [95]) already existed and one could determine whether a sequence passes these tests or not, the pseudorandom properties of the sequence were not classified. We also mention that by using these tests it was possible to test a sequence after generating it (*a posteriori testing*), but we did not have any *a priori* result which guaranteed the applicability of the sequence before generating it. There are two fundamental problems with a posteriori testing. Firstly, it could be quite lengthy to check whether or not a sequence passes these tests and it is much faster if certain properties of the construction guarantee that these tests are always passed for certain theoretical reasons (*a priori testing*). Secondly, in the case of a posteriori testing we always test only one certain, very special property of the sequence and nothing is known about the other pseudorandom properties. By using the pseudorandom measures of Mauduit and Sárközy it is possible to control several pseudorandom properties of sequences and it is also possible to *measure* their quality. In [118] Rivat and Sárközy estimated the outcome of certain basic statistical tests by the pseudorandom measures W and C_ℓ (see Section 2 below; the precise definitions of these tests can be found, for example, in [95]). In [122] Sárközy gave a survey of this new constructive theory of pseudorandomness. In the present survey we will focus mostly on pseudorandom measures; we will study the most important properties of these measures and their connections with other cryptographic tools.

2 Definition of the Pseudorandom Measures

In [86] Mauduit and Sárközy introduced the following pseudorandom measures in order to study the pseudorandom properties of *finite* binary sequences:

Definition 2.1. For a binary sequence $E_N = (e_1, \dots, e_N) \in \{-1, +1\}^N$ of length N, write

$$U(E_N, t, a, b) = \sum_{j=0}^{t} e_{a+jb} \, .$$

Then the *well-distribution measure of* E_N is defined as

$$W(E_N) = \max_{a,b,t} \left| U(E_N, t, a, b) \right| = \max_{a,b,t} \left| \sum_{j=0}^{t} e_{a+jb} \right| ,$$

where the maximum is taken over all a, b, t such that $a, b, t \in \mathbb{N}$ and $1 \le a \le a + tb \le N$.

The well-distribution measure studies how close are the frequencies of the $+1$'s and -1's in arithmetic progressions (for a binary sequence with strong pseudorandom properties these two quantities are expected to be very close). But often it is also necessary to study the connections between certain elements of the sequence. For example, if the subsequence $(+1,+1)$ occurs much more frequently than the subsequence $(-1,-1)$, it may cause problems in the applications, and we cannot say that our sequence has strong pseudorandom properties. In order to study connections of this type Mauduit and Sárközy [86] introduced the correlation and normality measures:

Definition 2.2. For a binary sequence $E_N = (e_1,\dots,e_N) \in \{-1,+1\}^N$ of length N and for $D = (d_1,\dots,d_\ell)$ with non-negative integers $0 \le d_1 < \cdots < d_\ell$, write

$$V(E_N,M,D) = \sum_{n=1}^{M} e_{n+d_1}\dots e_{n+d_\ell}.$$

Then the *correlation measure of order ℓ of E_N* is defined as

$$C_\ell(E_N) = \max_{M,D} |V(E_N,M,D)| = \max_{M,D} \left| \sum_{n=1}^{M} e_{n+d_1}\dots e_{n+d_\ell} \right|,$$

where the maximum is taken over all $D = (d_1,\dots,d_\ell)$ and M such that $0 \le d_1 < \cdots < d_\ell < M + d_\ell \le N$.

Definition 2.3. For a binary sequence $E_N = (e_1,\dots,e_N) \in \{-1,+1\}^N$ of length N and for $X = (x_1,\dots,x_\ell) \in \{-1,+1\}^\ell$ write

$$T(E_N,M,X) = \left|\{n : 0 \le n < M,\ (e_{n+1},e_{n+2},\dots,e_{n+\ell}) = X\}\right|.$$

Then the *normality measure of order ℓ of E_N* is defined as

$$N_\ell(E_N) = \max_{M,X} \left| T(E_N,M,X) - M/2^\ell \right|,$$

where the maximum is taken over all $X = (x_1,\dots,x_\ell) \in \{-1,+1\}^\ell$, and M such that $0 < M \le N - \ell + 1$.

We remark that *infinite* analogs of the functions U, V and T have been studied before (see, for example, [19, 66] and [111]), but the quantitative analysis of pseudorandom properties of *finite* sequences started with the work of Mauduit and Sárközy [86].

The *combined* (well-distribution correlation) pseudorandom measure [86] is a common generalization of well-distribution and correlation measures. This measure has an important role in the multidimensional extension of the theory of pseudorandomness (see Section 9).

Definition 2.4. For a binary sequence $E_N = (e_1, \dots, e_N) \in \{-1, +1\}^N$ of length N and for $D = (d_1, \dots, d_\ell)$ with non-negative integers $0 \le d_1 < \cdots < d_\ell$ write

$$Z(E_N, a, b, t, D) = \sum_{j=0}^{t} e_{a+jb+d_1} \dots e_{a+jb+d_\ell}\,.$$

Then the *combined (well-distribution correlation) measure of order ℓ of E_N* is defined as

$$Q_\ell(E_N) = \max_{a,b,t,D} \left| Z(E_N, a, b, t, D) \right| = \max_{a,b,t,D} \left| \sum_{j=0}^{t} e_{a+jb+d_1} \dots e_{a+jb+d_\ell} \right|,$$

where the maximum is taken over all a, b, t and $D = (d_1, \dots, d_\ell)$ such that all the subscripts $a + jb + d_i$ belong to $\{1, 2, \dots, N\}$.

When introducing their quantitative pseudorandom measures, the starting point of Mauduit and Sárközy was to balance the requirements possibly optimally. They decided to introduce functions that are real-valued and positive, and the pseudorandom properties of the sequence are characterized by the sizes of the values of these functions. It was also an important requirement that one should be able to present constructions for which these measures can be estimated well. It turned out that the measures W and C_ℓ do not only satisfy these criteria, but later Rivat and Sárközy [118] showed that if the values of W and C_ℓ are "small", then the outcome of many (previously used a posteriori) statistical tests is guaranteed to be (nearly) positive.

Although by W, C_ℓ, N_ℓ and Q_ℓ many pseudorandom properties of the sequence can be characterized, obviously not all of them can. For example, in [45] the symmetry measure was introduced in order to study symmetry properties of finite binary sequences (later the symmetry measure was generalized by Sziklai [125]). In [135] Winterhof gave an excellent survey on different pseudorandom measures and certain constructions. This is a fast developing area and many papers have been published; there are too many to list all of them here. However, introducing more and more pseudorandom measures, can make it quite lengthy to handle all these measures. Thus it is important to determine a not too large set of certain basic pseudorandom measures, which can guarantee the adequate security in the applications. The present research shows that the measures described in this section satisfy these criteria. The most studied measures are W and C_ℓ, and many papers use only these measures.

In the next section we will show that for a *random-type* sequence (i.e. for a sequence with strong pseudorandom properties) the well-distribution and correlation measures are expected to be small.

3 Typical Values of Pseudorandom Measures

In [16] Cassaigne, Ferenczi, Mauduit, Rivat and Sárközy formulated the following principle: "The sequence E_N is considered a 'good' pseudorandom sequence if these measures $W(E_N)$ and $C_\ell(E_N)$ (at least for 'small' ℓ) are 'small'." Indeed, the security of many cryptographic schemes is based on the property that the frequencies of the -1's and $+1$'s are about the same in certain "regular" subsequences of the used pseudorandom binary sequence $E_N \in \{-1,+1\}^N$.

In [18] Cassaigne, Mauduit and Sárközy proved that for the majority of the sequences $E_N \in \{-1,+1\}^N$ the measures $W(E_N)$ and $C_\ell(E_N)$ are around $N^{1/2}$ (up to some logarithmic factors). Later Alon, Kohayakawa, Mauduit, Moreira and Rödl [5] improved on these bounds:

Theorem 3.1. *Suppose that we choose each $E_N \in \{-1,+1\}^N$ with probability $1/2^N$. For all $\varepsilon > 0$ there exist $N_0 = N_0(\varepsilon)$ and $\delta = \delta(\varepsilon) > 0$ such that for $N > N_0$ we have*

$$P\left(\delta\sqrt{N} < W(E_N) < \tfrac{1}{\delta}\sqrt{N}\right) > 1 - \varepsilon\,.$$

Theorem 3.2. *Suppose that we choose each $E_N \in \{-1,+1\}^N$ with probability $1/2^N$. Then for all $0 < \varepsilon < 1/16$ there is a constant $N_0 = N_0(\varepsilon)$ such that for $N > N_0$ we have*

$$P\left(\tfrac{2}{5}\sqrt{N\log\binom{N}{\ell}} < C_\ell(E_N) < \tfrac{7}{4}\sqrt{N\log\binom{N}{\ell}}\right) > 1 - \varepsilon\,.$$

We remark that while it is important that for a binary sequence with strong pseudorandom properties these measures should be "small", lower bounds are not required (this will be justified by the results of Section 4, where the minimum values of these measures are studied). In many applications it is enough to guarantee that $W(E_N)$ and $C_\ell(E_N)$ are $o(N)$, but for the best constructions $E_N \in \{-1,+1\}^N$ *it is proved* that $W(E_N) \ll N^{1/2}\log N$, $C_\ell(E_N) \ll N^{1/2}(\log N)^{A_\ell}$ (see Section 6).

4 Minimum Values of Pseudorandom Measures

Write

$$m(N) = \min_{E_N\in\{-1,+1\}^N} W(E_N)\,, \qquad M_\ell(N) = \min_{E_N\in\{-1,+1\}^N} C_\ell(E_N)\,.$$

The estimate of $m(N)$ is a classical problem. In 1964 Roth [119] proved that $m(N) \gg N^{1/4}$. Upper bounds for $m(N)$ were given by Sárközy [32] and Beck [9]. Finally Matoušek and Spencer [78] showed that $m(N) \ll N^{1/4}$.

The value of $M_\ell(N)$ depends on the value of the order ℓ. Cassaigne, Mauduit and Sárközy [18] proved that $M_\ell(E_N) \ll (\ell N\log N)^{1/2}$. The results of [5] improved the implied constant factor (see Theorem 3.2 in the previous section). On the other hand, first Cassaigne, Mauduit and Sárközy [18] proved that $M_\ell(N) \gg \log(N/\ell)$ for even ℓ. This was improved considerably by Alon, Kohayakawa, Mauduit, Moreira and Rödl in [4] and [67], where the best lower bound is the following:

Theorem 4.1. *If ℓ is even then*

$$M_\ell(N) \geq \sqrt{\frac{1}{2}\left[\frac{N}{\ell+1}\right]}.$$

The proof of the theorem used deep linear algebraic tools, and later Anantharam [7] simplified the proof, but he obtained a slightly (by a constant factor) weaker result.

Cassaigne, Mauduit and Sárközy [18] noticed that the minimum values of correlation of odd order can be very small. Namely, for the sequence $E_N = (-1, +1, -1, +1, \dots) \in \{-1, +1\}^N$ we have $C_\ell(E_N) = 1$ for odd ℓ, since

$$e_{n+1+d_1} \cdots e_{n+1+d_\ell} = (-e_{n+d_1}) \cdots (-e_{n+d_\ell}) = (-1)^\ell e_{n+d_1} \cdots e_{n+d_\ell}.$$

Thus

$$\left| \sum_{n=1}^{M} e_{n+d_1} \cdots e_{n+d_\ell} \right| = |1 - 1 + 1 - 1 + \cdots| = \begin{cases} 1 & \text{if } M \text{ is odd,} \\ 0 & \text{if } M \text{ is even.} \end{cases}$$

So $C_\ell(E_N) = 1$ and thus $M_\ell(N) = 1$ for odd ℓ. Cassaigne, Mauduit and Sárközy [18] also observed that although for the sequence $E_N = (-1, +1, -1, +1, \dots)$, $C_3(E_N)$ is 1, the correlation measure of order 2 is large: $C_2(E_N) = \lceil \frac{N}{2} \rceil$. By solving problems of Cassaigne, Mauduit and Sárközy [18] and Mauduit [79], in [48] I proved that $C_2(E_N)C_3(E_N) \gg N^{2/3}$ always holds. Later Anantharam [8] proved that $C_2(E_N)C_3(E_N) \gg N$. By the methods of the proofs it is possible to compare correlation measures of odd and even order. With Mauduit we proved the following sharp result in [51]:

Theorem 4.2. *There is a constant $c_{k,\ell}$ depending only on k and ℓ such that if*

$$C_{2k+1}(E_N) < c_{k,\ell} N^{1/2},$$

then

$$C_{2k+1}(E_N)^{2\ell} C_{2\ell}(E_N)^{2k+1} \gg N^{2k+1},$$

where the implied constant factor depends only on k and ℓ.

This theorem has the following consequences:

Corollary 4.3. *If $C_{2k+1}(E_N) = O(1)$, then $C_{2\ell}(E_N) \gg N$, where the implied constant factor depends on k and ℓ.*

Corollary 4.4.

$$C_{2k+1}(E_N) C_{2\ell}(E_N) \gg N^{c(k,\ell)}$$

where the implied constant factor depends only on k and ℓ and where

$$c(k,\ell) = \begin{cases} 1 & \text{if } k \geq \ell, \\ \frac{1}{2} + \frac{2k+1}{4\ell} & \text{if } k < \ell. \end{cases}$$

The minimum of the normality measure was studied in [4] and [67], but there is a huge gap between the lower and upper bounds.

5 Connection between Pseudorandom Measures

It is a problem of basic importance to study the connections between the different pseudorandom measures. For example, Mauduit and Sárközy [86] proved that the normality measure can be bounded by the maximum of correlation measures:

Theorem 5.1.

$$N_\ell(E_N) \le \max_{1\le t\le \ell} C_t(E_N)\,.$$

Since the normality measures can be estimated by the correlation measures, most of the papers do not handle the normality measures separately, just they give non-trivial upper bounds for the well-distribution and correlation measures.

Cassaigne, Mauduit and Sárközy [18] compared correlation measures of different orders:

Theorem 5.2. *Suppose that* $2 \le k \mid \ell$ *and* $E_N \in \{-1,+1\}^N$. *Then*

$$C_k(E_N) \ll N^{1-k/\ell}\left(C_\ell(E_N)\right)^{k/\ell}\,.$$

If $k \nmid \ell$, it is possible to construct a sequence E_N for which $C_k(E_N)$ is large but $C_\ell(E_N)$ is small:

Theorem 5.3. *Suppose that* $2 \le k, \ell$ *and* $k \nmid \ell$. *Then there is a sequence* $E_N \in \{-1,+1\}^N$ *for which*

$$C_k(E_N) > \frac{N}{k} - 1 - 54k^2\log N\,,$$

$$C_\ell(E_N) < 27k^2\ell N^{1/2}\log N\,.$$

Indeed in [18], Theorem 5.2 and Theorem 5.3 were proved in a sharper form.

The well-distribution measure can be estimated by the correlation measures of even order. In [92] Mauduit and Sárközy proved that for all sequences $E_N \in \{-1,+1\}^N$ we have

$$W(E_N) \le \sqrt{NC_2(E_N)}\,.$$

Later in [42] and [44] this inequality was generalized by me to correlation measures of any even order.:

Theorem 5.4. *For all sequences* $E_N \in \{-1,+1\}^N$ *we have*

$$W(E_N) \ll N^{1-1/(2\ell)}\left(C_{2\ell}(E_N)\right)^{1/(2\ell)}\,. \tag{5.1}$$

In [42] I also proved that (5.1) is sharp apart from the implied constant factor.

6 Constructions

First Mauduit and Sárközy [86] studied the well-distribution and correlation measures of a finite binary sequence. Their construction was the following:

Construction 6.1. *Let p be a prime number, $N = p - 1$ and define the Legendre-sequence $E_N = (e_1, e_2, \dots, e_N) \in \{-1, +1\}^N$ by*

$$e_n = \left(\frac{n}{p}\right),$$

where $(\frac{\cdot}{p})$ denotes the Legendre symbol.

Then by Theorem 1 in [86] for the sequence E_N defined in Construction 6.1 we have

$$W(E_N) \ll N^{1/2} \log N \quad \text{and} \quad C_\ell(E_N) \ll N^{1/2} \log N\,.$$

After their first paper [86] on pseudorandomness, Mauduit and Sárközy continued with a series of papers ([16–18, 87–89]) in which they tested several constructions. Since then numerous constructions have been given, see, for example, [21, 23, 26, 28, 29, 36, 41, 71–73, 75, 82, 109, 112, 113, 116, 121]. We remark that the majority of these constructions are of modular type. It would be interesting to give a construction which is not of modular type, but (nearly) optimal bounds can be proved for its pseudorandom measures.

First for fixed N most constructions produced only a single sequence of length N; however, in many applications one needs many pseudorandom binary sequences. In 2004 Goubin, Mauduit and Sárközy [40] succeeded in constructing large families of pseudorandom binary sequences based on the Legendre symbol. Their construction was the following:

Construction 6.2. *Let $K \in \mathbb{N}$, p be a prime number and denote by $\mathcal{P}$ the set of polynomials $f(x) \in \mathbb{F}_p[x]$ of degree k, where $0 < k \le K$ and which have no multiple zero in $\overline{\mathbb{F}}_p$ (=the algebraic closure of $\mathbb{F}_p$). For $f \in \mathcal{P}$ define the binary sequence $E_p(f) = (e_1, \dots, e_p)$ by*

$$e_n = \begin{cases} \left(\frac{f(n)}{p}\right) & \text{for } (f(n), p) = 1\,, \\ +1 & \text{for } p \mid f(n)\,. \end{cases} \tag{6.1}$$

Let $\mathcal{F} = \{E_p(f) : f \in \mathcal{P}\}$.

Clearly $\mathcal{F}$ is a large family of pseudorandom binary sequences. Goubin, Mauduit and Sárközy [40] proved that, under some not too restrictive conditions on the polynomials f, the sequences $E_p(f)$ have strong pseudorandom properties:

Theorem 6.3. *Let p, $\mathcal{P}$ and $\mathcal{F}$ be defined as in Construction 6.2 and for $f \in \mathcal{P}$ define $E_p = E_p(f) \in \mathcal{F}$ by (6.1). Let k be the degree of f. Then*

$$W(E_p) \ll k p^{1/2} \log p\,.$$

Moreover, assume that for $\ell \in \mathbb{N}$ one of the following assumptions holds:
(i) $\ell = 2$;
(ii) $\ell < p$ *and 2 is a primitive root modulo p;*
(iii) $(4k)^{\ell} < p$.
Then we also have

$$C_\ell(E_p) \ll k\ell p^{1/2} \log p \, .$$

We remark that several important a posteriori tests (indicated by the 1.4-sts. package of the National Institute of Standards and Technology) were checked by Rivat and Sárközy [118] by computer for many sequences generated by Construction 6.2. In each case they obtained that the sequence passes all these tests.

The next construction was based on the discrete logarithm [43]:

Construction 6.4. *Let $K \in \mathbb{N}$, p be an odd prime number, and denote by $\mathcal{P}'$ the set of polynomials $f(x) \in \mathbb{F}_p[x]$ of degree k, where $0 < k \le K$. Let g be a primitive root modulo p and define $\operatorname{ind} n$ by $n \equiv g^{\operatorname{ind} n} \pmod p$ and $1 \le \operatorname{ind} n \le p-1$. For $f \in \mathcal{P}'$ define the binary sequence $E_{p-1}(f) = (e_1, \dots, e_{p-1})$ by*

$$e_n = \begin{cases} +1 & \text{if } 1 \le \operatorname{ind} f(n) \le (p-1)/2 \\ -1 & \text{if } (p+1)/2 \le \operatorname{ind} f(n) \le p-1 \text{ or } p \mid f(n) \, . \end{cases}$$

Let $\mathcal{F}' = \{E_p(f) : f \in \mathcal{P}'\}$.

This construction is nearly as good as Construction 6.2, the only problem is that it is slow to compute e_n, since no fast algorithm is known to compute $\operatorname{ind} n$. In [44] this construction was slightly modified such that the sequences in the new construction can be generated faster. Since then many other constructions of large families of pseudorandom sequences have been given (see, for example, [22, 24, 34, 35, 40, 43, 44, 59, 69, 74, 81, 84, 96–98, 117, 123, 127]).

Most constructions use finite fields and character sums over it (see the survey paper [127] for the most frequently used character sum estimates). One of the main tools in estimating the pseudorandom measures is Weil's theorem [133]:

Lemma 6.5. *Suppose that $\mathbb{F}_q$ is a finite field, χ is a non-principal character of order d over it, $f \in \mathbb{F}_q[x]$ has s distinct roots in $\overline{\mathbb{F}}_q$ and it is not a constant multiple of the d-th power of a polynomial over $\mathbb{F}_q$. Then:*

$$\left| \sum_{n \in \mathbb{F}_q} \chi(f(n)) \right| \le (s-1)p^{1/2} \, .$$

More precisely, the proofs of Theorem 6.3 and several other theorems (involving estimates of pseudorandom measures of different modular type constructions) are based on incomplete sums of multiplicative and additive characters. Such results can be derived from Weil's theorems on complete character sums (see, e.g. Lemma 6.5) by using a method of Vinogradov [131] (see also [64, 114, 126]).

Although many constructions exist, Construction 6.2 is one of the best: we have optimally good bounds for the pseudorandom measures and the elements of the sequences can be generated fast. In the next section we will analyze structural properties of large families of pseudorandom binary sequences.

7 Family Measures

In many applications it is not enough if our family $\mathcal{F}$ is large. For example, if $\mathcal{F}$ contains many sequences but they differ only in the last few bits, then one cannot use more than one sequence from the family. So it is very important to guarantee that the family $\mathcal{F}$ has a "rich", "complex" structure, there are many "independent" sequences in it which are "far apart." Thus one needs quantitative measures to study the structural properties of families of binary sequences. The first family measure was introduced by Ahlswede, Khachatrian, Mauduit and Sárközy in [1]:

Definition 7.1. Suppose that $\mathcal{F}$ is a family of binary sequences $E_N = (e_1, e_2, \dots, e_N) \in \{-1, +1\}^N$ and $(\varepsilon_1, \varepsilon_2, \dots, \varepsilon_j) \in \{-1, +1\}^j$ is a fixed binary sequence of length j (for some $j \le N$), and let $1 \le i_1 < i_2 < \cdots < i_j \le N$. If we consider binary sequences $E_N = (e_1, e_2, \dots, e_N) \in \{-1, +1\}^N$ with

$$e_{i_1} = \varepsilon_1 , \quad e_{i_2} = \varepsilon_2 , \quad \dots , \quad e_{i_j} = \varepsilon_j , \tag{7.1}$$

then (7.1) is said to be a *specification of length* j (of the binary sequence E_N).

Definition 7.2. The *family complexity* or briefly *f-complexity* of a family $\mathcal{F}$ of binary sequences $E_N \in \{-1, +1\}^N$ is defined as the greatest integer j such that for any specification (7.1) (of length j) there is at least one $E_N \in \mathcal{F}$ which satisfies it. The f-complexity of $\mathcal{F}$ is denoted by $\Gamma(\mathcal{F})$. (If there is no $j \in \mathbb{N}$ with the property above, we set $\Gamma(\mathcal{F}) = 0$.)

Note that an easy consequence of the definition is

Proposition 7.3.

$$\Gamma(\mathcal{F}) \le \frac{\log |\mathcal{F}|}{\log 2} . \tag{7.2}$$

Ahlswede, Khachatrian, Mauduit and Sárközy [1] showed that for the family $\mathcal{F}$ defined in Construction 6.2, the f-complexity $\Gamma(\mathcal{F})$ is large. Later Gyarmati [47] improved on their lower bound by showing that $\Gamma(\mathcal{F}) > c \log |\mathcal{F}|$ with some explicit constant c; we note that by (7.2), this estimate is best possible apart from the value of this constant c, and thus the f-complexity of this family is optimally large (apart from the constant factor). Since then the family complexity of many other constructions were also studied by several authors. In [85] Mauduit and Sárközy gave a survey paper on family complexity.

Another important tool for studying the pseudorandomness of families of binary sequences is the notion of *collision* (see, for example, [10, 95, 129, 130]):

Assuming that $N \in \mathbb{N}$, S is a given set (e.g. a set of certain polynomials or the set of all the binary sequences of a given length much less than N), to each $s \in S$ we assign a unique binary sequence

$$E_N = E_N(s) = (e_1, \ldots, e_N) \in \{-1, +1\}^N ,$$

and let $\mathcal{F} = \mathcal{F}(S)$ denote the family of the binary sequences obtained in this way:

$$\mathcal{F} = \mathcal{F}(S) = \{E_N(s) : s \in S\} . \tag{7.3}$$

Definition 7.4. If $s \in S$, $s' \in S$, $s \neq s'$ and

$$E_N(s) = E_N(s') , \tag{7.4}$$

then (7.4) is said to be a *collision* in $\mathcal{F} = \mathcal{F}(S)$. If there is no collision in $\mathcal{F} = \mathcal{F}(S)$, then $\mathcal{F}$ is said to be *collision free*.

In other words, $\mathcal{F} = \mathcal{F}(S)$ is collision free if we have $|\mathcal{F}| = |S|$. It turns out that in the best constructions, the families of pseudorandom binary sequences are collision free. If $\mathcal{F}$ is not collision free but the number of collisions is "small", then they may cause only minor problems in the applications. A good measure of the number of collisions is the following:

Definition 7.5. The *collision maximum* $M = M(\mathcal{F}, S)$ is defined by

$$M = M(\mathcal{F}, S) = \max_{E_N \in \mathcal{F}} |\{s : s \in S,\ E_N(s) = E_N\}|$$

(i.e. M is the maximal number of elements of S representing the same binary sequence E_N, and $\mathcal{F} = \mathcal{F}(S)$ is collision free if and only if $M(\mathcal{F}, S) = 1$).

Another important family requirement is the avalanche effect (see, e.g. [10, 33, 65, 129, 130]) which studies that by changing a few bits of the *seed* how many elements of the output sequence will change.

Definition 7.6. If in (7.3) we have $S = \{-1, +1\}^\ell$, and for any $s \in S$, changing any element of s changes "many" elements of $E_N(s)$ (i.e. for $s \neq s'$ many elements of the sequences $E_N(s)$ and $E_N(s')$ are different), then we speak about an *avalanche effect*, and we say that $\mathcal{F} = \mathcal{F}(S)$ *possesses the avalanche property*. If $N \to \infty$ and for any $s \in S$, $s' \in S$, $s \neq s'$ at least $(\frac{1}{2} - o(1))N$ elements of $E_N(s)$ and $E_N(s')$ are different, then $\mathcal{F}$ is said to *possess the strict avalanche property*.

To study the avalanche property, one may introduce the following quantitative measure:

Definition 7.7. If $N \in \mathbb{N}$, $E_N = (e_1, \dots, e_N) \in \{-1, +1\}^N$ and $E'_N = (e'_1, \dots, e'_N) \in \{-1, +1\}^N \in \{-1, +1\}^N$, then the *distance* $d(E_N, E'_N)$ between E_N and E'_N is defined by

$$d(E_N, E'_N) = |\{n : 1 \le n \le N,\ e_n \ne e'_n\}|$$

(a similar notion is introduced in [10]; this is a variant of the Hamming distance). Moreover, if $\mathcal{F}$ is a family of the form (7.3), then the *distance minimum* $m(\mathcal{F})$ of $\mathcal{F}$ is defined by

$$m(\mathcal{F}) = \min_{\substack{s,s' \in \mathcal{S} \\ s \ne s'}} d(E_N(s), E_N(s')) .$$

Thus the family $\mathcal{F}$ in (7.3) is collision free if and only if $m(\mathcal{F}) > 0$, and $\mathcal{F}$ possesses the strict avalanche property if

$$m(\mathcal{F}) \ge \left(\tfrac{1}{2} - o(1)\right) N .$$

In [129] Tóth studied the Legendre symbol construction described in Construction 6.2 and she showed that a variant of the family defined there (she replaced the condition $\deg f(x) \le K$ by $\deg f(x) = K$) is collision free if $K < p^{1/2}/2$ and it possesses the strong avalanche effect for $p \to \infty$, $K = o(p^{1/2})$. In [130] she also studied a further construction using additive characters and she showed that there are many collisions in it, but a large subfamily of it possesses the strong avalanche property.

8 Linear Complexity

Cryptographic applications require pseudorandom sequences which are "unpredictable" in a certain sense. Kolmogorov [68] and Chaitin [20] introduced the notion of Kolmogorov complexity, which is roughly speaking the length of the shortest computer program which generates the given sequence in a fixed Turing machine. From this point of view, a sequence can be considered a bad pseudorandom sequence if its Kolmogorov complexity is "small". Unfortunately, in practice, it is usually hopeless to compute the Kolmogorov complexity for a fixed sequence, thus this definition cannot be used in the applications. In this section we analyze a related measure, linear complexity, which is a computable measure. Mainly we will study the connection between linear complexity and other pseudorandom measures.

Feedback shift registers, in particular linear feedback shift registers are used in many cryptographic stream ciphers (see, e.g. [95]). The linear feedback shift registers (LFSR) have many equivalent definitions, here I use one from [132]:

Definition 8.1. The *linear feedback shift register* is a sequence of 0–1 bits $(s_1, s_2, \dots, s_\ell, c_1, \dots, c_\ell)$ with $c_1 = 1$. The output of the LFSR is the infinite sequence $(s_1, s_2, \dots)$

where $s_i (\in \{0,1\})$ for $i > \ell$ is defined by the following equation:

$$s_i = \sum_{j=1}^{\ell} c_j s_{i-\ell-1+j} \pmod 2 .$$

An LFSR $L(s_1, s_2, \dots, s_\ell, c_1, \dots, c_\ell)$ is said to *generate* an infinite sequence $s = (s_1, s_2, \dots)$ if s is the output sequence of $L(s_1, s_2, \dots, s_\ell, c_1, \dots, c_\ell)$. The *linear complexity* of an infinite sequence s, denoted by $L(s)$, is defined as follows:

(1) If s is the zero sequence $(0, 0, 0, \dots)$, then $L(s) = 0$.
(2) If no LFSR generates s, then $L(s) = \infty$.
(3) Otherwise $L(s)$ is the length of the shortest LFSR that generates s.

For finite sequence $s \in \{0,1\}^N$, the linear complexity $L(s)$ is the length of the shortest LFSR that generates an infinite sequence whose first N bits form the finite sequence s.

The relationship between linear complexity and Kolmogorov complexity was studied in [13, 132]. The linear complexity is an important cryptographic characteristic of sequences (see the monographs and surveys [27, 93, 95, 102, 108, 128, 134]). An excellent historical survey on the linear complexity is given in [115]. Here I mention only some of the most important properties of the linear complexity: It is known [120] that the linear complexity of a truly random bit sequence $s = (s_1, s_2, \dots, s_N) \in \{0,1\}^N$ is $(1 + o(1))\frac{N}{2}$. Based on this fact a sequence with low linear complexity is usually considered a "bad" pseudorandom sequence.

Using the Berlekamp–Massey algorithm (which is due to Massey [77] and based on an earlier algorithm of Berlekamp [12]), it is possible to calculate the value of the linear complexity of a fixed finite sequence. The linear complexity is usually defined for $0-1$ sequences (note that it can be defined similarly in the case of sequences of elements of $\mathbb{F}_q$ or $\mathbb{Z}_m$), but in this survey we study mostly ± 1 sequences. This problem can be easily avoided: there is a natural bijection $\varphi : \{-1,+1\}^N \to \{0,1\}^N$. Namely, if the sequence $E_N \in \{-1,+1\}^N$ is given, then $\varphi(E_N)$ can be defined by

$$\varphi(E_N) = \varphi((e_1, e_2, \dots, e_N)) = S_N = (s_0, s_1, \dots, s_{N-1}) \in \{0,1\}^N$$
$$\text{with } s_i = \frac{1 - e_{i+1}}{2} \text{ (or equivalently } (-1)^{s_i} = e_{i+1}) \text{ for } i = 0, 1, \dots, N-1 .$$

Hence we may define the linear complexity of the binary sequence $E_N \in \{+1,-1\}^N$ by

$$L(E_N) = L(\varphi(E_N)) .$$

Brandstätter and Winterhof [14] showed that the linear complexity of a binary sequence E_N can be estimated in terms of the correlation measures of the sequence:

Theorem 8.2. *If $N \geq 2$ and E_N is a binary sequence then we have*

$$L(E_N) \geq N - \max_{1 \leq k \leq L(E_N)+1} C_k(E_N) .$$

Using this inequality they were able to give (in some cases quite strong) lower estimates for the linear complexity of binary sequences occurring in certain constructions. While this theorem may give quite good estimates for linear complexity, it has the disadvantage that it also uses correlations of high order which can be very difficult to estimate. Thus Andics [8] proved another inequality which uses the correlation of order 2 only (but it usually gives a weak lower bound):

Theorem 8.3. *If* $N \in \mathbb{N}$ *and* E_N *is a binary sequence then we have*

$$2^{L(E_N)} \geq N - C_2(E_N)\,.$$

Further results related to the pseudorandom measures and linear complexity can be found in several works (see, e.g. the papers of Winterhof and co-authors [6, 14, 15, 25, 37, 93, 94, 124, 128, 134]).

9 Multidimensional Theory

In the recent years, the one-dimensional theory of pseudorandomness has been extended to several dimensions. For example, when we would like to encrypt a digital map or image by the multidimensional analog of the Vernam cipher, then instead of a pseudorandom binary sequence we need a *two or more dimensional pseudorandom binary lattice* as a keystream. The multidimensional theory of pseudorandomness was developed by Hubert, Mauduit and Sárközy [62]. They introduced the following definitions:

Denote by I_N^n the set of n-dimensional vectors whose coordinates are integer numbers between 0 and $N-1$:

$$I_N^n = \{\mathbf{x} = (x_1, \dots, x_n) : x_1, \dots, x_n \in \{0, 1, \dots, N-1\}\}\,.$$

This set is called an *n-dimensional N-lattice* or briefly *N-lattice*. Next they extended this definition to more general lattices in the following way: Let $\mathbf{u_1}, \mathbf{u_2}, \dots, \mathbf{u_n}$ be n linearly independent vectors, where the i-th coordinate of $\mathbf{u_i}$ is a non-zero integer, and the other coordinates of $\mathbf{u_i}$ are 0, so $\mathbf{u_i}$ is of the form $(0, \dots, 0, z_i, 0, \dots, 0)$. Let $t_1, t_2, \dots, t_n$ be integers with $0 \leq t_1, t_2, \dots, t_n < N$. Then we will call the set

$$B_N^n = \{\mathbf{x} = x_1\mathbf{u_1} + \cdots + x_n\mathbf{u_n} : x_i \in \mathbb{N} \cup \{0\}\,,\ \ 0 \leq x_i\,|\mathbf{u_i}| \leq t_i (< N) \quad \text{for } i = 1, \dots, n\}$$

an *n-dimensional box N-lattice* or briefly a *box N-lattice*.

In [62] the definition of binary sequences is extended to more dimensions by considering functions of type

$$e_{\mathbf{x}} = \eta(\mathbf{x}) : I_N^n \to \{-1, +1\}\,.$$

If $\mathbf{x} = (x_1, \dots, x_n)$ so that $\eta(\mathbf{x}) = \eta((x_1, \dots, x_n))$ then we will slightly simplify the notation by writing $\eta(\mathbf{x}) = \eta(x_1, \dots, x_n)$. These functions are called *bina-*

ry N-lattices or briefly *binary lattices*. One may visualize a binary lattice as the lattice points of the N-lattice replaced by the two symbols $+$ and $-$.

In [62] Hubert, Mauduit and Sárközy introduced the following pseudorandom measure of binary lattices (here we will present the definition in a slightly modified but equivalent form):

Definition 9.1. Let

$$\eta\colon I_N^n \to \{-1,+1\}$$

be a binary lattice. Define the *pseudorandom measure of order ℓ of η* by

$$Q_\ell(\eta) = \max_{B,\mathbf{d}_1,\ldots,\mathbf{d}_k} \left| \sum_{\mathbf{x}\in B} \eta(\mathbf{x}+\mathbf{d}_1)\cdots\eta(\mathbf{x}+\mathbf{d}_\ell) \right| ,$$

where the maximum is taken over all distinct $\mathbf{d}_1,\ldots,\mathbf{d}_\ell \in I_N^n$ and box N-lattice B such that $B+\mathbf{d}_1,\ldots,B+\mathbf{d}_\ell \subseteq I_N^n$.

Then η is said to have strong pseudorandom properties, or briefly, it is considered a "good" pseudorandom lattice if for fixed n and ℓ and "large" N the measure $Q_\ell(\eta)$ is "small" (much smaller than the trivial upper bound N^n). This terminology is justified by the fact that, as it was proved in [62], for a truly random binary lattice defined on I_N^n and for fixed ℓ the measure $Q_\ell(\eta)$ is "small" (less than $N^{n/2}$ multiplied by a logarithmic factor).

Recently several multidimensional constructions have been given for lattices with strong pseudorandom properties, see, for example, [52, 60–62, 70, 83, 91, 99, 100].

Some one-dimensional theorems can be generalized to the multidimensional case. For example, we studied the properties of the multidimensional pseudorandom measures in [53–58]. In particular, in [58] we compared the one-dimensional pseudorandom measures with the two or more dimensional pseudorandom measures and we showed that the study of the multidimensional measures cannot be reduced to one-dimensional ones, so indeed it was necessary to develop the multidimensional theory. In [55–57] we introduced the multidimensional analog of the normality, correlation and symmetry measures. We studied the connection between multidimensional pseudorandom measures of different orders and we proved the multidimensional analog of Theorem 5.1. We also studied the minimal values of the multidimensional pseudorandom measures. In [46] further multidimensional pseudorandom measures were introduced. In [53] and [54] the notions of family complexity, collision and avalanche effect were extended and studied in the multidimensional case.

10 Extensions

Pseudorandom binary sequences have many further generalizations. For example, Mauduit and Sárközy [90], Ahlswede, Mauduit and Sárközy [2, 3], Bérczi [11], Marzouk and Winterhof [76] and Mérai [101] studied the case of sequences of k symbols.

Hubert and Sárközy [63] studied the case of p-pseudorandom binary sequences, i.e. the case when the binary sequences simulate the binomial distribution of parameter p. Niederreiter, Rivat and Sárközy [110] studied pseudorandom sequences of binary vectors. In [30] and [31] Dartyge and Sárközy started to study pseudorandom subsets of $\{1, 2, \ldots, N\}$ and $\mathbb{Z}_n$. In [49] and [50] we studied pseudorandom binary functions on rooted plane trees. The connection between pseudorandom binary and $[0, 1)$ sequences was analyzed in [80] by Mauduit, Niederreiter and Sárközy.

References

[1] R. Ahlswede, L. H. Khachatrian, C. Mauduit, and A. Sárközy, *A complexity measure for families of binary sequences*, Period. Math. Hungar. **46** (2003), 107–118.

[2] R. Ahlswede, C. Mauduit, and A. Sárközy, *Large families of pseudorandom sequences of k symbols and their complexity. I*, Lecture Notes in Comput. Sci. 4123, General theory of information transfer and combinatorics, pp. 293–307, Springer, Berlin, Heidelberg, 2006.

[3] R. Ahlswede, C. Mauduit, and A. Sárközy, *Large families of pseudorandom sequences of k symbols and their complexity. II*, Lecture Notes in Comput. Sci. 4123, General theory of information transfer and combinatorics, pp. 308–325, Springer, Berlin, Heidelberg, 2006.

[4] N. Alon, Y. Kohayakawa, C. Mauduit, C. G. Moreira, and V. Rödl, *Measures of pseudorandomness for finite sequences: minimal values*, Combin., Probab. Comput. **15** (2005), 1–29.

[5] N. Alon, Y. Kohayakawa, C. Mauduit, C. G. Moreira, and V. Rödl, *Measures of pseudorandomness for finite sequences: typical values*, Proc. London Math. Soc. **95** (2007), 778–812.

[6] H. Aly and A. Winterhof, *On the k-error linear complexity over $\mathbb{F}_p$ of Legendre and Sidelnikov sequences*, Des. Codes Cryptogr. **40** (2006), 369–374.

[7] V. Anantharam, *A technique to study the correlation measures of binary sequences*, Discrete Math. **308(24)** (2008), 6203–6209.

[8] Á. Andics, *On the linear complexity of binary sequences*, Annales Univ. Sci. Budapest. **48** (2005), 173–180.

[9] J. Beck, *Roth's estimate on the discrepancy of integer sequences is nearly sharp*, Combinatorica **1** (1981), 319–325.

[10] A. Bérczes, J. Ködmön, and A. Pethő, *A one-way function based on norm form equations*, Period. Math. Hungar. **49** (2004), 1–13.

[11] G. Bérczi, *On finite pseudorandom sequences of k symbols*, Period. Math. Hungar. **47(1–2)** (2003), 29–44.

[12] E. R. Berlekamp, *Algebraic Coding Theory*, McGraw Hill, New York, 1968.

[13] T. Beth and Z. D. Dai, *On the complexity of pseudo-random sequences – or: if you can describe a sequence it can't be random*, Advances in Cryptology – EUROCRYPT '89 (Houthalen 1989), Lecture Notes in Computer Science 434, pp. 533–543, Springer, Berlin, 1990.

[14] N. Brandstätter and A. Winterhof, *Linear complexity profile of binary sequences with small correlation measure*, Period. Math. Hungar. **52** (2006), 1–8.

[15] N. Brandstätter and A. Winterhof, *k-error linear complexity over $\mathbb{F}_p$ of subsequences of Sidelnikov sequences of period $(p^r - 1)/3$*, J. Math. Cryptol. **3** (2009), 215–225.

[16] J. Cassaigne, S. Ferenczi, C. Mauduit, J. Rivat, and A. Sárközy, *On finite pseudorandom binary sequences III: The Liouville function, I*, Acta Arith. **87** (1999), 367–384.

[17] J. Cassaigne, S. Ferenczi, C. Mauduit, J. Rivat, and A. Sárközy, *On finite pseudorandom binary sequences. IV*, Acta Arith. **95**, 343–359 (2000).

[18] J. Cassaigne, C. Mauduit, and A. Sárközy, *On finite pseudorandom binary sequences VII: The measures of pseudorandomness*, Acta Arith. **103** (2002), 97–118.

[19] J. W. S. Cassels, *On a paper of Niven and Zuckerman*, Pacific J. Math. **2** (1952), 555–557.

[20] G. J. Chaitin, *On the length of programs for computing finite binary sequences*, J. Assoc. Comput. Mach. **13** (1966), 547–569.

[21] Z.-X. Chen, *Elliptic curve analogue of Legendre sequences*, Monatsh. Math. **154** (2008), 1–10.

[22] Z. X. Chen, X. N. Du, and G. Z. Xiao, *Sequences related to Legendre/Jacobi sequences*, Inform. Sci. **177** (2007), 4820–4831.

[23] Z. Chen and S. Li, *Some notes on generalized cyclotomic sequences of length pq*, J. Comput. Sci. Technology **23** (2008), 843–850.

[24] Z. Chen, S. Li and G. Xiao, *Construction of pseudorandom binary sequences from elliptic curves by using the discrete logarithms*, in: Sequences and their applications – SETA 2006, LNCS 4086, pp. 285–294, Springer, 2006.

[25] Z. Chen and A. Winterhof, *Linear complexity profile of m-ary pseudorandom sequences with small correlation measure* Indag. Math. (N. S.) **20(4)** (2009), 631–640.

[26] Z. Chen, Zhixiong, A. Ostafe, and A. Winterhof, *Structure of pseudorandom numbers derived from Fermat quotients*, Lecture Notes in Comput. Sci., 6087, Arithmetic of finite fields, pp. 73–85, Springer, Berlin, 2010.

[27] T. W. Cusick, C. Ding, and A. Renwall, *Stream Ciphers and Number Theory*, revised ed., North-Holland Mathematical Library 66, Elsevier Science B. V., Amsterdam, 2004.

[28] H. Daboussi, *On pseudorandom properties of multiplicative functions*, Acta Math. Hungar. **98** (2003), 273–300.

[29] H. Daboussi, *On the correlation of the truncated Liouville function*, Acta Arith. **108** (2003), 61–76.

[30] C. Dartyge and A. Sárközy, *On pseudo-random subsets of the set of the integers not exceeding N*, Period. Math. Hung. **54(2)** (2007), 183–200.

[31] C. Dartyge and A. Sárközy, *On pseudo-random subsets of $\mathbb{Z}_n$*, Monatsh. Math. **157(1)** (2009), 13–35.

[32] P. Erdős and A. Sárközy, *Some solved and unsolved problems in combinatorial number theory*, Math. Slovaca **28** (1978), 407–421 (p. 415).

[33] H. Feistel, W. A. Notz, and J. L. Smith, *Some cryptographic techniques for machine-to-machine data communications*, Proc. IEEE **63** (1975), 1545–1554.

[34] J. Folláth, *Construction of pseudorandom binary sequences using additive characters over $GF(2^k)$*, Period. Math. Hungar. **57** (2008), 73–81.

[35] J. Folláth, *Construction of pseudorandom binary sequences using additive characters over $GF(2^k)$. II*, Period. Math. Hungar. **60** (2010), 127–135.

[36] E. Fouvry, P. Michel, J. Rivat, and A. Sárközy, *On the pseudorandomness of the signs of Kloosterman sums*, J. Australian Math. Soc. **77** (2004), 425–436.

[37] M. Z. Garaev, F. Luca, Florian, I. E. Shparlinski, and A. Winterhof, *On the lower bound of the linear complexity over $\mathbb{F}_p$ of Sidelnikov sequences*, IEEE Trans. Inform. Theory **52(7)** (2006), 3299–3304.

[38] S. Goldwasser, *Mathematical Foundations of Modern Cryptography: Computational Complexity Perspective*, ICM 2002, vol. I, 245–272.

[39] S. W. Golomb and G. Gong, *Signal Design for Good Correlation: For Wireless Communication, Cryptography, and Radar*, University Press, Cambridge, New York, Melbourne, Madrid, Cape Town, Singapore, São Paulo, 2005.

[40] L. Goubin, C. Mauduit, and A. Sárközy, *Construction of large families of pseudorandom binary sequences*, J. Number Theory **106** (2004), 56–69.

[41] E. Grant, J. Shallit, and T. Stoll, *Bounds for the discrete correlation of infinite sequences on k symbols and generalized Rudin–Shapiro sequences*, Acta Arith. **140** (2009), 345–368.

[42] K. Gyarmati, *An inequality between the measures of pseudorandomness*, Ann. Univ. Sci. Budapest. Eötvös Sect. Math. **46** (2003), 157–166.

[43] K. Gyarmati, *On a family of pseudorandom binary sequences*, Period. Math. Hungar. **49** (2004), 45–63.

[44] K. Gyarmati, *On a fast version of a pseudorandom generator*, Lecture Notes in Comput. Sci. 4123, General theory of information transfer and combinatorics, pp. 326–342, Springer, Berlin, Heidelberg, 2006.

[45] K. Gyarmati, *On a pseudorandom property of binary sequences*, Ramanujan J. **8** (2004), 289–302.

[46] K. Gyarmati, *On new measures of pseudorandomness of binary lattices*, Acta Math. Hung. **131** (2011), 346–359.

[47] K. Gyarmati, *On the complexity of a family related to the Legendre symbol*, Period. Math. Hungar. **58** (2009), 209–215.

[48] K. Gyarmati, *On the correlation of binary sequences*, Studia Sci. Math. Hungar. **42** (2005), 59–75.

[49] K. Gyarmati, P. Hubert, and A. Sárközy, *Pseudorandom binary functions on almost uniform trees*, J. Combin. Number Theory **2** (2010), 1–24.

[50] K. Gyarmati, P. Hubert, and A. Sárközy, *Pseudorandom binary functions on rooted plane trees*, J. Combin. Number Theory, to appear.

[51] K. Gyarmati and C. Mauduit, *On the correlation of binary sequences, II*, Discrete Math. **312** (2012), 811–818.

[52] K. Gyarmati, C. Mauduit, and A. Sárközy, *Constructions of pseudorandom binary lattices*, Uniform Distribution Theory **4** (2009), 59–80.

[53] K. Gyarmati, C. Mauduit, and A. Sárközy, *Measures of pseudorandomness of families of binary lattices, I (Definitions, a construction using quadratic characters.)*, Publ. Math. Debrecen **79** (2011), 445–460.

[54] K. Gyarmati, C. Mauduit, and A. Sárközy, *Measures of pseudorandomness of families of binary lattices, II (A further construction.)*, Publ. Math. Debrecen **80** (2012), 481–504.

[55] K. Gyarmati, C. Mauduit, and A. Sárközy, *Measures of pseudorandomness of finite binary lattices, I (The measures Q_k, normality.)*, Acta Arith. **144** (2010), 295–313.

[56] K. Gyarmati, C. Mauduit, and A. Sárközy, *Measures of pseudorandomness of finite binary lattices, II (The symmetry measures.)*, Ramanujan J. **25** (2011), 155–178.

[57] K. Gyarmati, C. Mauduit, and A. Sárközy, *Measures of pseudorandomness of finite binary lattices, III (Q_k, correlation, normality, minimal values.)*, Unif. Distrib. Theory **5** (2010), 183–207.

[58] K. Gyarmati, C. Mauduit, and A. Sárközy, *Pseudorandom binary sequences and lattices*, Acta Arith. **135(2)** (2008), 181–197.

[59] K. Gyarmati, A. Pethő, and A. Sárközy, *On linear recursion and pseudorandomness*, Acta Arith. **118** (2005), 359–374.

[60] K. Gyarmati, A. Sárközy, and C. L. Stewart, *On Legendre symbol lattices*, Uniform Distribution Theory **4** (2009), 81–95.

[61] K. Gyarmati, A. Sárközy, and C. L. Stewart, *On Legendre symbol lattices, II*, to appear.

[62] P. Hubert, C. Mauduit, and A. Sárközy, *On pseudorandom binary lattices*, Acta Arith. **125** (2006), 51–62.

[63] P. Hubert and A. Sárközy, *On p-pseudorandom binary sequences*, Period. Math. Hungar. **49** (2004), 73–91.

[64] H. Iwaniec and E. Kowalski, *Analytic Number Theory*, Colloquium Publications 53, American Mathematical Society, 2004.

[65] J. Kam and G. Davida, *Structured design of substitution-permutation encryption networks*, IEEE Transactions on Computers **28** (1979), 747–753.

[66] D. E. Knuth, *The Art of Computer Programming*, Vol. 2, 2nd ed., Addison-Wesley, Reading, Mass., 1981.

[67] Y. Kohayakawa, C. Mauduit, C. G. Moreira, and V. Rödl, *Measures of pseudorandomness for finite sequences: minimum and typical values*, Proceedings of WORDS'03, TUCS Gen. Publ. 27, Turku Cent. Comput. Sci., Turku, 2003, 159–169.

[68] A. N. Kolmogorov, *Three approaches to the definition of the concept "quantity of information"*, Problemy Inform. Transmission **1(1)** (1965), 3–7.

[69] H. N. Liu, *A family of pseudorandom binary sequences constructed by the multiplicative inverse*, Acta Arith. **130** (2007), 167–180.

[70] H. Liu, *A large family of pseudorandom binary lattices*, Proc. Amer. Math. Soc. **137** (2009), 793–803.

[71] H. Liu, *New pseudorandom sequences constructed by quadratic residues and Lehmer numbers*, Proc. Amer. Math. Soc. **135** (2007), 1309–1318.

[72] H. Liu, *New pseudorandom sequences constructed using multiplicative inverses*, Acta Arith. **125** (2006), 11–19.

[73] H. N. Liu and W. G. Zhai, *A note on the pseudorandomness of the Liouville function*, Acta Arith. **136** (2009), 101–121.

[74] H. N. Liu, T. Zhan, and X. Y. Wang, *On the correlation of pseudorandom binary sequences with composite moduli*, Publ. Math. Debrecen **74** (2009), 195–214.

[75] S. Louboutin, J. Rivat, and A. Sárközy, *On a problem of D. H. Lehmer*, Proc. Amer. Math. Soc. **135** (2007), 969–975.

[76] R. Marzouk and A. Winterhof, *On the pseudorandomness of binary and quaternary sequences linked by the Gray mapping*, Period. Math. Hungar. **60** (2010), 13–23.

[77] J. L. Massey, *Shift register synthesis and BCH decoding*, IEEE Transactions on Information Theory **15** (1969), 122–127.

[78] J. Matoušek and J. Spencer, *Discrepancy in arithmetic progressions*, J. Amer. Math. Soc. **9** (1996), 195–204.

[79] C. Mauduit, *Construction of pseudorandom finite sequences*, unpublished lecture notes to the conference, Information Theory, and Some Friendly Neighbours – ein Wunschkonzert, Bielefeld, 2003.

[80] C. Mauduit, H. Niederreiter, and A. Sárközy, *On pseudorandom* $[0, 1)$ *and binary sequences*, Publ. Math. Debrecen **71** (2007), 305–327.

[81] C. Mauduit, J. Rivat, and A. Sárközy, *Construction of pseudorandom binary sequences using additive characters*, Monatsh. Math. **141** (2004), 197–208.

[82] C. Mauduit, J. Rivat, and A. Sárközy, *On the pseudo-random properties of* n^c, Illinois J. Math. **46** (2002), 185–197.

[83] C. Mauduit, and A. Sárközy, *Construction of pseudorandom binary lattices by using the multiplicative inverse*, Monatsh. Math. **153** (2008), 217–231.

[84] C. Mauduit and A. Sárközy, *Construction of pseudorandom binary sequences by using the multiplicative inverse*, Acta Math. Hungar. **108** (2005), 239–252.

[85] C. Mauduit and A. Sárközy, *Family Complexity and VC-dimension*, Lecture Notes in Computer Science, Springer, to appear.

[86] C. Mauduit and A. Sárközy, *On finite pseudorandom binary sequences I: Measures of pseudorandomness, the Legendre symbol*, Acta Arith. **82** (1997), 365–377.

[87] C. Mauduit and A. Sárközy, *On finite pseudorandom binary sequences. II: The Champernowne, Rudin–Shapiro, and Thue–Morse sequences, a further construction*, J. Number Theory **73(2)** (1998), 256–276.

[88] C. Mauduit and A. Sárközy, *On finite pseudorandom binary sequences. V: On $(n\alpha)$ and $(n^2\alpha)$ sequences*, Monatsh. Math. **129(3)** (2000), 197–216.

[89] C. Mauduit and A. Sárközy, *On finite pseudorandom binary sequences. VI: On $(n^k\alpha)$ sequences*, Monatsh. Math. **130(4)** (2000), 281–298.

[90] C. Mauduit and A. Sárközy, *On finite pseudorandom binary sequences of k symbols*, Indag. Math **13** (2002), 89–101.

[91] C. Mauduit and A. Sárközy, *On large families of pseudorandom binary lattices*, J. Uniform Distribution Theory **2** (2007), 23–37.

[92] C. Mauduit and A. Sárközy, *On the measures of pseudorandomness of binary sequences*, Discrete Math. **271** (2003), 195–207.

[93] W. Meidl and A. Winterhof, *Linear complexity of sequences and multisequences*, in: G. Mullen, D. Panario (eds.), Handbook of Finite Fields, Boca Raton, London, New York, CRC Press, to appear.

[94] W. Meidl and A. Winterhof, *Some notes on the linear complexity of Sidel'nikov–Lempel–Cohn-Eastman sequences*, Des. Codes Cryptogr. **38(2)** (2006), 159–178.

[95] A. Menezes, P. C. van Oorschot, and S. Vanstone, *Handbook of Applied Cryptography*, CRS Press, Boca Raton, 1997.

[96] L. Mérai, *A construction of pseudorandom binary sequences using both additive and multiplicative characters*, Acta Arith. **139** (2009), 241–252.

[97] L. Mérai, *A construction of pseudorandom binary sequences using rational functions*, Unif. Distrib. Theory **4** (2009), 35–49.

[98] L. Mérai, *Construction of large families of pseudorandom binary sequences*, Ramanujan J. **18** (2009), 341–349.

[99] L. Mérai, *Construction of pseudorandom binary lattices based on multiplicative characters*, Period. Math. Hungar. **59** (2009), 43–51.

[100] L. Mérai, *Construction of pseudorandom binary lattices using elliptic curves*, Proc. Amer. Math. Soc. **139** (2011), 407–420.

[101] L. Mérai, *On finite pseudorandom lattices of k symbols*, Monatsh. Math. **161(2)** (2010), 173–191.

[102] H. Niederreiter, *Linear complexity and related complexity measures for sequences*, Progress in Cryptology – INDOCRYPT 2003, Lecture Notes in Computer Science, Vol. 2904, pp. 1–17, Springer, Berlin, 2003.

[103] H. Niederreiter, *On the distribution of pseudo-random numbers generated by the linear congruential method.*, Math. Comp. **26** (1972), 793–795.

[104] H. Niederreiter, *On the distribution of pseudo-random numbers generated by the linear congruential method. II*, Math. Comp. **28** (1974), 1117–1132.

[105] H. Niederreiter, *On the distribution of pseudo-random numbers generated by the linear congruential method. III*, Math. Comp. **30** (1976), 571–597.

[106] H. Niederreiter, *Quasi-Monte Carlo methods and pseudorandom numbers*, Bull. Amer. Math. Soc. **84** (1978), 957–1041.

[107] H. Niederreiter, *Random Number Generation and Quasi-Monte Carlo Methods*, CBMS-NSF Regional Conference Series in Applied Math., Vol. 63, Soc. Industr. Applied Math., Philadelphia, 1992.

[108] H. Niederreiter, *Some computable complexity measures for binary sequences*, Sequences and their Applications, Singapore, 1998, Springer Ser. Discrete Math. Theor. Comput. Sci., pp. 67–78, Springer, London, 1999.

[109] H. Niederreiter and J. Rivat, *On the correlation of pseudorandom numbers generated by inversive methods*, Monatsh. Math. **153** (2008), 251–264.

[110] H. Niederreiter, J. Rivat, and A. Sárközy, *Pseudorandom sequences of binary vectors*, Acta Arith. **133(2)** (2008), 109–125.

[111] I. Niven and H. S. Zuckerman, *On the definition of normal numbers*, Pacific J. Math. **1** (1951), 103–109.

[112] S.-M. Oon, *On pseudo-random properties of some Dirichlet characters*, Ramanujan J. **15** (2008), 19–30.

[113] S.-M. Oon, *Pseudorandom properties of prime factors*, Period. Math. Hungar. **49** (2004), 107–118.

[114] A. Ostafe and A. Winterhof, *Some applications of character sums*, in: G. Mullen and D. Panario (eds.), Handbook of Finite Fields, Boca Raton, London, New York, CRC Press, to appear.

[115] T. Ritter, *Linear Complexity: A Literature Survey*, http://www.ciphersbyritter.com/RES/LINCOMPL.HTM.

[116] J. Rivat, *On pseudo-random properties of* $P(n)$ *and* $P(n+1)$, Period. Math. Hungar. **43** (2001), 121–136.

[117] J. Rivat and A. Sárközy, *Modular constructions of pseudorandom binary sequences with composite moduli*, Period. Math. Hungar. **51** (2005), 75–107.

[118] J. Rivat and A. Sárközy, *On pseudorandom sequences and their application*, Lecture Notes in Comput. Sci. 4123, General theory of information transfer and combinatorics, pp. 343–361, Springer, Berlin, Heidelberg, 2006.

[119] K. F. Roth, *Remark concerning integer sequences*, Acta Arith. **9** (1964), 257–260.

[120] R. A. Rueppel, *Linear complexity and Random Sequences*, Proc. Advances in Cryptology – EUROCRYPT '85, Linz, Austria, April 9–12, 1985, LNCS 219, pp. 167–188.

[121] A. Sárközy, *A finite pseudorandom binary sequence*, Studia Sci. Math. Hungar. **38** (2001), 377–384.

[122] A. Sárközy, *On finite pseudorandom binary sequences and their applications in cryptography*, Tatra Mt. Math. Publ. **37** (2007), 123–136.

[123] A. Sárközy and C. L. Stewart, *On pseudorandomness in families of sequences derived from the Legendre symbol*, Period. Math. Hungar. **54** (2007), 163–173.

[124] I. E. Shparlinski and A. Winterhof, *On the discrepancy and linear complexity of some counter-dependent recurrence sequences*, Lecture Notes in Comput. Sci., 4086, pp. 295–303, Springer, Berlin, 2006 Sequences and their applications – SETA 2006.

[125] B. Sziklai, *On the symmetry of finite pseudorandom binary sequences*, Uniform Distribution Theory **6** (2011), 143–156.

[126] A. Tietäväinen, *Vinogradov's method and some applications*, in: C. Yildirim and S. A. Stepanov (eds.), Number theory and its applications, Lecture Notes in Pure and Applied Math., Vol. 204, Dekker, New York, 1998, 261–282.

[127] A. Topuzoğlu and A. Winterhof, *Pseudorandom numbers: Uniform distribution and exponential sums*, in: S. Boztas (ed.), CRC Handbook of Sequences, Codes, and Applications: Chapman and Hall/CRC Press, to appear.

[128] A. Topuzoğlu and A. Winterhof, *Pseudorandom sequences*, in: G. Mullen and D. Panario (eds.), Handbook of Finite Fields, Boca Raton, London, New York, CRC Press, to appear.

[129] V. Tóth, *Collision and avalanche effect in families of pseudorandom binary sequences*, Period. Math. Hungar. **55** (2007), 185–196.

[130] V. Tóth, *The study of collision and avalanche effect in a family of pseudorandom binary sequences*, Period. Math. Hungar. **59** (2009), 1–8.

[131] I. M. Vinogradov, *Elements of Number Theory*, Dover 1954.

[132] Y. Wang, *Linear Complexity versus Pseudorandomness: On Beth and Dai's result*, Advances in Cryptology – ASIACRYPT'99 (Singapore), Lecture Notes in Computer Science, Vol. 1716, pp. 288–298, Springer, Berlin, 1999.

[133] A. Weil, *Sur les courbes algébriques et les variétés qui s'en déduisent*, Act. Sci. Ind. 1041, Hermann, Paris, 1948.
[134] A. Winterhof, *Linear complexity and related complexity measures*, in: I. Woungang, S Misra, and S. C. Misra (eds.), Selected Topics in Information and Coding Theory, Vol. 7, Singapore: World Scientific, 2010, 3–40.
[135] A. Winterhof, *Measures of pseudorandomness*, in: S. Boztas, (ed.), CRC Handbook of Sequences, Codes and Applications: Chapman and Hall/CRC Press, to appear.

Sophie Huczynska

Existence Results for Finite Field Polynomials with Specified Properties

Abstract: In this survey article, we discuss what is currently known about the existence of finite field polynomials with certain desirable properties. The properties with which we will be chiefly concerned are primitivity, normality and specified coefficients. Work on obtaining existence results for polynomials simultaneously possessing several properties of this type has been ongoing for several decades (the well-known Primitive Normal Basis Theorem is an early example) and continues to be a thriving area of research. Such polynomials have useful applications, including efficient computation in finite fields, fast Fourier transform, coding theory and cryptography. The main approach used to establish such results employs character sum techniques and estimates; further tools such as sieving techniques and p-adic methods have also been applied to obtain more complete results. Here, we review the literature and describe the most up-to-date results known, discuss key themes in the methods used to obtain them, and describe current open problems.

Keywords: Finite Field, Polynomial, Primitive, Normal, Specified Coefficients, Character Sum

2010 Mathematics Subject Classifications: Primary: 12-02; Secondary: 11T06, 11T24, 11T30

Sophie Huczynska: University of St Andrews, St Andrews, Fife, United Kingdom,
e-mail: sophieh@mcs.st-and.ac.uk

1 Introduction

The purpose of this survey article is to assemble, for the convenience of the theorist and practitioner alike, the results currently known about the existence of finite field polynomials with certain desirable properties, and to discuss the methods and approaches used to obtain them. The properties with which we will be chiefly concerned are primitivity, normality and specified coefficients. Work on obtaining existence results for polynomials simultaneously possessing several properties of this type has been ongoing for several decades (the well-known Primitive Normal Basis Theorem is

---

I would like to thank my colleague Nik Ruškuc for his constructive feedback on the first draft and the anonymous reviewer for his/her helpful comments.

an early example) and continues to be a thriving area of research. Such polynomials have useful applications, including efficient computation in finite fields, fast Fourier transform, coding theory and cryptography. In terms of methodology, the essence of the proof technique (involving character sums) dates back to the work of Carlitz, and a continuous thread runs through all the existence proofs, from Carlitz up to the present day. However, the inherent limitations of this basic approach have been the stimulus for many ingenious and innovative new approaches, introduced at various stages by various authors: two examples (now widely-used) are sieving and p-adic techniques. For polynomial existence problems, the first type of result to be seen in the literature is typically an asymptotic result, with the small-parameter cases proving the most difficult to "pin down." The technique generally consists of obtaining a condition which is satisfied for almost all pairs (q, n), then progressively resolving the remaining (small) cases using a combination of more specialized theoretical arguments and computation.

This survey article owes much to those which have come before. Prior to this article, the last survey article on the topic of existence results for polynomials was that of Cohen in 2005 [13]. Since the appearance of that work, however, many notable advances have been made, including the complete resolution (by Cohen himself and his coauthors) of the Hansen–Mullen primitivity conjecture, new results on primitive normal polynomials with around half their coefficients prescribed, and new extensions of the primitive normal basis theorem.

The article is structured as follows: the first part consists of a summary of results on the existence of finite field polynomials with specified properties, from their earliest appearance in the literature to the current state-of-the-art. The second part outlines the proof approaches used to obtain such results, drawing out the common underlying philosophy while explaining the additional techniques necessary in order to obtain more complete results. This survey, while comprehensive, does not claim to be an exhaustive review of the field.

2 A Survey of Known Results

Throughout this paper, we let $\mathbb{F}_q$ be the finite field of order q, where q is a power of its characteristic p (p prime). We denote its degree n extension by $\mathbb{F}_{q^n}$ ($n \in \mathbb{N}$). We will write a typical monic polynomial $f(x) \in \mathbb{F}_q[x]$ of degree n as

$$f(x) = x^n + a_1 x^{n-1} + \cdots + a_k x^{n-k} + \cdots + a_n \,. \tag{2.1}$$

We call a_i the i-th coefficient of f, $\{a_1, \ldots, a_k\}$ the *first k coefficients* of f, and $\{a_n, \ldots, a_{n-m+1}\}$ the *last m coefficients* of f.

It is well known that the multiplicative group $\mathbb{F}_{q^n}^*$ of non-zero elements of $\mathbb{F}_{q^n}$ is cyclic; a generator γ of this group is called a *primitive element* of $\mathbb{F}_{q^n}$. The element γ

and its conjugates are the roots of a *primitive polynomial* $f(x) \in \mathbb{F}_q[x]$ of degree n, which is monic and irreducible.

A classical result in finite field theory is the Normal Basis Theorem. An element γ of $\mathbb{F}_{q^n}$ is called *normal* (or *free*) over $\mathbb{F}_q$ if γ and its conjugates form a basis of $\mathbb{F}_{q^n}$ over $\mathbb{F}_q$; such a basis is called a *normal basis*. The Normal Basis Theorem states that a normal basis for $\mathbb{F}_{q^n}$ over $\mathbb{F}_q$ exists for every q and n. The element γ and its conjugates are the roots of a *normal polynomial* (*free polynomial*) $f(x) \in \mathbb{F}_q[x]$ of degree n, which is monic and irreducible. In fact, this result tells us that the additive structure of $\mathbb{F}_{q^n}$ is (in a sense which we will discuss later) also cyclic; its generators are precisely the free elements of $\mathbb{F}_{q^n}$. Further information on the theory of finite fields may be found in [62].

Throughout, we will speak interchangeably of the existence of polynomials with certain properties (primitivity, freeness or specified coefficients) and of the existence of field elements having such polynomials as their minimal polynomial. Depending on the nature of the individual existence problem, it may be the case that the problem is expressible more naturally in one of these contexts than the other.

2.1 Normal Bases

Normal bases can be considered in the more general setting of Galois extensions of arbitrary fields; they were first used by Gauss [39] in his work on the constructibility of regular polygons. For finite fields, the Normal Basis Theorem was stated without proof by Eisenstein [26] in 1850; a proof in the $\mathbb{F}_p$ case was given by Schönemann [74], also in 1850, and the first complete proof was given by Hensel [52] in 1888. Hensel himself observed the computational advantages of the normal basis representation, a theme which later came to prominence as the area was applied to coding theory and cryptosystems. Such applications require the implementation of finite field arithmetic in hardware or software; multiplication schemes such as those in [63] and [70] exploit the normal basis representation. More details on normal bases can be found, for example, in [37].

It is not the case that a normal basis of $\mathbb{F}_{q^n}$ over $\mathbb{F}_q$ is necessarily a normal basis over an intermediate field, nor vice versa. A normal basis of $\mathbb{F}_{q^n}$ which is a normal basis over all intermediate fields is known as a *complete normal basis*; the existence of such a basis for any finite field follows from the work of Blessenohl and Johnsen [4] (who prove the result for any Galois extension L of a field K).

Theorem 2.1 (Complete Normal Basis Theorem for Finite Fields). *For any prime power q and $n \in \mathbb{N}$, there exists a complete normal basis of $\mathbb{F}_{q^n}$ over $\mathbb{F}_q$. Equivalently, there exists an element $\gamma \in \mathbb{F}_{q^n}$ which is normal over K for every intermediate field K of $\mathbb{F}_{q^n}/\mathbb{F}_q$.*

Further discussion of this topic, and related questions, in the setting of finite fields can be found in [43].

2.2 Primitive Normal Bases

For a finite field, a normal element can be viewed as an additive generator of the field, while a primitive element is a multiplicative generator, so it is natural on aesthetic grounds to ask about the existence of an element which is at once an additive and multiplicative generator, i.e. a primitive normal element. The first results in this direction were due to Carlitz [5, 6] who established the Primitive Normal Basis Theorem for sufficiently large q, and Davenport [25], who established the result for all q, n with q prime. Existence of such a basis for every extension was first proved by Lenstra and Schoof [60], and a computer-free proof of this result was produced by Cohen and the present author [18]. We state it below:

Theorem 2.2 (Primitive Normal Basis Theorem (PNBT)). *For any prime power q and $n \in \mathbb{N}$, there exists a primitive normal basis of $\mathbb{F}_{q^n}$ over $\mathbb{F}_q$. Equivalently, there exists a primitive element $\gamma \in \mathbb{F}_{q^n}$ such that $\{\gamma, \gamma^q, \ldots, \gamma^{q^{n-1}}\}$ is a basis of $\mathbb{F}_{q^n}$ over $\mathbb{F}_q$.*

One of the factors which gives this topic its unique flavor is that the multiplicative and the additive group of the extension field have very similar structures, yet there is little interplay between the two. For $n \leq 2$, a primitive element is automatically normal, but for $n > 2$ it is certainly not the case in general that being primitive implies being normal, or vice versa.

All of the above-mentioned proofs of the Primitive Normal Basis Theorem are nonconstructive in nature; work of a different flavor has been done towards finding the primitive normal elements promised by the PNBT. Hachenberger, in [42], presented an algorithmic method which (at least theoretically) determines all primitive free elements (note that, contrary to what is suggested in [68], this latter paper does not reprove the PNBT). Shparlinski and Stepanov [73] provided an upper bound for N such that for a fixed primitive $\gamma \in \mathbb{F}_{q^n}$, a primitive normal element is guaranteed to occur in the set $\{\gamma, \gamma^2, \ldots, \gamma^N\}$. A probabilistic polynomial-time algorithm for finding a primitive normal element was given by von zur Gathen and Giesbrecht [38]. The distribution of primitive normal elements is discussed, for example, in Menezes [64] and Shparlinksi [72].

For practitioners seeking collections of primitive normal polynomials to use in applications, Beard and West [3] gave a primitive normal polynomial of degree n over $\mathbb{F}_{p^d}$ for each p, d and n satisfying $p < 10^2$, $p^d < 10^3$ and $p^{dn} < 10^6$. Gulliver, Serra and Bhargava [40] gave lists of primitive normal polynomials of small degrees over $\mathbb{F}_q$ for $q \leq 19$, $q \neq 9$. The work of Morgan and Mullen in [68] extended the range of published tables for such polynomials over prime fields, by exhibiting a primitive normal polynomial of degree n over $\mathbb{F}_p$ for each $p \le 97$ with $p^n < 10^{50}$.

Having seen that both the Primitive Normal Basis Theorem and the Complete Normal Basis Theorem hold for finite fields, a very natural next step is to ask for the simultaneous strengthening of both. Morgan and Mullen [69] found examples of primitive completely free elements by computer search in all cases when $q \le 97$ and $q^n \le 2^{31}$, and conjectured that such an element exists for every q and n.

Conjecture 2.3. *Given q and n, does there necessarily exist an element of $\mathbb{F}_{q^n}$ which is primitive and completely free over $\mathbb{F}_q$?*

Hachenberger has made several key steps towards establishing the Primitive Complete Normal Basis Theorem in full generality. In [46] it was established for a large class of extensions called *regular* extensions ($\mathbb{F}_{q^n}$ is called regular over $\mathbb{F}_q$ if, for each prime $r \neq \text{char}(\mathbb{F}_q)$ dividing n, $\text{ord}_r q$ and n are coprime) and further exceptional cases were resolved in [48].

Another natural strengthening of the Primitive Normal Basis Theorem is to ask for the existence of an element $y \in \mathbb{F}_{q^n}$ such that both y and its inverse are primitive and normal over $\mathbb{F}_q$. A proof for the case when $n \ge 32$, by Tian and Qi, is given in [75]; in [20], Cohen and Huczynska establish that this holds in all cases aside from five specific extensions.

Theorem 2.4 (Strong Primitive Normal Basis Theorem (SPNBT)). *For every prime power q and $n \in \mathbb{N}$, there exists a primitive element y of $\mathbb{F}_{q^n}$, free over $\mathbb{F}_q$, such that its reciprocal $y^{-1} \in \mathbb{F}_{q^n}$ is also primitive and free over $\mathbb{F}_q$, unless the pair (q, n) is one of the (genuine) exceptions*

$$(2,3), (2,4), (3,4), (4,3), (5,4) .$$

We end this section with a recent generalization of the Primitive Normal Basis Theorem due to Hsu and Nan [54], which extends the PNBT to the setting of Carlitz modules. Let A be the polynomial ring $\mathbb{F}_q[x]$ and let $f \in A$ be a monic irreducible; set $\mathcal{P} = (f)$ and $\mathbb{F}_{\mathcal{P}} = A/\mathcal{P}$. Let $\mathbb{F}_{\mathcal{P}^n}$ be the unique extension of $\mathbb{F}_{\mathcal{P}}$ of degree n; then $\mathbb{F}_{\mathcal{P}^n}$ may be regarded as a finite Carlitz A-module (for more details, see Section 3.1, or [41]). As such a module, $\mathbb{F}_{\mathcal{P}^n}$ is cyclic; in the spirit of the PNBT, it is therefore natural to ask about the existence of an element which is simultaneously a primitive element of $\mathbb{F}_{\mathcal{P}^n}$ and generates $\mathbb{F}_{\mathcal{P}^n}$ as an A-module. In the case when $f = x$, we see that $\mathbb{F}_{\mathcal{P}} \cong \mathbb{F}_q$, $\mathbb{F}_{\mathcal{P}^n} \cong \mathbb{F}_{q^n}$, and α is a generator of the A-module $\mathbb{F}_{\mathcal{P}^n}$ if and only if α is a normal element of $\mathbb{F}_{\mathcal{P}^n}$ over $\mathbb{F}_{\mathcal{P}}$.

Theorem 2.5. *Let $A = \mathbb{F}_q[x]$ and let $f \in A$ be a monic irreducible polynomial. Let $\mathbb{F}_{\mathcal{P}^n}$ be the unique extension of $\mathbb{F}_{\mathcal{P}} = A/\mathcal{P}$ of degree n, where $\mathcal{P} = (f)$. Then, except in finitely many cases, there exists a primitive element of $\mathbb{F}_{\mathcal{P}^n}$ which is a generator of the finite Carlitz A-module $\mathbb{F}_{\mathcal{P}^n}$.*

The finitely many possible exceptions are listed by the authors in [54]; a genuine exception occurs when $q = 2$, $f = x^2 + x + 1$ and $n = 1$.

2.3 Prescribed Coefficients

Given that, for any $\mathbb{F}_{q^n}$, there is guaranteed to exist an element which is primitive, or free, or indeed both, we may wish to impose some additional conditions on this element. There are both theoretical and practical reasons for making this demand: as well as providing the theoretician with a greater understanding of how widely-occurring these properties are, the knowledge of extra properties can help the practitioner by reducing the search space when searching for such an element, and is useful in cryptographic or coding applications.

One very natural condition is to demand that our primitive or free element should possess prescribed norm or trace. This is of course equivalent to specifying the constant term or coefficient of x^{n-1} in the corresponding primitive or free polynomial. We observe that the polynomial formulation of the question naturally suggests more general conditions which may be imposed, in terms of specifying various subsets of the coefficients of the primitive or free polynomial.

Throughout, we shall assume any polynomial to be monic of degree n over $\mathbb{F}_q$, and of the form $f(x) = x^n + \sum_{i=1}^{n} a_i x^{n-i}$, as given in (2.1).

2.4 Primitive Polynomials: Prescribed Coefficients

The first problem of this type to be found in the literature concerns the existence of primitive polynomials with prescribed first coefficient a (equivalent to prescribed trace for a primitive element); the $q = 2$ case was proved by Davenport in [25]. Motivated by applications such as the design of Costas arrays, the general case was resolved in 1989 by Jungnickel and Vanstone in [56] for all but finitely many exceptional cases; the exceptions were independently resolved by Moreno in [67]. A single complete proof was given by Cohen in [9] (subsequently streamlined in [22]).

Theorem 2.6. *Let $n > 1$ and $a \in \mathbb{F}_q$. There exists a primitive polynomial of degree n with $a_1 = a$, except when $(q, n, a) = (2, q, 0)$ or $(4, 3, 0)$.*

Motivation for further work on the existence of a primitive polynomial with a single specified coefficient was provided by the Hansen–Mullen Conjecture (1992), which appeared in [51] and was based on computational evidence. It conjectured that, with the genuine exception of a few specific parameter sets, a primitive degree-n polynomial with m-th coefficient prescribed ($1 \leq m < n$) always exists. Assorted special cases of the conjecture were established by various authors. Of particular note is the 1997 proof by Han [50] that the second coefficient of a primitive polynomial may be arbitrarily prescribed as $a \in \mathbb{F}_q$, for any even prime power q and $(n, a) \neq (4, 0), (5, 0), (6, 0)$. In [50], Han introduced the idea of using p-adic methods to overcome the long-standing problem related to the characteristic. This p-adic approach was further developed and applied in a series of papers in 2004 by Han and his coauthor Fan ([28–32]), and has been instrumental in allowing significant progress to be made in this area. More details of this p-adic approach are given in the next section.

In 2004, the Hansen–Mullen conjecture was shown to hold asymptotically for large q as a function of n, by Fan and Han in [28], who also established an explicit result in the case when q is even and $n \geq 7$ odd [29]. The first comprehensive proof was given in 2006 by Cohen in [14] for $n \geq 9$, while the remaining cases were completed by Cohen and Prešern in [23] and [24].

Theorem 2.7. *Given integers m, n with $1 \leq m < n$ and $a \in \mathbb{F}_q$, there exists a monic primitive polynomial $f(x) \in \mathbb{F}_q[x]$ of degree n with $a_m = a$, with genuine exceptions when (q, n, m, a) take the following values:*

$$(q, 2, 1, 0), (4, 3, 1, 0), (4, 3, 2, 0) \text{ or } (2, 4, 2, 1)\,.$$

In some sense, the Hansen–Mullen Conjecture is a rather conservative description of the landscape, because it transpires that stronger results hold for primitive polynomials, in which a higher proportion of the coefficients may be specified. In fact, many cases of Theorem 2.7 may be deduced from results in the literature concerning the prescription of several coefficients.

We began this section with the question of prescribed trace. One may ask about the existence of a primitive polynomial with the coefficients of x^{n-1} and x both prescribed; this is equivalent to the existence of a primitive element y such that y and its inverse both have prescribed trace. The case when these prescribed coefficients are not both zero was dealt with by Cohen in [11], while the case when both of these coefficients are 0 was resolved in [8] by Chou and Cohen. In both cases, the result is established for $n \geq 5$; this represents a complete result in the latter case (since the answer is negative for $2 \leq n \leq 4$), but in the former case the question is still open for $2 \leq n \leq 4$.

Theorem 2.8. *Let $n \geq 5$ and $a, b \in \mathbb{F}_q$. Then there exists a primitive polynomial of degree n over $\mathbb{F}_q$ with $a_1 = a$ and $a_{n-1} = b$, except when $a = b = 0$ and $(q, n) = (2, 6), (3, 6)$ or $(4, 5)$.*

Another natural pair of coefficients to prescribe is that corresponding to the trace and norm of a primitive element (equivalent to prescribing the first and last coefficients of a primitive polynomial). Note that, in this case, the norm must itself be a primitive element of $\mathbb{F}_q$; we can also assume $n \geq 3$. An asymptotic result establishing this for q^n "sufficiently large" was proved by Chang and Lee in [7] in 2001. In the setting when the polynomial in question is both primitive and normal, a full result has now been proved (see the next section). From this, a near-complete result may be deduced for a primitive element with trace and norm prescribed; "near-complete" because the normality condition implies that the trace must be non-zero. The remaining zero trace case for $n = 3, 4$ was recently resolved by Cohen in [15], thereby yielding the following complete result:

Theorem 2.9. *Let $n \geq 3$. Let $a, b \in \mathbb{F}_q$, where b is a primitive element of $\mathbb{F}_q$. Then there exists a primitive polynomial of degree n over $\mathbb{F}_q$ with $a_1 = a$ and $a_n = (-1)^n b$ respectively, with genuine exceptions when $a = 0$ and $(q, n) = (4, 3)$ or $(7, 3)$.*

Note that, for primitive polynomials, prescribing norm and trace is equivalent to prescribing the last two coefficients (proceeding via the reciprocal polynomial); however this is no longer the case when the normality condition is added, since the monic reciprocal of a normal polynomial need not be normal.

One currently active direction of research is to consider the existence of primitive polynomials with the first m coefficients fixed (this is also being pursued in the primitive normal case). The case with $m = 2$, i.e. with the first two coefficients prescribed, was first considered by Han, who in 1996 proved the result for q odd and $n \geq 7$ [49]. In 2003, the cases $n = 5$ and 6 (also for q odd) were resolved by Cohen and Mills; their work also showed that, when $n = 4$, the result is true for sufficiently large (odd) q if the prescribed coefficients are not both zero. Some parts of the two-coefficient problem have now been subsumed, as the existence question for primitive polynomials with the first three coefficients fixed has been completely resolved for $n \geq 7$. In 2004, Fan and Han proved the result for $n \geq 8$ [30] and $n = 7$ with $p = 2, 3$ [31]; while Mills established the $n = 7$, $p > 3$ case in [65]. The finite number of remaining cases are resolved by examples given in [21].

Theorem 2.10. *Let $n \geq 7$. Then there exists a primitive polynomial of degree n over $\mathbb{F}_q$ with its first three coefficients arbitrarily prescribed.*

Hence, for the three-coefficient problem, the cases $n = 5$ and $n = 6$ remain to be dealt with (although this is expected to be routine). For the two-coefficient problem, there remain some unresolved cases for $4 \leq n \leq 7$.

It is natural to ask that, instead of taking a specific numerical value, the number m of prescribed coefficients should be some proportion of the total number n of coefficients. Various asymptotic results for primitive polynomials with their first m coefficients prescribed, where $m < \frac{n}{2}$, were produced between 1996 and 2001 (see, for example, [71]), all subject to the restriction that characteristic $p > m$. In 2004, Fan and Han [32] published an asymptotic result which is not subject to this restriction:

Theorem 2.11. *Let $n \in \mathbb{N}$. Then there exists a constant $C(n)$ such that, for $q > C(n)$, there exists a primitive polynomial of degree n over $\mathbb{F}_q$ with its first $\lfloor \frac{n-1}{2} \rfloor$ coefficients arbitrarily specified.*

This has now been strengthened, by Fan, Han and Feng, to a result (also asymptotic) concerning primitive normal polynomials, which will be discussed in the next section. An explicit result of this flavor is given by Cohen in [12]:

Theorem 2.12. *Let $m \leq \frac{n}{3}$ ($m \leq \frac{n}{4}$ when $q = 2$). Then there exists a primitive polynomial with its first m coefficients arbitrarily prescribed, with the (genuine) exception that there is no primitive cubic over $\mathbb{F}_4$ with first coefficient 0.*

It is possible, as suggested by Cohen in [13], to pose more general questions about primitive polynomials whose first k and last m coefficients are prescribed (subject to appropriate restrictions).

2.5 Primitive Normal Polynomials: Prescribed Coefficients

As in the primitive case, the first results in the literature concerning primitive normal polynomials with prescribed coefficients have involved norm and trace: note that, in this case, the norm must be primitive and the trace must be non-zero. In [16], Cohen and Hachenberger resolved the question of a primitive free element with prescribed (non-zero) trace:

Theorem 2.13. *Let $n \in \mathbb{N}$ and $a \neq 0 \in \mathbb{F}_q$. Then there exists a primitive normal polynomial of degree n over $\mathbb{F}_q$ with $a_1 = a$.*

In [17] the same authors studied the existence of primitive free elements with prescribed norm and obtained a complete result:

Theorem 2.14. *Let $n \in \mathbb{N}$ and $b \in \mathbb{F}_q$ with b a primitive element of $\mathbb{F}_q$. Then there exists a primitive normal polynomial of degree n over $\mathbb{F}_q$ with $a_n = (-1)^n b$.*

In this setting, the analog of the Hansen–Mullen Conjecture is to ask whether there always exists a primitive normal polynomial of degree n over $\mathbb{F}_q$ with m-th coefficient specified ($1 \leq m < n$), with an additional condition ensuring non-zero trace. This has been resolved for $n \geq 15$ by Fan and Wang in [36]; in theory their approach could be extended to cover smaller values. The authors conjecture that the result holds for $n \geq 2$ except in a small number of stated cases.

Theorem 2.15. *Let $n \geq 15$ and $1 \leq m < n$ and that $a \in \mathbb{F}_q$ (with $a \neq 0$ if $m = 1$). Then there exists a primitive normal polynomial with $a_m = a$.*

Conjecture 2.16. *Let $n \geq 2$ and $1 \leq m < n$. Let $a \in \mathbb{F}_q$ (with $a \neq 0$ if $m = 1$). Then there exists a primitive normal polynomial with $a_m = a$, with (genuine) exceptions when (q, n, m, a) take the following values:*

$$(2,3,2,1), (2,4,2,1), (2,4,3,1), (2,6,3,1), (3,4,2,2), (5,3,4,3), (4,3,2,1+\theta)$$

where $\mathbb{F}_4 = \mathbb{F}_2(\theta)$ with $\theta^2 + \theta + 1 = 0$.

Some work has been done on the existence of normal polynomials with a prescribed coefficient, i.e. without the requirement that the polynomial be primitive (as in the previous section). A complete existence result for a normal polynomial with prescribed last coefficient, is given by Cohen and Hachenberger in [17]. Here the only restriction is that the last coefficient be non-zero; this is more general than the case of a primitive polynomial (which has a primitivity condition on its last coefficient).

Theorem 2.17. *Let $n \in \mathbb{N}$ and $a \in \mathbb{F}_q^*$. Then there exists a normal polynomial of degree n with $a_n = a$.*

Concerning the existence of a primitive normal polynomial with multiple coefficients prescribed, Cohen and Hachenberger (in 2000) considered the constraint of both fixed norm and trace in [17], and resolved this problem for $n \geq 7$, except for a finite set of parameters. This work was improved up to $n \geq 5$ by Cohen in [10], and the remaining cases $n = 3$ and $n = 4$ were resolved by Cohen and Huczynska in [19] and [55], thereby completing the proof process.

Theorem 2.18. *Let $n \geq 3$. Let $a \in \mathbb{F}_q^*$ and let b be a primitive element of $\mathbb{F}_q$. Then there exists a primitive normal polynomial with $a_1 = a$ and $a_n = (-1)^n b$.*

We remark that this is particularly noteworthy in the case when $n = 3$, when it guarantees that we may take a cubic $x^3 + ax^2 + cx + b$ and fix a and b, with only c allowed to vary.

In 2007, the question of prescribed first and second coefficients was resolved in [34] for $n \geq 7$:

Theorem 2.19. *Let $n \geq 7$. Let $a, b \in \mathbb{F}_q$ with $a \neq 0$. Then there exists a primitive normal polynomial with $a_1 = a$ and $a_2 = b$.*

The case of the penultimate and last coefficients was established in [35] for $n \geq 5$.

Theorem 2.20. *Let $n \geq 5$. Let $a, b \in \mathbb{F}_q$ with b a primitive element of $\mathbb{F}_q$. Then there exists a primitive normal polynomial with $a_{n-1} = a$ and $a_n = (-1)^n b$.*

As in the previous section, we may ask for results when a certain proportion of the n coefficients are prescribed. For primitive normal polynomials, the following asymptotic results relating to the problem of specifying the first or last $\lfloor \frac{n}{2} \rfloor$ coefficients have been established. In 2007, it was shown in [33] that:

Theorem 2.21. *Let $n \geq 2$. Then there is a constant $C(n)$ such that for $q > C(n)$, there exists a primitive normal polynomial with its first $\lfloor \frac{n}{2} \rfloor$ coefficients arbitrarily prescribed (where a_1 must be non-zero).*

Observe that, compared to the result for primitive polynomials in [32], there is an increase in the number of coefficients specified, from $\lfloor \frac{n-1}{2} \rfloor$ to $\lfloor \frac{n}{2} \rfloor$. However, this improvement is obtained via a character sum estimate which uses the fact that the first coefficient is non-zero, so for polynomials with zero first coefficient, the previous result is not subsumed.

The last m coefficients are dealt with in [27] (2009):

Theorem 2.22. *Let $n \geq 2$. Then there is a constant $C(n)$ such that for $q > C(n)$, there exists a primitive normal polynomial with its last $\lfloor \frac{n}{2} \rfloor$ coefficients arbitrarily prescribed (where $a_n = (-1)^n b$ for some primitive element b of $\mathbb{F}_q$).*

Hachenberger has also investigated the question of imposing primitivity, normality and prescribed norm/trace constraints on elements of $\mathbb{F}_{q^n}$, in the context of intermediate fields. Let $\mathbb{F}_{q^k}$ be intermediate between $\mathbb{F}_{q^n}$ and $\mathbb{F}_q$ (i.e. $n = ke$ for some e, where $\mathbb{F}_{q^n}$ is the extension of $\mathbb{F}_{q^k}$ of degree e). Denote by $\mathcal{T}$ the set of triples (q, k, e) such that the following holds: for every $a \in \mathbb{F}_{q^k}$ which is normal over $\mathbb{F}_q$, there exists a primitive $\gamma \in \mathbb{F}_{q^n}$ which is normal over $\mathbb{F}_q$ and whose $(\mathbb{F}_{q^n}, \mathbb{F}_{q^k})$-trace is equal to a.

The result of Theorem 2.13 may be viewed as stating that $(q, 1, e) \in \mathcal{T}$ for all $e \geq 2$ and $q \geq 2$. In general, the question of determining whether $(q, k, e) \in \mathcal{T}$ (where $k, e \geq 2$) is found to be a difficult one. In [45] and [47], various results are established, including the following:

Theorem 2.23.
(i) $(q, k, 2) \in \mathcal{T}$ *if q is a power of 2;*
(ii) *with finitely many exceptions, $(q, k, e) \in \mathcal{T}$ whenever $e \geq 3$ and $q \geq 17$;*
(iii) *if $k, e(> 1)$ are powers of the same prime $r \geq 5$, then $(q, k, e) \in \mathcal{T}$;*
(iv) *if $a \geq 0$, $b \geq 1$ and q is congruent to 2, 4, 5 or 7 modulo 9, then $(q, 3^a, 3^b) \in \mathcal{T}$.*

This work has applications to *trace-compatible sequences* of primitive normal elements in towers of Galois fields. In [44] and [47], two intermediate fields are considered and a norm condition is added.

3 A Survey of Methodology and Techniques

In this section, we shall let E be the finite field of order q^n and F its unique subfield of order q. Let $\bar{F}$ be the algebraic closure of F.

Throughout the literature, it is noteworthy that all proofs establishing the existence of elements with these "desired" properties, proceed by obtaining an estimate of the number of such elements and showing this to be greater than zero. Hence, though the problems are stated in terms of finite field elements, their solutions are ultimately reached through the manipulation of inequalities in the rational numbers, a reformulation which allows the application of the results and techniques of number theory. While there are, of course, many individual differences across the papers in the literature, there is a uniform underlying approach: the estimate is obtained via the use of character sums and, in the case of primitivity and normality, originates in the Vinogradov criterion for primitive roots. We begin by explaining this basic approach; we then consider its limitations and how these have been overcome by the use of additional techniques, in particular sieving and the p-adic method.

3.1 Basic Approach

As suggested earlier, the approach is rooted in the fact that the multiplicative and additive groups of E over F are very similar. Both are finite cyclic modules over principal ideal domains. While there may be little interplay between them, their considerable structural analogy means that a unified approach may be adopted.

The field E is a cyclic Galois extension of degree n over F; a canonical generator of the Galois group is the Frobenius automorphism σ:

$$\sigma : E \to E, x \mapsto x^q .$$

We may consider σ as an F-linear mapping on E; then E carries the structure of a module over the polynomial ring $F[x]$ (with respect to σ) if we define a scalar multiplication

$$f \circ_\sigma \gamma := f^\sigma(\gamma) := \sum_{i=0}^{n} f_i \sigma^i(\gamma)$$

where $f = \sum_{i=0}^{n} f_i x^i \in F[x]$ and $\gamma \in E$. It transpires that many well-known properties of the multiplicative group have analogs for the additive group when considered as an $F[x]$-module.

For an element $\gamma \in \bar{F}^*$, $\gamma \in E$ if and only if $\gamma^{q^n-1} = 1$. The definition of the multiplicative order of γ as the smallest positive integer k such that $\gamma^k = 1$, is well known; γ is a primitive element of E precisely if $\operatorname{ord}(\gamma) = q^n - 1$. For $\gamma \in \bar{F}$, $\gamma \in E$ if and only if $(x^n - 1)^\sigma(\gamma) = 0$: its additive F-order $\operatorname{Ord}(\gamma)$ is defined to be the unique monic polynomial in $F[x]$ generating the annihilator of γ in $F[x]$ as an ideal, and γ is a free element precisely if $\operatorname{Ord}(\gamma) = x^n - 1$.

It is well known that the number of primitive elements of E is given by $\phi(q^n - 1)$, where ϕ is Euler's phi function. For a monic $f \in F[x]$, we define $\Phi(f) := |(F[x]/fF[x])^*|$ and $N(f) := |F[x]/fF[x]| = q^{\deg(f)}$. The additive analog M of the Möbius function may be defined in the natural way, setting $M(f) = (-1)^r$ if f is a product of r distinct monic irreducible factors, and $M(f) = 0$ if f is divisible by the square of such a factor. It can be shown that Φ behaves analogously to Euler's phi function and that, in particular, the number of free elements of E is given by $\Phi(x^n - 1)$. To complete the analogy between the multiplicative and additive groups of E considered as modules, observe that $E \cong F[x]/(x^n - 1)F[x]$ as $F[x]$-modules while $E^* \cong \mathbb{Z}/(q^n - 1)\mathbb{Z}$ as $\mathbb{Z}$-modules.

The following result is a variation on one of Vinogradov (see [59]). Let G be a finite cyclic group of order $|G|$. For any divisor m of $|G|$, we define $\gamma \in G$ to be m-free if

$\gamma = v^d$ (where $v \in G$ and $d|m$) implies $d = 1$; so in particular $\gamma \in G$ is $|G|$-free if and only if γ generates G.

Lemma 3.1. *For any divisor k of $|G|$, define the function V_k as follows:*

$$V_k(\gamma) = \sum_{d|k} \frac{\mu(d)}{\phi(d)} \sum_{\substack{\chi \in \hat{G} \\ \operatorname{ord}(\chi)=d}} \chi(\gamma)$$

for all $\gamma \in G$, where μ denotes the Möbius function and $\operatorname{ord}(\chi)$ *the order of χ in the group $\hat{G}$ of characters of G. Then*

$$V_k(\gamma) = \begin{cases} 0\,, & \text{if } \gamma \text{ is not } k\text{-free}, \\ \frac{k}{\phi(k)}\,, & \text{if } \gamma \text{ is } k\text{-free}. \end{cases}$$

In particular, when $k = |G|$, then $V_{|G|}(\gamma) \neq 0$ if and only if γ is a generator of G.

It is clear how this may immediately be applied to the multiplicative group E^* of our finite field, to yield a characteristic function for the set of k-free elements of E^*. For $k|q^n - 1$, define $V_k : E^* \to \mathbb{C}$ by

$$V_k(\gamma) = \sum_{d|k} \frac{\mu(d)}{\phi(d)} \sum_{\substack{\chi \in \widehat{E^*} \\ \operatorname{ord}(\chi)=d}} \chi(\gamma)\,;$$

then $\theta(k)V_k$ is the required characteristic function, where $\theta(k) := \frac{\phi(k)}{k}$.

An additive analog may be obtained as follows: for $g|x^n - 1$, we say that $\gamma \in E$ is g-free if $\gamma = h^\sigma(v)$, where $v \in E$ and h is an F-divisor of g, implies $h = 1$. In particular, an element is normal if and only if it is $x^n - 1$-free. To obtain a characteristic function for the set of g-free elements of E, define $V_g : E \to \mathbb{C}$ by

$$V_g(\gamma) = \sum_{f|g} \frac{M(f)}{\Phi(f)} \sum_{\substack{\chi \in \hat{E} \\ \operatorname{Ord}(\chi)=f}} \chi(\gamma)\,.$$

Then $\Theta(g)V_g$ is the required characteristic function, where $\Theta(g) := \frac{\Phi(g)}{N(g)}$. It is by combining these characteristic functions for appropriate k and g that the existence proofs will proceed.

Analogous character functions may be obtained in the Carlitz module setting (for more about Carlitz modules, see for example [41] or [1]; for more about this approach, see [54]). Let A be the polynomial ring $F[x]$ and let G_a be the additive group of A. The Carlitz action $\phi : A \to \operatorname{End}(G_a)$ of A upon G_a is defined to be the unique F-linear homomorphism such that $\phi(x) := (\alpha \mapsto x\alpha + \alpha^q) \in \operatorname{End}(G_a)$. The Carlitz module is a copy of G_a equipped with the Carlitz action of A. An alternative viewpoint is obtained by taking σ to be the q-th power Frobenius mapping; we may then identify the subring of $\operatorname{End}(G_a)$ comprising the F-linear endomorphisms, with the ring $A\{\sigma\}$

of polynomials in σ with coefficients in A equipped with the "twisted" multiplication law. Then the Carlitz action may be expressed as the F-algebra homomorphism $\phi : A \to A\{\sigma\}$ defined by $\phi(1) = \sigma^0, \phi(x) = x\sigma^0 + \sigma^1$.

A field L containing F is called an A-field if there is a morphism $\iota : A \to L$. Applying ι to the coefficients of $\phi(a)$ $(a \in A)$ yields elements of $L\{\sigma\}$; we can therefore consider the Carlitz module over an A-field L, by proceeding via ϕ. This yields the F-algebra homomorphism $\Phi : A \to L\{\sigma\}$; the A-module structure is given by $\alpha^a = \Phi(a)(\alpha)$.

Let $f \in A$ be a monic irreducible; set $\mathcal{P} = (f)$ and $\mathbb{F}_{\mathcal{P}} = A/\mathcal{P}$. Let $\mathbb{F}_{\mathcal{P}^n}$ be the unique (up to isomorphism) extension of $\mathbb{F}_{\mathcal{P}}$ of degree n. Then $\mathbb{F}_{\mathcal{P}^n}$ is an A-field via the map $A \to A/\mathcal{P} \hookrightarrow \mathbb{F}_{\mathcal{P}^n}$ and is an A-module via ϕ. As A-modules, $\mathbb{F}_{\mathcal{P}^n} \cong A/(f^n - 1)$, so in particular $\mathbb{F}_{\mathcal{P}^n}$ is cyclic. We may define a notion of order in this setting: for $\alpha \in \mathbb{F}_{\mathcal{P}^n}$, define its order $\mathrm{ord}(\alpha)$ to be the monic polynomial $h \in A$ of least degree such that $\alpha^h = 0$. Here, α is a generator of $\mathbb{F}_{\mathcal{P}^n}$ precisely if $\mathrm{ord}(\alpha) = f^n - 1$. Using the Vinogradov approach, a characteristic function may now be obtained for the set of elements which are generators of the finite Carlitz A-module $\mathbb{F}_{\mathcal{P}^n}$, and this may be used to obtain PNBT-type results.

Returning to the main discussion, one may formulate characteristic functions for the set of $\gamma \in E$ whose minimal polynomials have prescribed coefficients. These characteristic functions also employ character sums. For example, the set of $\gamma \in E$ with norm $N_{E/F}(\gamma) = b$ has characteristic function given by

$$\frac{1}{q-1} \sum_{\chi \in \hat{F^*}} \chi(N_{E/F}(\gamma) b^{-1}),$$

where $\hat{F^*}$ denotes the group of multiplicative characters of F^*, while the characteristic function for the set of $\gamma \in E$ with trace $Tr_{E/F}(\gamma) = a$ is given by

$$\frac{1}{q} \sum_{c \in F} \lambda(c(Tr_{E/F}(\gamma) - a)),$$

where λ is the canonical additive character of F.

3.2 A p-adic Approach to Coefficient Constraints

Specifying the trace of a root γ of an irreducible polynomial f is equivalent to specifying the first coefficient of f. Higher coefficients may be expressed as $\pm\sigma_k$, where σ_k is the k-th symmetric function of the roots of f; specifically,

$$f(x) = x^n - \sigma_1 x^{n-1} + \cdots + (-1)^m \sigma_m x^{n-m} + \cdots + (-1)^n \sigma_n .$$

Unfortunately, these functions are not easy to work with; a more satisfactory approach is to use the connection, via Newton's identities, between the set of functions

σ_k and the set of functions s_k (defined as the trace of γ^k). However, the drawback of this reformulation is that, while the values of $\{s_1, \dots, s_k\}$ are determined by those of the $\{\sigma_1, \dots, \sigma_k\}$, the converse is true only for $p = \mathrm{char}(F) > k$.

A breakthrough in the attempt to overcome this restriction, has been provided by the application of p-adic techniques (first introduced by Han and Fan, and subsequently developed by Cohen). The approach is to translate the problem into one over finite Galois rings, thereby eliminating the difficulty regarding the characteristic. Character sum estimates over Galois rings may then be applied (see Li [61]), and the traditional proof strategy pursued. This technique has facilitated the proof of various new results, such as estimates for the number of primitive polynomials whose first m coefficients are arbitrarily prescribed. The following exposition is based mainly on that of Cohen; for more details see (for example) [14], while for more information on the p-adic background, see a text such as [58]. To streamline the discussion, we concentrate on the case when a single coefficient is specified (say the m-th).

We begin by briefly outlining the necessary p-adic theory. Let $\mathbb{Q}_p$ be the completion of $\mathbb{Q}$ with respect to the usual p-adic metric, and C_p the usual completion of its algebraic closure. Let K_n be the splitting field (in C_p) of $x^{q^n} - x$ over $\mathbb{Q}_p$, and let Γ_n be the roots of this polynomial, i.e. $\Gamma_n = \{\zeta \in K_n : \zeta^{q^n} - \zeta = 0\}$. This set of roots is called the set of *Teichmüller points* of K_n; note that Γ_n^* is cyclic of order $q^n - 1$. The set Γ_n lies in the ring of integers R_n of K_n. R_n is a local ring with unique maximal ideal pR_n, and $R_n/pR_n \cong \mathbb{F}_{q^n}$. The ring R_n has the form

$$R_n = \left\{ \sum_{i=0}^{\infty} p^i \gamma_i \colon \gamma_i \in \Gamma_n \right\}.$$

Next, we take congruence classes modulo p^e where $e \geq 1$. Given the Teichmüller points Γ_n, define $\Gamma_{n,e}$ to be the set of classes of Γ_n which are congruent mod p^e. We have $\gamma^{q^n} = \gamma$ for $\gamma \in \Gamma_{n,e}$, and $\Gamma_{n,1} \cong \mathbb{F}_{q^n}$. Next, we define the *Galois ring*. For any positive integer e, the Galois ring $R_{n,e}$ is defined to be $R_n/p^e R_n$:

$$R_{n,e} = \left\{ \sum_{i=0}^{e-1} p^i \gamma_i \colon \gamma_i \in \Gamma_{n,e} \right\}.$$

Observe that $R_{n,1} = \Gamma_{n,1} \cong \mathbb{F}_{q^n}$. Every $\gamma \in \Gamma_{n,1}$ has a unique lift to each $\Gamma_{n,e}$ and to Γ_n. Multiplicative order is preserved in this process; in particular the lift of a primitive element is also a primitive element. If we consider the extension structure, we find that K_1 is a subfield of K_n, and the Galois group of K_n/K_1 is isomorphic to the Galois group of $\mathbb{F}_{q^n}/\mathbb{F}_q$, i.e. cyclic of order n and generated by the Frobenius automorphism. Hence, in this new setting we can ensure that normal elements correspond to normal elements, and that trace and norm may naturally be defined.

We now introduce the process of *lifting*. Let $f(x)$ be a monic irreducible over $\mathbb{F}_q$; equivalently, we may consider f over $R_{1,1}$:

$$f(x) = (x - \gamma)(x - \gamma^q) \cdots (x - \gamma^{q^{d-1}}) = x^d - \sigma_1 x^{d-1} + \cdots + (-1)^d \sigma_d\,,$$

where $\gamma \in \Gamma_{n,1}$ and each $\sigma_j \in \Gamma_{1,1} = R_{1,1}$. The polynomial f may be lifted to a unique polynomial (of the same form) over each $R_{1,e}$ (and over R_1). Now the root γ lies in $\Gamma_{n,e}$ (or Γ_n), while the coefficients σ_j now lie in $R_{1,e}$ (or R_1), but need not be in $\Gamma_{1,e}$ (or Γ_1). Moreover, from the discussion above, we can ensure that a finite field polynomial which is primitive or normal, is lifted to a corresponding primitive or normal polynomial (with respect to its new setting). The advantage of having performed the lifting process, is that there is a result applicable to irreducible polynomials over R_1 which generalizes the Newton identities, without any undesirable caveats regarding the characteristic. Hence we have that the values of the (lifted) σ_k are determined by those of the (lifted) s_i (more specifically, the so-called "s-components"). We first work over R_1, then reduce to $R_{1,e}$ for a suitable choice of e (all the preceding theory translates to $R_{1,e}$). The transition from R_1 to $R_{1,e}$ will enable us to employ character sums over finite sets. It transpires that, to ensure the m-th coefficient is specified, it suffices to prove that at most $\frac{m}{2}$ conditions on the s-components are satisfied (plus perhaps one additional condition on the norm). This reduction is crucial to the subsequent stages of the proof since $\frac{m}{2} + 1$ is, in general, much less than $\frac{n}{2}$, allowing terms of order $O(q^{n/2})$ to be dealt with in the inequalities. Hence, in summary, if we wish to prove the existence of a finite field polynomial with specified properties, the conditions for specifying coefficients up to the m-th can be replaced by (significantly fewer than m) trace equations in $R_{1,e}$ (for appropriate choice of e) where the "Newton's identity" relationship holds without restrictions; other properties (primitivity, normality) can also be translated into conditions in the Galois ring setting, and we can proceed to pursue the standard character sum/estimate approach with Galois rings replacing Galois fields.

A useful stratagem in tackling problems concerning the existence of primitive polynomials with the m-th coefficient, or last m coefficients, specified, is to consider the reciprocal polynomial. When $a_n \neq 0$, the monic reciprocal of polynomial f is given by

$$f^*(x) = \frac{x^n}{a_n} f\left(\frac{1}{x}\right) = x^n + \cdots + \frac{a_m}{a_n} x^m + \cdots + \frac{1}{a_n}.$$

It can be checked that f is primitive precisely if f^* is primitive. Then, to estimate the number of primitive polynomials with last m coefficients prescribed, it suffices to estimate the number of primitive polynomials whose first $m - 1$ coefficients and trace are prescribed. For the problem of a primitive polynomial f of degree n with specified m-th coefficient, where $m \leq \frac{n}{2}$, considering the reciprocal polynomial allows the $m > \frac{n}{2}$ case to be reduced to the $m < \frac{n}{2}$ case, now with the extra condition that the norm of f must also be specified. Note that this trick cannot be played if we impose the condition of normality upon f, since the reciprocal of a normal polynomial need not itself be normal.

There are various other reductions and simplifications which may be made to the problem in certain cases. For example, in the case where the "specified properties" are simply to be both primitive and normal, it is enough to show the existence of a normal element which is $Q(q,n)$-free, where $Q(q,n)$ is the square-free part of $\frac{q^n-1}{(q-1)\gcd(n,q-1)}$. However, this reduction may not be valid when additional requirements are imposed.

3.3 The Sieving Technique

We now discuss the type of estimates obtainable from character sums, their limitations, and how the concept of sieving can help overcome these limitations. Taking the characteristic functions of the sets of elements in E possessing each of the desired properties (k-free with $k|q^n-1$, g-free with $g|x^n-1$, or having coefficients specified in the minimal polynomial), and combining them, one obtains an expression for the number of elements of E possessing all such properties simultaneously. Call this quantity $N(k,g)$. Clearly our existence problem will be resolved if we can show that $N(k,g) > 0$. The expression for $N(k,g)$ is in terms of exponential sums; these are typically Gauss sums, but other types such as Jacobi or Kloosterman sums may also arise. Standard results on such sums may be applied to obtain lower bounds, which are frequently of the order $O(q^{n/2})$.

We illustrate with the case of primitive free elements. Define $W(k) = 2^{\omega(k)}$ to be the number of square-free divisors of k (with $\omega(k)$ being the number of distinct primes in k) and similarly define $W(g) = 2^{\omega(g)}$ to be the number of square-free divisors of g (where $\omega(g)$ counts the number of distinct monic irreducibles in g). In this case, $N(k,g)$ denotes the number of elements of E which are both k-free and g-free, so the number of primitive free elements is given by $N(q^n-1, x^n-1)$. We wish to show that $N(q^n-1, x^n-1)$ is positive for all q and n; in fact, as discussed above, it suffices to show that $N(Q, x^n-1) > 0$. A lower bound is immediately given by

$$N(k,g) \geq \theta(k)\Theta(g)(q^n - \epsilon_g - (W(k)-1)(W(g)-1)q^{n/2}) .$$

While the quantities $W(Q)$ and $W(x^n-1)$ may be bounded accurately enough for "q sufficiently large," it is difficult to estimate them with the precision necessary to establish the result for smaller q and n. This is an inevitable consequence of the "unpredictability" of factorization, particularly over the integers. For certain existence problems, by combining these direct character sum estimates with technical bounds on the number of prime/irreducible factors, analysis of specific prime decompositions and computational work, complete or near-complete results may be obtained (as in the Lenstra–Schoof proof of the PNBT, or the analogous Hsu–Nan proof in the Carlitz module case). However, for a general polynomial existence problem it is desirable (indeed, often necessary) to employ an approach which overcomes these limitations, by working instead with the divisors of q^n-1 and x^n-1. This motivates the use of *sieving*.

A sieve such as the following Brun-type sieve, can usefully be applied to yield a more flexible criterion. Given a set of factors $\{k_1, \ldots, k_r\}$ of k, we call this a set of *complementary divisors* of k with common divisor k_0 if $\operatorname{lcm}\{k_1, \ldots, k_r\} = k$ and for any distinct pair (i, j), $\gcd(k_i, k_j) = k_0$ (and similarly for g).

Lemma 3.2. *For $k | q^n - 1$ and $g | x^n - 1$, given complementary divisor pairs*

$$\{(k_1, g_1), \ldots, (k_r, g_r)\}$$

with common divisor pair (k_0, g_0),

$$N(k, g) \geq \left(\sum_{i=1}^{r} N(k_i, g_i) \right) - (r - 1)N(k_0, g_0) . \tag{3.1}$$

There is strong motivation for applying a sieving process. In passing from the original quantities ($q^n - 1$ for the primitive criterion and $x^n - 1$ for the normal criterion) to sets of their divisors, we work with smaller, simpler quantities which may be selected so that they possess more amenable properties. In particular, we can work with divisors d of $q^n - 1$ or $x^n - 1$ whose factorization into primes or irreducibles is more easily understood than those of the original quantities; hence $W(d)$ (the number of square-free divisors of d) can be more accurately estimated. For each individual problem, we can tailor our sieving approach. It may be helpful to subdivide a problem into cases, and apply a different sieving set for each (for example, conditions such as $n | q - 1$ or n prime, place restrictions on the factorization of $x^n - 1$, yielding a set of known factors of lower degree which may be used in the sieve). To resolve the case of general q and n, it can be helpful to choose a specific factorization of $x^n - 1$, for which specialized estimates may then be derived. For example in the proof of the PNBT in [18], $x^n - 1$ is factorized as $G_1 \ldots G_r g$ where the G_i are the irreducible factors of $x^n - 1$ of maximal degree $\operatorname{ord}_n q$; bounds on the number of irreducible factors of g (i.e. the number of irreducible factors of $x^n - 1$ of non-maximal degree) are separately established via an inductive proof. To resolve the general case of the PNBT, the sieve is applied with divisors $G_i g$ $(i = 1, \ldots, r)$.

In applying a sieving process, the challenge is to balance the potential gains and losses. For example, it may seem appealing to decompose each quantity into its constituent primes and irreducibles and apply the sieve with these, avoiding the need to estimate $W(d)$ for divisors d. However this must be balanced against increasing the size of other terms in the sieving inequality; it transpires that this approach would be rendered invalid if the r primes and irreducibles involved failed to satisfy $\sum_{i=1}^{r} \frac{1}{|p_i|} < 1$ (where $|p|$ denotes p if p prime and $q^{\deg p}$ if p irreducible).

The sieving approach has been extensively used in the work of Cohen and his coauthors; most recently in the form of "core-atom decomposition." Here, the formal product $m = kg$ (where $k | q^n - 1$ and $g | x^n - 1$) is decomposed as a "core" m_0 together with "atoms" $p_1, \ldots, p_r$ which are prime divisors of k or irreducible divisors of g not in m_0; the sieve is applied using the quantities $m_0 p_i$ $(i = 1, \ldots, r)$, and it is essential that the core and atoms are selected so that $\delta := 1 - \sum_{i=1}^{r} \frac{1}{|p_i|}$ is positive. It can then be shown (see, for example, [20]) that the original condition for $N(m)$ to be positive, namely

$$q > (2W(m))^{2/n},$$

can be replaced by the criterion

$$q > (2W(m_0)\Delta)^{2/n},$$

where $\Delta = \frac{r-1}{\delta} + 2$. Observe that this core-atom approach systematizes the approach often seen, in a more *ad hoc* context, in earlier papers (such as that of [18] outlined above).

A further advantage of sieving, is that it enables a separate treatment of the multiplicative and additive components; this is valuable for several reasons. We observe that the emphasis on $x^n - 1$ (rather than $q^n - 1$) in the preceding discussion of factorization is not coincidental: for an existence problem in which the desired properties include primitivity and normality, it is often preferable to sieve on the additive rather than the multiplicative part. Factorization of $x^n - 1$ is easier to predict and control than factorization of $q^n - 1$, since the finite field structure causes certain constraints and conditions on the polynomial (as noted above), whereas integer factorization is more "erratic".

Furthermore, this ability to separate the additive and multiplicative strands is useful in order to apply character sum estimates which hold for multiplicative or additive characters alone (but not for "mixed" character sums). The challenge of bounding quantities sufficiently well is a key challenge in this type of work. While the usual bounds can be surprisingly effective over a range of values, to establish the most delicate cases it can be beneficial, even necessary, to employ bounds derived in other areas of number theory. For example, results of Katz ([57]) arising from Soto–Andrade sums play a crucial role in establishing the quartic and cubic results in [19] and [55]; these hold only for multiplicative character sums. We note that, since [19] and [55] appeared, the results in [57] have been slightly improved and rendered more accessible by Moisio in [66]; use of these results would have marginally simplified the proofs in [19] and [55].

4 Conclusion

What does the future hold for the area of polynomial existence problems? Although the area has its roots in work which began in the 1930s, significant progress continues to be made, with particular momentum gained in the last few years as the fusion of p-adic techniques with the sieving method has rendered a new vein of problems tractable.

It is clear from the foregoing that there are challenges on various levels for researchers in this field. One such level is certainly filling in the gaps in our current knowledge, resolving various small cases to obtain complete results. These small cases are often the most challenging and can require very artful arguments. In particular, the analog of the Hansen–Mullen primitivity conjecture has been resolved for $n \geq 15$; a complete resolution would be satisfying. Some striking recent results of an asymptotic nature have been mentioned in this survey, particularly for polynomials with first (or last) m coefficients prescribed, when m is around $\frac{n}{2}$. It would be desirable to see these asymptotic results converted into complete results. More ambitiously, it would be of great interest to see new techniques developed which allow us to increase the number of specified coefficients beyond $\frac{n}{2}$.

The existence of a primitive completely free polynomial for all q and n is another key conjecture currently awaiting resolution. Combining the requirements to be primitive and normal over some (or all) intermediate fields, with various coefficient prescriptions, offers further avenues of investigation. This topic involves considerable structure theory of modules as well as character sums, suggesting another interesting emphasis.

The recent generalization of the Primitive Normal Basis Theorem in the Carlitz module context suggests a range of new problems. Firstly, since the result was proved in [54] without sieving, there is scope for an improved version of this result in which sieving (in particular, additive sieving) is employed. More generally, there are various ways in which the property of being a generator in the Carlitz module sense, could be combined with other properties described in this article to produce new existence questions. It would be good to see further interplay between the type of problems surveyed in this survey article, and the area of function fields; there are interesting questions about Carlitz module generators (see, for example, [53], concerning an analog of Artin's conjecture) which could inspire new research directions here.

Finally we remark that, while grounded in algebraic theory, this area has ongoing links with applications, particularly to coding theory and cryptography. New applications of the results continue to be developed (see for example [2], where work involving permutation groups and error-correcting codes employs existence results including the PNBT), and conversely applications generate new existence questions. From the research of Davenport onwards, the needs of practitioners have provided stimuli for research directions in polynomial existence problems, and we envisage this as a continuing theme.

References

[1] G. W. Anderson and D. S. Thakur, *Tensor powers of the Carlitz module and zeta values*, Ann. of Math. **132** (1990), 159–191.

[2] R. Bailey, *Uncoverings-by-bases for base-transitive permutation groups*, Designs, Codes and Cryptography **41** (2006), 153–176.

[3] J. T. Beard and K. I. West, *Some primitive polynomials of the third kind*, Math.Comp. **28** (1974), 1166–1167.

[4] D. Blessenohl and K. Johnsen, *A sharpening of the normal basis theorem*, J. Algebra **103** (1986), 141–159.

[5] L. Carlitz, *Primitive roots in a finite field*, Trans. Amer. Math. Soc. **73** (1952), 373–382.

[6] L. Carlitz, *Some problems involving primitive roots in a finite field*, Proc. Nat. Acad. Sci. U. S.A. **38** (1952), 314–318.

[7] S. Chang and J. B. Lee, *Some primitive polynomials over finite fields*, Acta Math. Sci. Ser. B Engl. Ed. **21** (2001), 412–416.

[8] W-S. Chou and S. D. Cohen, *Primitive elements with zero traces*, Finite Fields Appl. **7** (2001), 125–141.

[9] S. D. Cohen, *Primitive elements and polynomials with arbitrary trace*, Discrete Math. **83** (1990), 1–7.

[10] S. D. Cohen, *Gauss sums and a sieve for generators of Galois fields*, Publ. Math. Debrecen **56** (2000), 293–312.

[11] S. D. Cohen, *Kloosterman sums and primitive elements in Galois fields*, Acta Arithmetica **94** (2000), 173–201.

[12] S. D. Cohen, *Primitive polynomials over small fields*, Finite Fields and Applications, Seventh International Conference, Tolouse, 2003, Lecture Notes in Computer Science, vol 2948, pp. 197–214, Springer, Berlin, 2004.

[13] S. D. Cohen, *Explicit theorems on generator polynomials*, Finite Fields Appl. **11** (2005), 337–357.

[14] S. D. Cohen, *Primitive polynomials with a prescribed coefficient*, Finite Fields Appl. **12** (2006), 425–491.

[15] S. D. Cohen, *Primitive cubics and quartics with zero trace and prescribed norm*, Finite Fields Appl. (2012), http://dx.doi.org/10.1016/j.ffa.2012.09.008

[16] S. D. Cohen and D. Hachenberger, *Primitive normal bases with prescribed trace*, Appl. Algebra. Engrg. Comm. Comp. **9** (1999), 383–403.

[17] S. D. Cohen and D. Hachenberger, *Primitivity, freeness, norm and trace*, Discrete Math. **214** (2000), 135–144.

[18] S. D. Cohen and S. Huczynska, *The primitive normal basis theorem – without a computer*, J. London Math. Soc **67** (2003), 41–56.

[19] S. D. Cohen and S. Huczynska, *Primitive free quartics with specified norm and trace*, Acta Arith. **109** (2003), 359–385.

[20] S. D. Cohen and S. Huczynska, *The Strong Primitive Normal Basis Theorem*, Acta Arith. **143**, (2010), 299–332.

[21] S. D. Cohen and C. King, *The three fixed coefficient primitive polynomial problem*, JP J. Algebra Number Theory Appl. **4** (2004), 79–87.

[22] S. D. Cohen and M. Prešern, *Primitive finite field elements with prescribed trace*, Southeast Asian Bull. Math. **29** (2005), 283–300.

[23] S. D. Cohen and M. Prešern, *Primitive polynomials with prescribed second coefficient*, Glasg. Math. J. **48** (2006), 281–307.

[24] S. D. Cohen and M. Prešern, *The Hansen–Mullen primitive conjecture: completion of proof*, Number theory and polynomials, London Math. Soc. Lecture Note Ser. 352, pp. 89–120, Cambridge Univ. Press, Cambridge, 2008.

[25] H. Davenport, *Bases for finite fields*, J. London Math. Soc. **43** (1968), 21–39; **44** (1969), 378.

[26] G. Eisenstein, *J. reine angew. Math.* **39** (1850), 180–182; Math. Werke, vol. 2, Chelsea, New York, 1975, pp. 620–622.

[27] S. Fan, *Primitive normal polynomials with the last half coefficients prescribed*, Finite Fields Appl. **15** (2009), 604–614.

[28] S. Fan and W. Han, *p-adic formal series and primitive polynomials over finite fields*, Proc. Amer. Math. Soc. **132** (2004), 15–31.

[29] S. Fan and W. Han, *Primitive polynomials over finite fields of characteristic two*, Appl. Algebra Engrg. Comm. Comput. **14** (2004), 381–395.

[30] S. Fan and W. Han, *Character sums over Galois rings and primitive polynomials over finite fields*, Finite Fields Appl. **10** (2004), 36–52.

[31] S. Fan and W. Han, *Primitive polynomials with three coefficients prescribed*, Finite Fields Appl. **10** (2004), 506–521.

[32] S. Fan and W. Han, *p-adic formal series and Cohen's problem*, Glasg. Math. J. **46** (2004), 47–61.

[33] S. Fan, W. Han, and K. Feng, *Primitive normal polynomials with multiple coefficients prescribed: an asymptotic result*, Finite Fields Appl. **13** (2007), 1029–1044.

[34] S. Fan, W. Han, K. Feng, and X. Zhang, *Primitive normal polynomials with the first two coefficients prescribed: a revised p-adic method*, Finite Fields Appl. **13** (2007), 577–604.

[35] S. Fan and X. Wang, *Primitive normal polynomials with the specified last two coefficients*, Discrete Math. **309** (2009), 4502–4513.

[36] S. Fan and X. Wang, *Primitive normal polynomials with a prescribed coefficient*, Finite Fields Appl. **15** (2009), 682–730.

[37] S. Gao, PhD thesis, University of Waterloo, 2003.

[38] J. von zur Gathen and M. Giesbrecht, *Constructing normal bases in finite fields*, J. Symbolic Comput. **10** (1990), 547–570.

[39] C. F. Gauss, *Disquisitiones Arithmeticae*, Braunschweig, 1801.

[40] T. A. Gulliver, M. Serra, and V. K. Bhargava, *The generation of primitive polynomials in $GF(q)$ with independent roots and their applications for power residue codes, VLSI testing and finite field multipliers using normal basis*, Internat. J. Electron. **71** (1991), 559–576.

[41] D. Goss, *Basic structures of function field arithmetic*, Springer-Verlag, 1996.

[42] D. Hachenberger, *On primitive and free roots in a finite field*, Appl. Alg. Eng. Comm. Computing **3** (1992), 139–150.

[43] D. Hachenberger, *Finite fields. Normal bases and completely free elements*, The Kluwer International Series in Engineering and Computer Science, 390, Kluwer Academic Publishers, Boston, MA, 1997.

[44] D. Hachenberger, *Universal generators for primary closures of Galois fields*, Finite Fields and Applications, Augsburg, 1999, Springer, Berlin, 2001, pp. 208–223.

[45] D. Hachenberger, *Primitive normal bases for towers of field extensions*, Finite Fields Appl. **5** (1999), 378–385.

[46] D. Hachenberger, *Primitive complete normal bases for regular extensions*, Glasg. Math. J. **43** (2001), 383–398.

[47] D. Hachenberger, *Generators for primary closures of Galois fields*, Finite Fields Appl. **9** (2003), 122–128.

[48] D. Hachenberger, *Primitive complete normal bases: existence in certain 2-power extensions and lower bounds*, Discrete Math. **310** (2010), 3246–3250.

[49] W-B. Han, *The coefficients of primitive polynomials over finite fields*, Math. Comp. **65** (1996), 331–340.

[50] W-B. Han, *On two exponential sums and their applications*, Finite Fields Appl. **3** (1997), 115–130.

[51] T. Hansen and G. L. Mullen, *Primitive polynomials over finite fields*, Math. Comput. **59** (1992), 639–643.

[52] K. Hensel, *Über die Darstellung der Zahlen eines Gattungsbereiches für einen beliebigen Primdivisor*, J. Reine Angew. Math. **103** (1888), 230–237.

[53] C-N. Hsu, *On Artin's conjecture for the Carlitz module*, Compositio Math. **106** (1997), 247–266.

[54] C-N. Hsu and T-T. Nan, *A generalization of the Primitive Normal Basis Theorem*, J. Number Theory **131** (2011), 146–157.

[55] S. Huczynska and S. D. Cohen, *Primitive free cubics with specified norm and trace*, Trans. Amer. Math. Soc. **355** (2003), 3099–3116.

[56] D. Jungnickel and S. Vanstone, *On primitive polynomials over finite fields*, J. Algebra **124** (1989), 337–353.

[57] N. Katz, *Estimates for Soto–Andrade sums*, J. reine. angew. Math **438** (1993), 143–161.

[58] N. Koblitz, *p-adic numbers, p-adic analysis and zeta-functions*, Springer, New York, 1984.

[59] E. Landau, *Vorlesungen über Zahlentheorie II*, Hirzel, Leipzig, 1927.

[60] H. W. Lenstra, Jr. and R. J. Schoof, *Primitive normal bases for finite fields*, Mathematics of Computation, **48** (1987), 217–231.

[61] W-C. W. Li, *Character sums over p-adic fields*, J. Number Theory **74** (1999), 181–229.

[62] R. Lidl and H. Niederreiter, *Finite fields*, 2nd ed., Cambridge University Press, Cambridge, 1997.

[63] J. L. Massey and J. K. Omura, *Computational method and apparatus for finite field arithmetic*, U. S. patent no. 4,587,627, May 1986.

[64] A. J. Menezes (ed.), *Applications of finite fields*, Kluwer, Dordrecht, 1993.

[65] D. Mills, *Existence of primitive polynomials with three coefficients prescribed*, JP J. Algebra Number Theory Appl. **4** (2004), 1–22.

[66] M. Moisio, *Kloosterman sums, elliptic curves, and irreducible polynomials with prescribed trace and norm*, Acta Arith. **132** (2008), 329–350.

[67] O. Moreno, *On the existence of a primitive quadratic of trace 1 over GF(pm)*, J. Combin. Theory Ser. A **51** (1989), 104–110.

[68] I. Morgan and G. L. Mullen, *Primitive normal polynomials over finite fields*, Mathematics of Computation **63** (1994), 759–765.

[69] I. Morgan and G. L. Mullen, *Completely normal primitive basis generators of finite fields*, Utilitas Math. **49** (1996), 21–43.

[70] I. M. Onyszchuk, R. C. Mullin, and S. A. Vanstone, *Computational method and apparatus for finite field multiplication*, U. S. patent no. 4,745,568, May 1988.

[71] D-B. Ren, *On the coefficients of primitive polynomials over finite fields*, Sichuan Daxue Xuebao **38** (2001), 33–36.

[72] I. E. Shparlinski, *Computational and algorithmic problems in finite fields*, Math. Appl., Kluwer, Dordrecht, 1992.

[73] S. A. Stepanov and I. E. Shparlinski, *On the construction of primitive elements and primitive normal bases in a finite field*, Computational number theory (Debrecen, 1989), pp. 1–14, de Gruyter, Berlin, 1991.

[74] T. Schönemann, *Über einige von Herrn Dr. Einsentein aufgestellte Lehrsätze*, J. reine angew. Math. **40** (1850), 185–187.

[75] T. Tian and W-F. Qi, *Primitive normal element and its inverse in finite fields* (Chinese), Acta Math. Sinica (Chin. Ser.) **49** (2006), 657–668.

Dieter Jungnickel

Incidence Structures, Codes, and Galois Geometries

Abstract: It is the aim of this survey article to give a self-contained exposition of a recent, surprisingly tight, connection between incidence structures, linear codes, and Galois geometry: in joint work with Vladimir Tonchev [35], we have introduced new invariants for finite simple incidence structures $\mathcal{D}$, which admit both an algebraic and a geometric description. More precisely, there is one such invariant associated with the isomorphism class of $\mathcal{D}$ for each prime power q.
This approach was motivated by our study of the longstanding Hamada conjecture, which concerns a possible coding theoretic characterization of the classical designs formed by the points and d-subspaces of a finite projective or affine space. We will give a self-contained exposition of the resulting new theory, including the necessary background from coding theory, and discuss applications as well as open problems.

Keywords: Incidence Structure, Design, Configuration, Projective Space, Affine Space, Galois Closed Code, Simplex Code, Reed–Muller Code, Embedding Theorems

2010 Mathematics Subject Classifications: 51E20, 11T71, 94B27, 05B05

Dieter Jungnickel: Lehrstuhl für Diskrete Mathematik, Optimierung, und Operations Research, Universität Augsburg, Augsburg, Germany, e-mail: jungnickel@math.uni-augsburg.de

1 Introduction

As is well known, there is a close and fruitful interaction between design theory and coding theory. Indeed, the codes defined by incidence matrices of designs – or, from the opposite point of view, designs supported by codewords of a given weight – have been studied extensively for a long time now. We refer the reader to the monograph by Assmus and Key [2] for a systematic treatment of this topic.

There are also close connections between coding theory and Galois geometry. For instance, many interesting codes may be viewed as systems of points in finite projective spaces. Again, these connections have been extensively studied, and we refer the

Acknowledgments: As the general theory presented in this survey is joint work with Vladimir Tonchev, my first thanks go to him: as always, our collaboration was both fruitful and pleasant. I am also grateful to several colleagues for helpful discussions and for pointing out relevant citations regarding the research presented here: Jürgen Bierbrauer, Tor Helleseth, Relinde Jurrius, Michel Lavrauw, Tim Penttila, Alexander Pott, and Henning Stichtenoth.

reader to the recent survey by Landjev and Storme [39] for more information on this topic. For our purposes, the fundamental 1972 paper of Delsarte [13] and the subsequent, highly influential 1988 survey by Calderbank and Kantor [6] on the geometry of two-weight codes are of particular interest, as our results can be viewed as a far-reaching generalization of their approach. We shall give a self-contained exposition of a recent, surprisingly tight, connection between quite general incidence structures, linear codes, and Galois geometry.

In joint work with Vladimir Tonchev [35], we introduced new invariants for a large class of finite incidence structures $\mathcal{D}$, which admit both an algebraic and a geometric description. More precisely, there is one such invariant associated with the isomorphism class of $\mathcal{D}$ for each prime power q. This invariant may either be described geometrically in terms of embeddings of $\mathcal{D}$ into finite projective spaces over $GF(q)$, or algebraically in terms of codes over extension fields $GF(q^t)$ which are associated with the complementary incidence structure $\mathcal{D}^*$ of $\mathcal{D}$.

We now give a brief overview of what follows. The next two sections will deal with the necessary background from coding theory going beyond what is generally taken for granted: on the one hand, we will discuss trace codes and Galois closed codes in general, and on the other hand, we will present the recent results of Jurrius [36] on extension codes of the simplex codes and the first-order Reed–Muller codes.

Then, in Sections 4 and 5, we outline our new theory, by first introducing the q-dimension of an incidence structure, and then studying embeddings of incidence structures satisfying certain mild (and natural) restrictions into low-dimensional projective spaces $PG(n,q)$. In this way, we will obtain the new algebraic and geometric invariants for the specified prime power q, and we will see that these invariants essentially – namely, up to a term -1 – agree. We will also discuss the possibility of embeddings into affine spaces $AG(n,q)$.

In Section 6, we will give a first concrete example for our general theory by considering designs with classical parameters, that is, designs with the same parameters as some design formed by points and subspaces of a specified dimension in either projective space $PG(n,q)$ or affine space $AG(n,q)$. In particular, our q-invariant may be used to characterize the classical examples (at least, in most cases) – the original motivation for our investigations.

After this, we return to the general theory in Section 7 by briefly considering the connections to the geometry of two-weight codes. The subsequent two sections discuss further examples for the new theory by looking at some classes of Steiner systems and of configurations, respectively. The concluding section contains a selection of open problems for further research.

We assume that the reader is familiar with basic facts and terminology from the three areas we are concerned with. As general references, we recommend the following books: for design theory, [3]; for coding theory, [5, 40]; and for Galois geometry (that is, finite projective spaces), [27–29].

2 Galois Closed Codes

In this section, we review some facts from coding theory which are probably not as well known as they deserve to be. Throughout, let C denote a code over some extension field $E = GF(q^t)$ of $F = GF(q)$. There are two natural ways of associating with C a code over the ground field F. First, the *subfield subcode* C_F of C simply consists of all those words in C which have coordinates in F only. The second construction is a little more involved: the *trace code* $\mathrm{Tr}(C)$ of C is the code one obtains by applying the trace function $\mathrm{Tr} = \mathrm{Tr}_{E/F}$ from E to F, that is, the function defined by

$$\mathrm{Tr}(\xi) = \xi + \xi^q + \xi^{q^2} + \cdots + \xi^{q^{t-1}},$$

coordinate-wise to the words of C. The fundamental result on trace codes is the *Delsarte duality theorem* [14], which states that the dual of the trace code is the subfield subcode of the dual code; see also [5, Theorem 12.14]. For the convenience of the reader, we will include the simple proof. (Note that the two duals occurring in Delsarte's theorem are taken with respect to different vector spaces!)

Theorem 2.1. *Let C be an arbitrary code over E, and $\mathrm{Tr}(C)$ its trace code over F. Then*

$$(\mathrm{Tr}(C))^\perp = (C^\perp)_F.$$

Proof. A vector $\mathbf{x} = (\xi_1, \ldots, \xi_n) \in F^n$ belongs to $(\mathrm{Tr}(C))^\perp$ if and only if

$$\mathrm{Tr}\left(\sum_{i=1}^{n} \xi_i c_i\right) = \sum_{i=1}^{n} \xi_i \,\mathrm{Tr}(c_i) = 0$$

holds for all $\mathbf{c} \in C$. Let us assume $\mathrm{Tr}(\alpha) = 0$, but $\alpha \neq 0$, for such a vector $\mathbf{x}$ and some codeword $\mathbf{c} \in C$, where we write $\alpha = \sum_{i=1}^{n} \xi_i c_i \in E$. Note that we also get $\mathrm{Tr}(\alpha\lambda) = 0$ for all $\lambda \in E$, by considering the code words $\lambda\mathbf{c}$ with $\lambda \in E$. But then the trace function from E to F would be identically 0, which is absurd. Therefore, $\mathbf{x} \in (\mathrm{Tr}(C))^\perp$ holds if and only if $\sum_{i=1}^{n} \xi_i c_i = 0$ for all $\mathbf{c} \in C$, that is, iff $\mathbf{x} \in (C^\perp)_F$. □

In general, there is no further direct link between the trace code and the subfield subcode; however, for an important class of codes over E, both associated codes over F actually coincide. We need another definition: C is called *Galois closed* if it is invariant under applying the Frobenius automorphism $\xi \mapsto \xi^q$ of E over F (again coordinate-wise). We do not know where this terminology was first introduced (it appears in the books of Huffman and Pless [30] and Bierbrauer [5]), but the concept was already used in 1990 in a paper by Stichtenoth [42].

Galois closed codes have particularly nice properties; see [5, Theorem 12.17]. Again, we will include the short proof.

Theorem 2.2. *For any Galois closed code C over E, the subfield subcode C_F over F coincides with the trace code* $\mathrm{Tr}(C)$*; moreover, the dimension of C over E equals the dimension of $C_F = \mathrm{Tr}(C)$ over F.*

Proof. As C is Galois closed, $\mathrm{Tr}(\mathbf{c}) = \mathbf{c} + \mathbf{c}^q + \cdots + \mathbf{c}^{q^{t-1}} \in C$ for each codeword $\mathbf{c}$, so that $\mathrm{Tr}(C) \subseteq C_F$. Thus it suffices to show that both codes have the same dimension. By elementary linear algebra, linearly independent vectors in F^n stay linearly independent when considered as vectors in E^n, and therefore

$$\dim_F C_F \le \dim_E C \quad \text{and} \quad \dim_F (C^\perp)_F \le \dim_E C^\perp .$$

We may now apply Delsarte's Theorem 2.1 as follows:

$$\begin{aligned} \dim_E C = n - \dim_E C^\perp \ &\le \ n - \dim_F (C^\perp)_F \\ = n - \dim_F (\mathrm{Tr}(C))^\perp \ &= \ \dim_F \mathrm{Tr}(C) \ \le \ \dim_F C_F . \end{aligned}$$

This proves that all previous estimates have to hold with equality, establishing the theorem. □

Actually, in the situation of Theorem 2.2, the codes C and C_F also have the same minimum weight and the same strength, but we will not require these facts.

Later, we shall make essential use of the *Galois closure* $\overline{C}$ of an arbitrary code C over E: this is the smallest Galois closed code over E containing C. It may be obtained from C by taking the span of all images of some set of generators of C under the Frobenius automorphism and its powers. The following result stated in [5, Theorem 12.16] is an immediate consequence of the definitions and therefore left to the reader; nevertheless, it is crucial for our later proofs.

Proposition 2.3. *Let C be an arbitrary code over E, and $\overline{C}$ its Galois closure. Then the trace codes* $\mathrm{Tr}(C)$ *and* $\mathrm{Tr}(\overline{C})$ *coincide.*

Finally, we also need the following special case of a recent result of Giorgietti and Previtali [19] dealing with general Galois extensions, not only finite fields. It shows that Galois closed codes C over E and extension codes $S \otimes E$ of codes S over F are the same objects. We note that the explicit assertion on bases below is not included in [19], but has been stated much earlier in [42]. As this result is also crucial for our approach, we present the rather simple proof given in [34].

Theorem 2.4. *Let C be an arbitrary code over E. Then C is Galois closed if and only if it is the extension code $C = S \otimes E$ of some code S over F. In this case, S is the subfield subcode C_F of C, and every basis of C_F over F is also a basis of C over E.*

Proof. We first show that the extension $C = S \otimes E$ of an arbitrary code S over F is Galois closed: just choose a basis $B = \{\mathbf{b}_1, \ldots, \mathbf{b}_k\}$ of S and consider any element of C, that is, a linear combination $\mathbf{v} = \sum_i \lambda_i \mathbf{b}_i$ with scalars λ_i in E. Then

$$\mathbf{v}^q = \Big(\sum_i \lambda_i \mathbf{b}_i\Big)^q = \sum_i \lambda_i^q \mathbf{b}_i^q = \sum_i \lambda_i^q \mathbf{b}_i \in C\,,$$

where $\mathbf{b}_i^q = \mathbf{b}_i$ as $\mathbf{b}_i$ has all its coordinates in F.

Conversely, let C be any Galois closed code over E. By Theorem 2.2, the subfield subcode C_F satisfies $\dim_F C_F = \dim_E C$. Consider the extension code $C' = C_F \otimes E$ of C' and observe $C' \subseteq C$. By the first part of the proof, C' is Galois closed, and another application of Theorem 2.2 gives $\dim_E C' = \dim_F C_F = \dim_E C$, and thus $C = C'$ is indeed the extension code of its subfield subcode C_F. This also establishes the claim on bases. □

3 Extension Codes of Simplex and First-Order Reed–Muller Codes

In this section, we consider two particular classes of Galois closed codes, namely arbitrary extension codes of the simplex codes and the first-order Reed–Muller codes. The structure of these codes was recently investigated by Jurrius [36] who managed to compute their weight enumerators and even to determine the supports of all code words. Her results form the basis of our general theory, and therefore we will not only quote them but also sketch her proofs, which rely on a general approach that we cannot present in its entirety here. While we will outline the necessary results, we refer the reader to the expository articles by Jurrius and Pellikaan [37] and by Tsfasman and Vlădut [49] for more details and further references.

Throughout, we will only deal with *projective codes*, that is, with codes C whose dual code $C^\perp$ has minimum distance at least 3. In other words, no two columns of a generator matrix G of C can be linearly dependent. Therefore the projective codes of length N and dimension k over $GF(q)$ correspond – up to monomial equivalence – to equivalence classes of sets of N points in the projective space $PG(k-1,q)$, not all of which lie in a hyperplane. For the purposes of this paper, we call such a set of points a *projective system*. (In the literature, this term usually also allows using multisets of points.) The codes of interest to us provide particularly simple examples for this approach.

Example 3.1. The *q-ary simplex code S* of dimension k and length $N = q^{k-1} + \cdots + q + 1$ over $F = GF(q)$ is the monomially unique code over F whose generator matrix consists of N column vectors of length k which form a system of representatives for the 1-dimensional subspaces of $GF(q)^k$. Thus the corresponding projective system simply consists of *all* points of $\Pi = PG(k-1,q)$.

Example 3.2. Similarly, the *first-order q-ary Reed–Muller code R* of dimension k and length $N = q^{k-1}$ uses only the points of the affine space $AG(k-1,q)$; that is, R is the monomially unique code over F for which the corresponding projective system consists of the points of $\Pi = PG(k-1, q)$ contained in the complement A of some fixed hyperplane H of Π. Explicitly, up to monomial equivalence over F, we may use for H the hyperplane with the equation $x_1 = 0$ and normalize the first coordinates of the points in A to 1, so that a generator matrix G for R consists of the all-one row as first row, while the other positions in the columns of G contain all possible vectors in F^{k-1}.

Now let C be any code of length N and dimension k over $F = GF(q)$. We may be interested in the weights – and perhaps even in the possible supports of codewords – of an (arbitrary) extension code $C \otimes GF(q^t)$. In order to deal with this problem for all values of t simultaneously, one introduces a polynomial in three variables as follows; this generalizes the well-known (homogeneous) weight enumerator of C. It seems that weight enumerators of extension codes were first considered by Kløve in 1978, although in different form.

Definition 3.3. Let C be any code of length N and dimension k over $F = GF(q)$. The *extended weight enumerator* of C is the polynomial

$$W_C(X, Y, T) = \sum_{w=0}^{N} A_w(T) X^{N-w} Y^w \,, \tag{3.1}$$

where the $A_w(T)$ are polynomials in T for which $A_w(q^t)$ is the number of codewords of weight w in $C \otimes GF(q^t)$. It can be shown that the $A_w(T)$ are indeed integral polynomials of degree at most k.

Using the approach via projective systems, it is advisable to deal with another problem first, namely the determination of weights and supports of subcodes U of C of a given dimension r. Here the *support* of U is the set $\operatorname{supp} U$ of positions for which at least one word in U has a non-zero entry, and the *weight* of U is the cardinality $\operatorname{wt}(U)$ of $\operatorname{supp} U$. We remark that the smallest weight for which a subcode of dimension r exists is usually called the *r-th generalized Hamming weight* of C. Again, it is convenient to record the weight distribution for subcodes of dimension r, where $0 \le r \le k$, via a homogeneous polynomial:

Definition 3.4. Let C be any code of length N and dimension k over $F = GF(q)$. The *r-th generalized weight enumerator* of C is the polynomial

$$W_C^{(r)}(X, Y) = \sum_{w=0}^{N} A_w^{(r)} X^{N-w} Y^w \,, \tag{3.2}$$

where $A_w^{(r)}$ denotes the number of subcodes U of C with dimension r and weight w.

Generalized weight enumerators were first introduced by Helleseth, Kløve and Mykkeltveit [25] in 1979. It turns out that the two types of weight enumerators just defined determine each other. We state here only one of the two formulas, which we will need later.

Theorem 3.5. *The extended weight enumerator and the generalized weight enumerators of C satisfy the identity*

$$W_C(X,Y,T) = \sum_{r=0}^{k} \left(\prod_{j=0}^{r-1} (T - q^j) \right) W_C^{(r)}(X,Y) .$$

We now explain how the numbers $A_w^{(r)}$ may – at least in principle – be determined when C is given via a projective system in $\Pi = PG(k-1,q)$. This rests on a canonical correspondence between subspaces W of codimension r of Π and subcodes of dimension r of C. Explicitly, we may represent the points of W via the solutions of the linear system $A\mathbf{w}^T = \mathbf{0}$, where A is a suitable $r \times k$ matrix. Then AG is an $r \times n$ matrix of rank r, and therefore the rows of AG form a basis for a subcode U of dimension r of C. This gives the desired correspondence, which is actually independent of the choice of the matrix A describing W and the generator matrix G of C. Moreover, it allows one to determine the support of U via the following simple result.

Theorem 3.6. *Let U be a subcode of dimension r of C, and let W be the corresponding subspace of codimension r of Π. Then a coordinate j is contained in the support of U if and only if the associated point p_j of the projective system $\mathcal{P}$ defining C does not belong to W. In particular, the weight of U equals $N - |\mathcal{P} \cap W|$.*

Proof. Note that j does not belong to $\operatorname{supp} U$ if and only if all elements in a basis for U have a zero in this position. But this means that the j-th column of G is in the nullspace of A, that is, the point $p_j \in \mathcal{P}$ belongs to W. □

Finally, we need one further simple result which connects the supports of words in an extension code of C with the supports of certain subcodes of C.

Theorem 3.7. *Let C^t denote the vector space formed by the $t \times n$ matrices over F all of whose rows belong to C. Then there is an isomorphism α from the extension code $C \otimes E$ to C^t, where $E = GF(q^t)$.*

Now fix a word $\mathbf{c} \in C \otimes E$, and let U be the subcode of C generated by $M = \alpha(\mathbf{c})$. Then $\operatorname{supp} \mathbf{c} = \operatorname{supp} U$, and therefore $\operatorname{wt}(\mathbf{c}) = \operatorname{wt}(U)$.

Proof. Choose some basis B of E over F, so that the elements of E can be identified with the column vectors of length t representing them with respect to B. Doing so for all coordinates associates a matrix $\alpha(\mathbf{c})$ with any given code word $\mathbf{c} \in C \otimes E$, and it is easily checked that all rows of this matrix are codewords, that is, $\alpha(\mathbf{c}) \in C^t$. Clearly, α is an injective linear mapping from $C \otimes E$ to C^t and hence an isomorphism, since these two vector spaces have the same cardinality. (Note that this isomorphism is, of course, not canonical but depends on the choice of B.) The second assertion is clear, since a component of $\mathbf{c}$ is 0 if and only if the associated column of $\alpha(\mathbf{c})$ is $\mathbf{0}$. □

We now follow Jurrius [36] and apply the geometric approach to the simplex codes. First, we determine their generalized weight enumerators. In the corresponding formula, we need the *Gaussian coefficients* $\left[{m \atop k} \right]_q$ which give the number of k-dimensional subspaces of an m-dimensional vector space over $GF(q)$ (and constitute a q-analog of the binomial coefficients). Explicitly, one has

$$\begin{bmatrix} m \\ k \end{bmatrix}_q = \frac{(q^m - 1)(q^{m-1} - 1) \cdots (q^{m-k+1} - 1)}{(q^k - 1)(q^{k-1} - 1) \cdots (q - 1)}.$$

Theorem 3.8. *The generalized weight enumerators of the q-ary simplex code S of dimension k and length $N = q^{k-1} + \cdots + q + 1$ are given by*

$$W_S^{(r)}(X, Y) = \begin{bmatrix} k \\ r \end{bmatrix}_q X^{(q^{k-r}-1)/(q-1)} Y^{(q^k - q^{k-r})/(q-1)},$$

for $0 \le r \le n$.

Proof. We need to determine the weights of all subcodes of S. For this, let U be any subcode of dimension r of C, corresponding to the subspace W of codimension r of $\Pi = PG(k-1, q)$. Since the projective system $\mathcal{P}$ defining S is simply the set of all points of Π, the weight of U equals

$$N - |W| = \frac{q^k - 1}{q - 1} - \frac{q^{k-r} - 1}{q - 1},$$

by Theorem 3.6. As there are precisely $\left[{k \atop r} \right]_q$ subcodes of dimension r, the desired formula follows. □

Now a simple application of Theorem 3.5 gives the extended weight enumerator of the simplex codes:

Theorem 3.9. *The extended weight enumerator of the q-ary simplex code S of dimension k and length $N = q^{k-1} + \cdots + q + 1$ is given by*

$$W_S(X, Y, T) = \sum_{r=0}^{k} \left(\prod_{j=0}^{r-1} (T - q^j) \right) \begin{bmatrix} k \\ r \end{bmatrix}_q X^{(q^{k-r}-1)/(q-1)} Y^{(q^k - q^{k-r})/(q-1)}.$$

After these preparations, it is a simple matter to obtain the following structural result for arbitrary extension codes of the simplex codes:

Theorem 3.10. *Let S denote the q-ary simplex code of dimension k and length $N = q^{k-1} + \cdots + q + 1$, and let $S \otimes E$ be the Galois closed extension code of S over $E = GF(q^t)$. Then:*

(i) *The non-zero weights of $S \otimes E$ are exactly the cardinalities of the complements of subspaces of codimension r of $\Pi = PG(k-1, q)$, where $1 \le r \le t$.*

(ii) *The support of any word of weight $(q^k - q^{k-r})/(q-1)$ corresponds to the points not contained in some subspace of codimension r of Π, and every subspace of codimension r occurs in this manner.*

Proof. Assertion (i) is immediate from Theorem 3.9. For (ii), consider any word $\mathbf{c}$ of weight $(q^k - q^{k-r})/(q-1)$ in $S \otimes E$. Fix an isomorphism α as described in Theorem 3.7. Then $\operatorname{supp} \mathbf{c} = \operatorname{supp} U$, where U is the subcode of S generated by $M = \alpha(\mathbf{c})$. By Theorem 3.6, U is associated with a subspace W of codimension r of Π, and $\operatorname{supp} U$ is the complement of W in Π.

Finally, let W be an arbitrary subspace of codimension r of Π, where $1 \leq r \leq t$. By Theorem 3.6, W is associated with a subcode of dimension r of S. As $r \leq t$, this subcode U has a generator matrix M with t rows, and Theorem 3.7 yields a word $\mathbf{c} = \alpha^{-1}(M)$ in $S \otimes E$ with $\operatorname{supp} \mathbf{c} = \operatorname{supp} U$. This proves the second assertion in (ii). (Alternatively, this can also be established via a simple counting argument, see Lemma 2.6 of [34].) □

Jurrius [36] has also obtained a structural description for the extension codes of a first-order Reed–Muller code R. This situation can be dealt with in complete analogy to the case of simplex codes, and therefore we will just state the result, but omit its proof. We merely remark that the situation is a little more involved, as one needs to distinguish two possibilities when determining the generalized weight enumerators, which is due to the fact the projective system $\mathcal{P}$ used now is the set of points not contained in a given hyperplane H of Π. Namely, either the subspace W associated with a subcode U of R is contained in H, in which case the support of U is the entire complement A of H; or W intersects H in a subspace of codimension $r+1$ of Π, in which case the support of U is a subspace of codimension r of the affine space induced on A.

Theorem 3.11. *Let R denote the first-order q-ary Reed–Muller code of dimension k and length q^{k-1}, and let $R \otimes E$ be the Galois closed extension code of R over $E = GF(q^t)$. Then:*

(i) *The non-zero weights of C are q^{k-1} and the cardinalities of the complements of subspaces of codimension r of $\Sigma = AG(k-1, q)$, where $1 \leq r \leq t$.*

(ii) *The support of any word of weight $q^{k-1} - q^{k-1-r}$ corresponds to the points not contained in some subspace of codimension r of Σ, and every subspace of codimension r occurs in this manner.*

4 Simple Incidence Structures and Their Codes

In this section, we describe the new coding theoretic invariants for incidence structures which were recently introduced by Jungnickel and Tonchev [35]. We will only consider finite incidence structures. Throughout, we use the following (non-standard) terminology introduced in [35].

Definition 4.1. An incidence structure $\mathcal{D}$ will be called *simple* provided that $\mathcal{D}$ has neither repeated blocks nor repeated points and that each point belongs to some block, but not to all blocks, and dually.

While the notion of an incidence structure without repeated blocks is standard (and usually such incidence structures are called "simple"), our notion of simplicity asks for a bit more than usual: first, we add the analogous requirement that there should be no two points which are incident with exactly the same set of blocks. In addition, we assume that there are neither isolated points nor empty blocks, and that the *complementary incidence structure* $\mathcal{D}^*$ of $\mathcal{D}$ – that is, the incidence structure obtained by replacing every block of $\mathcal{D}$ with its complement – satisfies the same condition.

In terms of incidence matrices, all this just means that the incidence matrix A of $\mathcal{D}$ is not allowed to have two identical rows or columns, and that each row or column of A should contain both an entry 0 and an entry 1. In this connection we remark that, throughout this survey, incidence matrices will have their rows indexed by blocks and their columns indexed by points.

As we want to study, among other things, embeddings of incidence structures into projective or affine spaces, the somewhat strengthened notion of simplicity in Definition 4.1 is an entirely natural requirement.

We can now give the algebraic description of the new invariants; this vastly generalizes an idea introduced in [34] in connection with a coding theoretic characterization of the classical geometric designs.

Definition 4.2. Let A be the incidence matrix of some simple incidence structure $\mathcal{D}$, and let E be some finite field. Then any matrix obtained by replacing the entries 1 of A with arbitrary non-zero elements from E is called a *generalized incidence matrix* of $\mathcal{D}$ over E, or simply an *E-incidence matrix* of $\mathcal{D}$.

Now fix a prime power q and consider some generalized incidence matrix M with entries from $E = GF(q^t)$ for $\mathcal{D}$, where t may be any positive integer. We let $C(M)$ denote the code spanned by the rows of M over E. There is a natural way to obtain also an associated code over $F = GF(q)$, namely the trace code $\mathrm{Tr}(C(M))$. With this setup, the new invariant is defined as follows.

Definition 4.3. Let $\mathcal{D}$ be a simple incidence structure $\mathcal{D}$. Then the *q-dimension* $\dim_q \mathcal{D}$ of $\mathcal{D}$ is the smallest dimension of any trace code $\mathrm{Tr}(C(M))$ which can be obtained in the manner just described: that is,

$$\dim_q \mathcal{D} = \min \{ \dim \mathrm{Tr}(C(M)) : \\ M \text{ is a } GF(q^t)\text{-incidence matrix of } \mathcal{D} \text{ for some } t \} .$$

We remark that Definition 4.3 modifies and generalizes the definition of the q-dimension of a design as introduced by Tonchev [46], where only incidence matrices over $GF(q)$ were considered, while extension fields played no role at all.

It should be emphasized that the minimum is taken over a huge collection of codes associated with $\mathcal{D}$: a priori, one needs to consider all finite extension fields $E = GF(q^t)$ of $GF(q)$, and then all E-incidence matrices of $\mathcal{D}$. As we shall see via the equivalent geometric approach which will be the topic of the next section, there is actually a natural bound on t: it always suffices to restrict attention to $t \leq v - 1$, where v denotes the number of points of $\mathcal{D}$, and usually less trivial bounds on t can be achieved. Still, the new invariants would, in general, seem rather difficult to determine. Indeed, there is no known subexponential time algorithm for computing the q-dimension of an arbitrary simple incidence structure. Nevertheless, we are convinced of the theoretical importance of these new invariants, even though we are far less optimistic with respect to their practical applicability.

Finally, let us recall that the *q-rank* of an incidence structure $\mathcal{D}$, denoted by $r_q(\mathcal{D})$, is defined as the rank of its usual $(0, 1)$-incidence matrix over $GF(q)$. Trivially, the q-rank gives an upper bound for the q-dimension; of course, one would expect this bound to be rather weak in general.

Corollary 4.4. *Let $\mathcal{D}$ be a simple incidence structure. Then* $\dim_q \mathcal{D} \leq r_q(\mathcal{D})$.

In the case of simple 2-(v, k, λ)-designs, the bound given in Corollary 4.4 is interesting only for those prime powers $q = p^e$ for which the prime p divides the *order* $r - \lambda$ of $\mathcal{D}$: as is well known, in all other cases $r_q(\mathcal{D}) \in \{v - 1, v\}$.

5 Embedding Theorems

In this section, we consider embeddings of simple incidence structures in Galois geometries. Let us first define formally what we mean by an "embedding".

Definition 5.1. Let $\mathcal{D}$ be a simple incidence structure, q a prime power, and $\Pi = PG(n, q)$ some projective geometry over $GF(q)$. We say that $\mathcal{D}$ is *embedded* in Π if its point set V consists of points of Π and if each block X of $\mathcal{D}$ is induced by some subspace W of Π, that is, $X = V \cap W$. Note that there is always a unique smallest subspace W with this property; we will consider this subspace as *associated* with X.

An embedding is called *strong* if V spans Π, that is, if V contains $n + 1$ points in general position. We call $\mathcal{D}$ *(strongly) embeddable* if an isomorphic copy of $\mathcal{D}$ is (strongly) embedded in Π.

Finally, we speak of an *affine embedding* if the point set V of the embedding is contained in the complement of some hyperplane H of Π, since we can then view $\mathcal{D}$ as embedded into the corresponding affine space $\Sigma \cong AG(n, q)$.

At this point, we can explain one reason why we do not consider non-simple incidence structures: when trying to embed such a structure, repeated blocks would have to be induced by the same associated subspace. This would be contrary to the standard notions of embedding used in geometry, where all objects always are embedded via injections.

Note that any simple incidence structure on v points admits a trivial embedding into $PG(v-1,q)$, by using for V a set of v points of Π in general position. Thus it is natural to consider embeddings into projective spaces of as small a dimension as possible. In this context, we introduce the following terminology proposed in [18].

Definition 5.2. Let $\mathcal{D}$ be a simple incidence structure and q a prime power. Then the smallest integer n for which $\mathcal{D}$ can be embedded into the projective geometry $\Pi = PG(n,q)$ is called the *geometric q-dimension* of $\mathcal{D}$ and will be denoted by $\operatorname{gdim}_q \mathcal{D}$.

Using this notation, the trivial bound for the geometric dimension of a simple incidence structure $\mathcal{D}$ on v points reads $\operatorname{gdim}_q \mathcal{D} \le v-1$. If $\mathcal{D}$ has constant block size k, one has a considerably more interesting bound noted in [12]:

Proposition 5.3. *Any simple incidence structure on v points with constant block size k admits a strong embedding into $PG(k,q)$ for all $q \ge v-1$. That is,*

$$\operatorname{gdim}_q \mathcal{D} \le k \quad \textit{for } q \ge v-1\,. \tag{5.1}$$

Proof. As is well known, there exists a $(q+1)$-arc in $PG(k,q)$ (that is, $q+1$ points no $k+1$ of which are in a hyperplane) whenever $q \ge k+1$; see, for instance, [29]. Hence we may select v (arbitrary) points of such an arc A to embed $\mathcal{D}$ whenever $q \ge v-1$. Note that the defining property of an arc guarantees that every block X of $\mathcal{D}$ is induced from the hyperplane spanned by the k points of A determined by X. □

The main interest of the theory developed in [35] lies in the fact that the geometric q-dimension of any simple incidence structure $\mathcal{D}$ is closely related to the q-dimension of its complementary incidence structure $\mathcal{D}^*$. We shall use the notation introduced in the previous section and begin by showing that any generalized incidence matrix M of $\mathcal{D}^*$ with a trace code of dimension k can be used to obtain an embedding of $\mathcal{D}$ into $PG(k-1,q)$.

Proposition 5.4. *Let $\mathcal{D}$ be a simple incidence structure, q a prime power, and $E = GF(q^t)$ any extension field of $F = GF(q)$. Moreover, let M be some E-incidence matrix for the complementary incidence structure $\mathcal{D}^*$ and assume that the trace code $\operatorname{Tr}(C(M))$ associated with M has dimension k. Then $\mathcal{D}$ is strongly embeddable into the projective geometry $PG(k-1,q)$.*

Proof. In view of Proposition 2.3, we may replace C with its Galois closure $\overline{C}$, as we are not interested in the code associated with M itself, but merely in its trace code

$\mathrm{Tr}(C(M))$. By Theorem 2.2, this trace code coincides with the subfield subcode of $\overline{C}$:

$$\overline{C}_F = \mathrm{Tr}(\overline{C}) = \mathrm{Tr}(C)\,.$$

Let us choose a basis of $\overline{C}_F$ over F, say $\mathbf{b}_1, \ldots, \mathbf{b}_k$; by Theorem 2.4, this is also a basis of $\overline{C}$ over E. In particular, all rows of the given generalized incidence matrix M are linear combinations of these basis vectors with coefficients from E. Let us write $\mathbf{b}_1, \ldots, \mathbf{b}_k$ as the rows of an $(k \times v)$-matrix B, so that the rows of M lie in the row space of B over E.

Now consider the columns of B. Clearly, B cannot contain a column $\mathbf{0}$, since otherwise the corresponding column of M would consist of entries 0 only, and so the associated point of $\mathcal{D}^*$ would not be contained in any blocks at all, contradicting the simplicity of $\mathcal{D}^*$. Similarly, B cannot contain two linearly dependent non-zero columns; otherwise, for any linear combination of $\mathbf{b}_1, \ldots, \mathbf{b}_k$ with coefficients from E, we would obtain 0 in either both or none of the two associated positions. In particular, the points corresponding to these two columns would be on exactly the same set of blocks of $\mathcal{D}^*$, again contradicting the simplicity of the incidence structure.

Hence B consists of v column vectors of length k over F, no two of which are linearly dependent, and therefore the columns of B give us a set V of v distinct points in $\Pi = PG(k-1, q)$. Note that V is a projective system, as B has (row and column) rank k, so that V contains k points in general position. We now identify the point set of $\mathcal{D}^*$ with this projective system in the natural way (via the corresponding columns of M and hence of B), and claim that this identification gives the desired strong embedding of $\mathcal{D}$ into Π. In particular, each block of $\mathcal{D}^*$ – and therefore also each block of the complementary incidence structure $\mathcal{D}$ of $\mathcal{D}^*$ – is now identified with some subset of the point set V in Π.

Consider an arbitrary block X^* of $\mathcal{D}^*$ and let $\mathbf{x}_1, \ldots, \mathbf{x}_c$ be points of V contained in the complementary block $X = V \setminus X^*$ of $\mathcal{D}$, so that the incidence matrix M has entries 0 in the row corresponding to X^* for all positions indexed by the points $\mathbf{x}_1, \ldots, \mathbf{x}_c$. As X^* is a linear combination of the rows of B with coefficients from E, we necessarily have entries 0 in row X^* of M in *all* columns which correspond to a point in V given by some linear combination of the columns of B associated with $\mathbf{x}_1, \ldots, \mathbf{x}_c$. In other words, if a block X^* has entries 0 in all positions indexed by some points $\mathbf{x}_1, \ldots, \mathbf{x}_c$, then it has entry 0 in *all* positions corresponding to a point in the intersection of V with the subspace of Π generated by these points. In terms of the incidence structure $\mathcal{D}$, this observation simply means that the block X of $\mathcal{D}$ is closed under intersections with subspaces of Π. Therefore, $X = V \cap W$, where W is the subspace of Π spanned by the points of X. (Note that W is the subspace associated with X, in the sense of Definition 5.1.) □

Next, we prove the following partial converse of Proposition 5.4:

Proposition 5.5. *Let $\mathcal{D}$ be any simple incidence structure embedded in the projective geometry $\Pi = PG(k-1, q)$ and let $E = GF(q^t)$ be any extension field of $GF(q)$ satisfy-*

ing $t \geq k-1-d$, where d is the smallest dimension of a subspace of Π associated with some block of $\mathcal{D}$ in the given embedding. Then there exists an E-incidence matrix M for the complementary incidence structure $\mathcal{D}^*$ of $\mathcal{D}$ such that the trace code $\mathrm{Tr}(C(M))$ associated with M has dimension at most k.

Proof. By definition, the point set of $\mathcal{D}$ is identified with a subset V of the point set of Π. We choose a coordinate vector for each of the $q^{k-1}+\cdots+q+1$ points of Π and write all these vectors as the columns of a matrix B. Thus B is a generator matrix for the simplex code S of dimension k over $F = GF(q)$. Let $C = S \otimes E$ denote the Galois closed extension code of S over E; we will now use Theorem 3.10. By hypothesis, $t \geq k-1-d$, so that all complements of subspaces of Π with codimension at most $k-1-d$ (and hence dimension at least d) arise as supports of codewords in C. Selecting such a word for each of these subspaces gives an E-incidence matrix $\tilde{M}$ for the incidence structure formed by the points of Π and all complements of subspaces of Π with dimension at least d.

We claim that the submatrix of $\tilde{M}$ which consists of the columns associated with the points in V contains an E-incidence matrix M for $\mathcal{D}^*$. To see this, let X be any block of $\mathcal{D}$ and $X^* = V \setminus X$ its complementary block in $\mathcal{D}^*$. As $\mathcal{D}$ is embedded in Π, there is a unique subspace W of Π associated with X. Then $X = V \cap W$ and hence $X^* = V \cap W^*$, where W^* denotes the complement of the subspace W in Π. By hypothesis, the dimension of W is at least d, and thus $\tilde{M}$ contains a row associated with W, which is a word $\mathbf{c} \in C$ with support W^*. Therefore, $\mathbf{c}$ has a non-zero entry in every position associated with a point in X^* and an entry 0 in every position associated with a point in X. Thus the restriction of $\mathbf{c}$ to the positions indexed by V is indeed an E-incidence vector for X^*, which verifies the claim.

Therefore, the restriction $C' = C|V$ of C to the positions associated with the common point set V of $\mathcal{D}$ and $\mathcal{D}^*$ contains an E-incidence matrix M of $\mathcal{D}^*$, as all rows of $\tilde{M}$ are, by construction, words in the E-extension C of the simplex code S. By Theorem 2.4, C is the E-extension of its trace code $\mathrm{Tr}(C) = S$, and thus C' is the E-extension of its trace code $\mathrm{Tr}(C') = S|V$. Since the E-code $C(M)$ generated by M is contained in the Galois closed code C', its trace code $\mathrm{Tr}(C(M))$ is contained in $\mathrm{Tr}(C') = S|V$. Trivially, $S|V$ has dimension at most k, and the assertion follows. □

It seems plausible that the trace code $\mathrm{Tr}(C(M))$ associated with M actually has exactly dimension k provided that $\mathcal{D}$ is strongly embedded in Π. Clearly, the code $S|V$ then has exactly dimension k, but it is not clear whether or not $\mathrm{Tr}(C(M))$ and $S|V$ necessarily coincide. In general, this is still an open problem. Nevertheless, it is easy to see that equality indeed holds if k is chosen minimally in Proposition 5.5:

Theorem 5.6. *Let $\mathcal{D}$ be a simple incidence structure and q a prime power. Then the q-dimension of $\mathcal{D}^*$ is one more than the smallest integer n for which $\mathcal{D}$ can be* *embedded into the projective geometry* $\Pi = PG(n, q)$*:*

$$\dim_q \mathcal{D}^* = \operatorname{gdim}_q \mathcal{D} + 1 . \tag{5.2}$$

Proof. By Proposition 5.4, $\mathcal{D}$ can be embedded into $PG(k-1, q)$ whenever $\mathcal{D}^*$ admits an E-incidence matrix M with $\dim \operatorname{Tr}(C(M)) = k$; in particular, this holds for $k = \dim_q \mathcal{D}^*$, so that $n = \operatorname{gdim}_q \mathcal{D} \leq \dim_q \mathcal{D}^* - 1$. On the other hand, by Proposition 5.5, there exists a generalized incidence matrix M for $\mathcal{D}^*$ for which $\dim \operatorname{Tr}(C(M)) \leq n + 1$. If we had strict inequality, say $\dim \operatorname{Tr}(C(M)) = k \leq n$, another application of Proposition 5.4 would also provide an embedding of $\mathcal{D}$ into $PG(k-1, q)$, contradicting the minimality of n. □

The next result of [35] settles the problem when an embedding of the smallest possible dimension can be affine.

Theorem 5.7. *Let* $\mathcal{D}$ *be a simple incidence structure, let* q *be a prime power, and write* $\operatorname{gdim}_q \mathcal{D} = n$*. Then* $\mathcal{D}$ *actually admits an affine embedding into* $AG(n, q)$ *if and only if* $\mathcal{D}^*$ *has an* E*-incidence matrix* M *with* $\dim \operatorname{Tr}(C(M)) = n + 1$ *for which the trace code* $\operatorname{Tr}(C(M))$ *contains a word with full support* V*.*

Proof. We need to consider the existence of embeddings into $\Pi = PG(n, q)$ where the point set V used is disjoint to some hyperplane H of Π. Assume first that the criterion stated in the assertion is satisfied, so that we have an E-incidence matrix M for $\mathcal{D}^*$ with $\dim \operatorname{Tr}(C(M)) = n + 1$ for which $C(M)$ contains a word with full support V. In the proof of Proposition 5.4, we may choose such a word as the first basis vector $\mathbf{b}_1$, so that all columns of the matrix B constructed there correspond to points of Π in the complement A of the hyperplane H with equation $x_1 = 0$. Then $\mathcal{D}$ is indeed embedded in the affine geometry $\Sigma = \Pi \setminus H$.

Conversely, assume the existence of an embedding of $\mathcal{D}$ into $\Pi = PG(n, q)$ for which V is disjoint to some hyperplane H of Π. The construction of the desired generalized incidence matrix is similar to the one used in the proof of Proposition 5.5, but relies on Theorem 3.11 instead of Theorem 3.10. Therefore, we will merely sketch the arguments.

Recall that the complement A of the hyperplane H is a projective system for the q-ary first-order Reed–Muller code R of dimension $n + 1$, see Example 3.2. Let t be any integer satisfying $t \geq n - d$, where d is the smallest dimension of a subspace of Π associated with some block of $\mathcal{D}$ in the given embedding and consider the extension field $E = GF(q^t)$. By Theorem 3.11, we may construct an E-incidence matrix $\tilde{M}$ for the incidence structure formed by the points in A and all complements of (affine) subspaces with dimension at least d of the affine space Σ determined on A, where all rows of $\tilde{M}$ are words of the extension code $C = R \otimes E$ of R. Again, the restriction of $\tilde{M}$ to the columns corresponding to the points of V contains an E-incidence matrix M for $\mathcal{D}^*$. Since $C' = C|V$ is the E-extension of its trace code $\operatorname{Tr}(C') = R|V$, the trace code $\operatorname{Tr}(C(M))$ is contained in $R|V$. As in the proof of Theorem 5.6, actually $\mathrm{Tr}(C(M)) = R|V$, since $n = \mathrm{gdim}_q \mathcal{D}$. Therefore, the restriction of any word of full weight q^n in R to V is indeed a word in the trace code $\mathrm{Tr}(C(M))$ with full support V. □

We mention the following interesting consequence of Theorem 5.7:

Corollary 5.8. *Assume that $\mathcal{D}$ is resolvable with parallel classes of size 2 (that is, the complement of every block is again a block), so that $\mathcal{D}^* = \mathcal{D}$. Then*

$$\dim_q \mathcal{D} = \mathrm{gdim}_q \mathcal{D} + 1 ,$$

and $\mathcal{D}$ has an affine embedding into $AG(n, q)$, where $n = \mathrm{gdim}_q \mathcal{D}$.

Finally, we present two new results. The first of these shows that – at least in principle – it is also possible to determine the smallest possible extension degree t for which $\mathcal{D}^*$ can be supported by a $GF(q^t)$-incidence matrix leading to a trace code of dimension $\dim_q \mathcal{D}^*$ geometrically, and to relate the associated Galois closed codes to extensions of simplex codes. This generalizes a result of [34] for the classical designs $PG_d(n, q)$.

Theorem 5.9. *Let $\mathcal{D}$ be a simple incidence structure and q a prime power, and write $n = \mathrm{gdim}_q \mathcal{D}$. Define D as the maximum of the smallest dimension d of a subspace of Π associated with some block of $\mathcal{D}$ in a given embedding of $\mathcal{D}$ into $\Pi = PG(n, q)$, taken over all such embeddings.*

Then $\mathcal{D}^*$ has a $GF(q^t)$-incidence matrix M satisfying $\dim \mathrm{Tr}(C(M)) = n + 1$ if and only if $t \geq n - D$. Also, for any such matrix M, the Galois closure of $C(M)$ is a truncation of the extension code $S \otimes GF(q^t)$ of the simplex code S of dimension $n + 1$ over $GF(q)$.

Proof. The existence of a generalized incidence matrix M for $\mathcal{D}^*$ over $GF(q^t)$ satisfying $\dim \mathrm{Tr}(C(M)) = n + 1$ is guaranteed for every $t \geq n - D$ by Proposition 5.5 and Theorem 5.6. Conversely, let M be any $GF(q^t)$-incidence matrix for $\mathcal{D}^*$ satisfying $\dim \mathrm{Tr}(C(M)) = n + 1$. Let us return to the proof of Proposition 5.4. Clearly, the matrix B constructed there can be enlarged to a generator matrix for the simplex code S of dimension $n + 1$ over $GF(q)$, by adjoining suitable further columns. Hence the trace code $\mathrm{Tr}(C(M))$ is a truncation of S, and therefore the Galois closure $\overline{C}$ of $C(M)$ is indeed a truncation of the extension code $S \otimes GF(q^t)$. In particular, all rows of M are restrictions of codewords in C to the set V used in the embedding constructed from M in the proof of Proposition 5.4. Using Theorem 3.10, we conclude that all blocks of $\mathcal{D}$ are associated with subspaces of codimension at most t and thus of dimension at least $n - t$ of Π. By the definition of D, we have $D \geq n - t$, which proves the assertion. □

Our second new result is an analog of Theorem 5.9 for affine embeddings, which generalizes a result for the classical designs $AG_d(n, q)$ obtained in [34]. We will omit the proof, as it is in complete analogy with that of Theorem 5.9, taking into account Theorem 5.7 and using Theorem 3.11 instead of Theorem 3.10.

Theorem 5.10. *Let $\mathcal{D}$ be a simple incidence structure and q a prime power, write $n = \operatorname{gdim}_q \mathcal{D}$, and assume that $\mathcal{D}$ admits an affine embedding into $\Sigma = AG(n, q)$. Define D' as the maximum of the smallest dimension d of a subspace of Σ associated with some block of $\mathcal{D}$, taken over all such affine embeddings.*

Then $\mathcal{D}^*$ has a $GF(q^t)$-incidence matrix M satisfying $\dim \operatorname{Tr}(C(M)) = n+1$ for which $\operatorname{Tr}(C(M))$ contains a word of full support if and only if $t \geq n - D'$. Also, for any such matrix M, the Galois closure of $C(M)$ is a truncation of the extension code $R \otimes GF(q^t)$ of first-order Reed–Muller code R of dimension $n+1$ over $GF(q)$.

6 Designs with Classical Parameters

In this section, we illustrate the general theory which we have presented by applying it to a particularly interesting class of simple incidence structures, namely simple designs with classical parameters. Recall that the *classical* or *geometric* designs are the designs $PG_d(n, q)$ and $AG_d(n, q)$ formed by the points and d-spaces in some projective or affine geometry $PG(n, q)$ or $AG(n, q)$, respectively, where $1 \leq d \leq n-1$, over a finite field $GF(q)$. Any design which has the same parameters as some geometric design is said to be a *design with classical parameters.*

As is well known, the parameters of $PG_d(n, q)$ are as follows:

$$v = \frac{q^{n+1}-1}{q-1},\ k = \frac{q^{d+1}-1}{q-1},\ \lambda = \begin{bmatrix} n-1 \\ d-1 \end{bmatrix}_q; \tag{6.1}$$

this design has $b = \begin{bmatrix} n+1 \\ d+1 \end{bmatrix}_q$ blocks, and each point is in $r = \begin{bmatrix} n \\ d \end{bmatrix}_q$ blocks. Similarly, $AG_d(n, q)$ has parameters

$$v = q^n,\ k = q^d,\ \lambda = \begin{bmatrix} n-1 \\ d-1 \end{bmatrix}_q; \tag{6.2}$$

in this case, there are $b = q^{n-d}\begin{bmatrix} n \\ d \end{bmatrix}_q$ blocks, and each point is in $r = \begin{bmatrix} n \\ d \end{bmatrix}_q$ blocks.

It is known [8, 33] that the number of simple designs with classical parameters grows exponentially with linear growth of n if one fixes either the dimension d or the codimension $n-d$. Therefore, it is a natural problem to try to characterize the geometric designs among the vast number of simple designs with the same parameters. There are two main approaches to this problem: one may use either combinatorial properties (e.g. the correct line[1] size), or one may try to give a coding theoretic characterization. We refer the reader to [31] for a recent survey on designs with classical

1 Recall that the *line* determined by two points of a design is defined as the intersection of all blocks containing the specified points.

parameters, with particular emphasis on the characterization problem. We remark that there are rather satisfactory results for the combinatorial approach.

In contrast, the coding theoretic approach has met with remarkably little success for a long time. The seminal work in this direction is due to Hamada [22] who gave a general – albeit very involved and hard to handle – formula for the p-ranks of the incidence matrices of the geometric designs $PG_d(n,q)$ and $AG_d(n,q)$, where q is a power of the prime p and where $1 \leq d \leq n-1$. He also conjectured that the geometric designs always have the smallest p-rank among all designs with the same parameters; later, in a joint paper with Ohmori [23], he proposed an even stronger conjecture, namely that the classical designs can be characterized among all designs with the same parameters as those of minimum p-rank.

While this strong version of *Hamada's conjecture* has been established in a few cases [16, 23, 44], a first counterexample was already contained in a paper by Goethals and Delsarte [20] well before the conjecture was made! A handful of further sporadic counterexamples were discovered later [24, 45], and only recently infinite families of counterexamples were constructed [7, 32]. In contrast, the original (weak) version of the conjecture is still wide open: not even a single counterexample is known, and the only cases established are those for which actually the strong version of the conjecture holds. The reader may find more details on the status of Hamada's conjecture in [31, 48].

The quest for an alternative coding theoretic characterization of the geometric designs among all simple designs with the same parameters was the starting point for our investigations in [34, 35] which led to the general theory presented here. In the projective case, we have the following result first proved in [34] (with a somewhat different formulation).

Theorem 6.1. *Let $\mathcal{D}$ be a simple design with the parameters of $PG_d(n,q)$. Then $\dim_q \mathcal{D}^* \geq n+1$, with equality if and only if $\mathcal{D}$ is the classical design.*

Moreover, $\mathcal{D}^*$ admits a $GF(q^t)$-incidence matrix M with $\dim \mathrm{Tr}(C(M)) = n+1$ if and only if $t \geq n-d$. Also, for any such matrix M, the Galois closure of $C(M)$ is the extension code $S \otimes GF(q^t)$ of the simplex code S of dimension $n+1$ and length $q^n + \cdots + q + 1$ over $GF(q)$.

Proof. The first assertion is a trivial consequence of Theorem 5.6. The remaining assertions hold by Theorem 5.9, since in the special case considered here the matrix B constructed in the proof of Proposition 5.4 is already a generator matrix for the simplex code S. □

We remark that Tonchev's characterization [46] of the classical point-hyperplane designs $PG_{n-1}(n,q)$ is essentially the special case $d = n-1$ of Theorem 6.1. Note, however, that Tonchev only allowed $GF(q)$-incidence matrices. Then the trace is trivial, namely the identity map; so the trace code is just the code itself; and C itself is vacuously Galois closed.

We also note that recognizing a code monomially equivalent (with respect to E) to an extended simplex code $S \otimes E$ seems to be a rather hard problem, which also makes it difficult to decide whether or not an arbitrary E-incidence matrix M for a design with the parameters of $PG_d(n, q)$ actually belongs to the classical design. In [34], there is a small example for this phenomenon; we refer to this paper for more details.

Example 6.2. Consider the binary simplex code S of dimension 4 and length 15. Let $\mathcal{D} = PG_1(3, 2)$, and let M be a "nice" $GF(4)$-incidence matrix for $\mathcal{D}^*$ constructed from the $GF(4)$-extension C of S, as in Proposition 5.5. It is easily seen that C coincides with $C(M)$ in this case. Now let $k \in \{5, 6, 7, 8\}$. Then:
- There exists a $GF(4)$-incidence matrix M_k for $\mathcal{D}^*$ which is monomially equivalent to M and has $\dim \mathrm{Tr}(C(M_k)) = k$.
- For $k = 5, 6$, the Galois closure of $C(M_k)$ contains the simplex code in its subfield subcode.
- For $k = 7, 8$, the Galois closure of $C(M_k)$ does not contain the simplex code in its subfield subcode.

Next, we state a recent result due to Ghinelli, Jungnickel, and Metsch [18].

Theorem 6.3. *There exists a simple design $\mathcal{D}$ with the same parameters as the classical design $PG_d(n, q)$, where $2 \le d \le n - 1$, such that* $\mathrm{gdim}_q\, \mathcal{D}^* = n + 1$.

The proof proceeds by showing that there are examples with the desired geometric dimension among the *distorted designs* with the specified parameters constructed in [33]. The authors of [18] also give some further results on the dimension of distorted designs in general and pose a conjecture about the possible embeddings of lowest dimension for such designs. We refer the reader to the original paper.

We now turn our attention to the affine case, which is considerably more difficult and solved completely only for a little more than half the possible values of d. Let us begin with the following result.

Theorem 6.4. *Let $\mathcal{D}$ be a simple design with the parameters of $AG_d(n, q)$. Then* $\dim_q \mathcal{D}^* \ge n + 1$, *and equality holds for the classical design.*

Now assume that $\mathcal{D}$ is classical. Then $\mathcal{D}^*$ admits a $GF(q^t)$-incidence matrix M with $\dim \mathrm{Tr}(C(M)) = n + 1$ *if and only if $t \ge n - d$. Also, for any such matrix M, the Galois closure of $C(M)$ is the extension code $R \otimes GF(q^t)$ of the first-order Reed–Muller code R of dimension $n + 1$ and length q^n over $GF(q)$.*

Proof. The lower bound stated in the assertion and the fact that the classical design satisfies this bound with equality are trivial consequences of Theorem 5.6. Then the assertions about the possible generalized incidence matrices of dimension $n + 1$ for the complementary design of the classical example hold by Theorem 5.10. □

In [34], we have conjectured the following coding theoretic characterization of the classical designs $AG_d(n, q)$, in analogy to the projective case:

Conjecture 6.5. *Let $\mathcal{D}$ be a simple design with the parameters of $AG_d(n,q)$. Then* $\dim_q \mathcal{D}^* = n+1$ *holds if and only if $\mathcal{D}$ is the classical design.*

By Theorem 5.7, Conjecture 6.5 is equivalent to the following geometric conjecture:

Conjecture 6.6. *Let $\mathcal{D}$ be a simple design with the parameters of the classical design $AG_d(n,q)$ and assume that $\mathcal{D}$ can be embedded into $PG(n,q)$. Then $\mathcal{D}$ is classical.*

Using a mixture of geometric and combinatorial arguments, the following result was obtained in [34]; all cases not covered by this result are still open.

Theorem 6.7. *Conjecture* 6.6 *– and hence also Conjecture* 6.5 *– holds provided that $d = 1$ or $d > (n-2)/2$.*

We refer to the original paper for a proof of Theorem 6.7 and conclude this section with a final remark. The strong Hamada conjecture would have provided an elegant and computationally simple characterization of the classical geometric designs in terms of the p-rank of their incidence matrices – probably its most attractive feature. The complexity of computing the rank of a matrix is a cubic polynomial in the number of rows (or columns), while the complexity of finding isomorphisms between block designs is as hard as the notoriously difficult graph isomorphism problem; see [9, Remark VII.6.6]. In contrast, the coding theoretic characterization of the geometric designs discussed here is certainly of considerable theoretical interest, but it seems to be of no help in deciding whether or not a design with the appropriate parameters is actually classical, since it is not at all clear how one could actually compute its q-dimension.

7 Two-Weight Codes

It is the purpose of this brief section to discuss how the well-known connection between two-weight codes and projective (N,k,h_1,h_2)-sets fits into the general theory discussed in the present survey. Let us recall the necessary definitions.

Definition 7.1. A code C is called a *two-weight code* if the weight of every non-zero codeword is one of two possible integers, which we will denote by w_1 and w_2.

Definition 7.2. Let V be a set of N points in the projective space $\Pi = PG(k-1,q)$ which spans Π, so that V is not contained in some hyperplane of Π. Following [6], V is said to be an (N,k,h_1,h_2)*-set* if every hyperplane H of Π intersects P in either h_1 or h_2 points. In Galois Geometry, such a set is usually called a *two-character set* with *characters* h_1, h_2.

The following fundamental result is due to Delsarte [13].

Theorem 7.3. *The existence of a projective (N, k, h_1, h_2)-set in $PG(k-1, q)$ is equivalent to that of a two-weight code of length N and dimension k with weights $w_1 = N - h_1$ and $w_2 = N - h_2$ over $GF(q)$.*

Delsarte's theory actually yields a lot more, as it also contains the equivalence to certain strongly regular graphs and certain examples for a type of difference sets now usually called "partial difference sets". We refer the reader to the seminal survey by Calderbanck and Kantor [6] for a detailed treatment of this topic, including all then known examples. A nice recent survey of two-character sets from the geometric point of view is given by De Clerck and Durante [11].

We now show how that part of Delsarte's theory stated in Theorem 7.3 is related to a very special case of the general approach presented here. Of course, Theorem 7.3 admits a simple direct proof, so that there is no need to appeal to the general theory. Moreover, we also have to assume certain restrictions in order to be able to apply our results. For instance, neither of the two characters h_1 and h_2 should be 0. However, such cases also provide interesting examples for our theory (see the next section) and could be dealt with by modifying our general arguments. In spite of these qualifications, it is certainly interesting to point out that a close connection exists.

First let V be a projective (N, k, h_1, h_2)-set in $PG(k-1, q)$ and consider the incidence structure $\mathcal{D}$ formed by the points of V and all intersections of V with hyperplanes as blocks. Let us assume that $\mathcal{D}$ is simple. Then Proposition 5.5 shows that the complementary incidence structure $\mathcal{D}^*$ is supported by a code C of length N and dimension at most n over $GF(q)$; for reasons of cardinality, the dimension here has to be exactly k. As the blocks of $\mathcal{D}^*$ have cardinality $w_1 = N - h_1$ or $w_2 = N - h_2$, the code C is indeed a two-weight code.

Conversely, let C be a two-weight code of length N and dimension k with weights w_1 and w_2 over $GF(q)$. Denote the incidence structure formed by the supports of all codewords by $\mathcal{D}^*$. Let us assume that $\mathcal{D}^*$ is simple. Then Proposition 5.4 shows that $\mathcal{D}$ can be embedded in $PG(k-1, q)$. Clearly, we can view C as a truncation of the simplex code S of dimension n over $GF(q)$, and therefore the blocks of $\mathcal{D}$ are induced by intersecting the point set V of the embedding with the complements of hyperplanes. As C is a two-weight code, the blocks of $\mathcal{D}^*$ have cardinalities w_1 or w_2, and thus the blocks of $\mathcal{D}$ have cardinality $h_1 = N - w_1$ or $h_2 = N - w_2$, so that V is the desired projective set.

8 Steiner Systems

In this section, we illustrate the general theory for some Steiner systems. These examples are taken from [35], and the first two are of particular interest. But first we state a general lower bound on the geometric dimension of Steiner systems and, more generally, t-designs taken from [12]; we shall omit the rather simple proof.

Proposition 8.1. *Let $\mathcal{D}$ be any simple t-(v, k, λ) design and let q be any prime power. Then*

$$\operatorname{gdim}_q \mathcal{D} \geq \begin{cases} t & \text{if } \lambda = 1\,, \\ t+1 & \text{if } \lambda \neq 1\,. \end{cases} \tag{8.1}$$

Example 8.2. Consider the ternary Golay code C, that is, the unique $[11, 6, 5]$-code over $GF(3)$. The set of 66 distinct supports of the 132 words of C with (minimum) weight 5 form the unique Steiner system $S(4, 5, 11)$. We take this Steiner system as our simple incidence structure $\mathcal{D}$. Note that C also contains 132 words of weight 6, which constitute the supports of the blocks of the complementary structure $\mathcal{D}^*$.

Moreover, all words of weight 6 have parity check sum 0, which is easily seen by comparing the weight distributions of C and of its parity check extension C', the extended ternary Golay code. Thus $\mathcal{D}^*$ is supported by the $[11, 5, 5]$-subcode C_0 of C which consists of all words with parity check sum 0. This shows $\dim_3 \mathcal{D}^* \leq 5$; on the other hand, also $\dim_3 \mathcal{D}^* - 1 = \operatorname{gdim}_3 \mathcal{D} \geq 4$, by Proposition 8.1.

By Theorem 5.6, $\mathcal{D}$ can be embedded into $PG(4, 3)$, and this is the smallest possible embedding into a projective space over $GF(3)$. Such an embedding is already contained in the paper of Tallini [43]; its point set is actually the (unique) smallest complete cap in $PG(4, 3)$. See also Hirschfeld [26] for an explicit construction, which takes some effort, and further references. The general theory yields this example in a very simple way.

Example 8.3. Consider the extended ternary Golay code C', that is, the unique [12,6,6]-code over $GF(3)$; this is the parity check extension of the code C considered in Example 8.2. Here the set of 132 distinct supports of the 264 words of C' with (minimum) weight 6 form the unique Steiner system $S(5, 6, 12)$. We now choose this Steiner system as our simple incidence structure $\mathcal{D}$. Note that $\mathcal{D}$ is resolvable, so that $\mathcal{D} = \mathcal{D}^*$.

As $\mathcal{D}$ is supported by C', we have $\dim_3 \mathcal{D} \leq 6$; on the other hand, also $\dim_3 \mathcal{D} - 1 = \operatorname{gdim}_3 \mathcal{D} \geq 5$, by Proposition 8.1. By Corollary 5.8, $\mathcal{D}$ can be embedded into $AG(5, 3)$, and this is the smallest possible embedding into a projective space over $GF(3)$. We remark that an embedding into $PG(5, 3)$ has been known for a long time: Coxeter [10] gave an explicit, considerably more involved construction. As in Example 8.2, the new general theory yields this example in a nearly trivial manner.

Next, we mention some well-known examples associated with two-weight codes.

Example 8.4. Recall that a *Möbius plane* (or *inversive plane* in the terminology of Dembowski [15]) of order q is a one-point extension of an affine plane of order q, that is, a Steiner system $S(3, q+1, q^2+1)$. The classical (or *Miquelian*) examples of Möbius planes admit several distinct constructions. For our purposes, it is best to define them in the following geometric way: as points, one may take the points of a non-degenerate elliptic quadric Q in $PG(3, q)$, and as blocks the intersections of Q with all secant planes of Q. As this already describes an embedding of our Steiner system, we conclude that the complementary 3-design $\mathcal{D}^*$ of a classical Möbius plane $\mathcal{D}$ of order q always has q-dimension 4. By Theorem 5.6, $\mathcal{D}^*$ is supported by an associated code C over $GF(q)$. This code is a two-weight code of type TF3 in the notation of [6], and a $4 \times (q^2+1)$ generator matrix G is obtained by taking as columns a set of q^2+1 vectors representing the projective coordinates of the points of Q.

Example 8.5. A *maximal* 2^s*-arc* in $PG(2, 2^t)$, where $1 \le s \le t-1$, is a set of $2^s(2^t - 2^{t-s}+1)$ points that meets every line in either none or 2^s points; a maximal arc with $s = 1$ is a *hyperoval*. Any maximal 2^s-arc defines a two-weight code C over $GF(2^t)$ with length $n = 2^s(2^t - 2^{t-s}+1)$, dimension 3, and non-zero weights $2^t(2^s - 1)$ and $2^s(2^t - 2^{t-s}+1)$, which is a code of type TF2 in the terminology of [6]. The points of a maximal 2^s-arc P and the intersections of P with secant lines considered as blocks form an $S(2, 2^s, 2^s(2^t - 2^{t-s}+1))$, say $\mathcal{D}$, and the complementary design $\mathcal{D}^*$ is supported by the minimum weight vectors of the related two-weight code.

In particular, a hyperoval defines a two-weight code C over $GF(2^t)$ with length $n = 2^t + 2$, dimension $k = 3$, weights $w_1 = 2^t$, $w_2 = n = 2^t + 2$. Note that $w_1 = 2^t = n-k+1$, hence C is an MDS-code in this case. The codewords of weight 2^t support the complete design on 2^t+2 points having as blocks all subsets of P of size 2^t. Thus, for every $t \ge 1$, the dimension of the trivial 2-$(2^t+2, 2^t, 2^{t-1}(2^t-1))$ design over $GF(2^t)$ is 3, in accordance with a result by Tonchev [47] concerning the connection between the q-dimension of complete designs and MDS codes.

Example 8.6. A *unital* in $PG(2, q^2)$ is a set of q^3+1 points that meets every line in either one or $q+1$ points. It defines a two-weight code C over $GF(q^2)$ with length $n = q^3+1$, dimension 3, and non-zero weights $q^3 - q$ and q^3. The q^3+1 points of a unital together with the line intersections of size $q+1$ form an $S(2, q+1, q^3+1)$, say $\mathcal{D}$, and again the blocks of $\mathcal{D}^*$ are supported by the minimum weight vectors of C.

9 Configurations

In this section, we illustrate the general theory for some small (mainly symmetric) configurations. Recall that a *configuration* v_k is a simple 1-design on v points, with k points per block and k blocks per point, and no two points on more than one block; hence one usually speaks of *lines* instead of *blocks* in this situation. We refer the reader to the book of Grünbaum [21] for a systematic treatment of configurations.

We begin with three famous examples which arise in the axiomatic foundation of projective geometry; see, for instance, [4] for background. The first two of these share the remarkable property that the q-dimension of their complementary incidence structures basically does not depend on the choice of q, with only $q = 2$ being an accidental exception.

Example 9.1. The well-known *Desargues configuration*, a configuration 10_3, is used to characterize the (not necessarily finite) projective planes which can be coordinatized over a skewfield (and, more generally, to derive the standard algebraic representation for projective spaces if one starts with a synthetic definition via the famous Veblen–Young axioms). Hence this configuration $\mathcal{D}$ is embedded in $PG(2,q)$, so that $\mathrm{gdim}_q\, \mathcal{D} = 2$, for every prime power $q \geq 3$. For $q = 2$, we have a sort of accident: $\mathcal{D}$ cannot possibly live in $PG(2,2)$, as it just has too many points. Here one easily checks $\mathrm{gdim}_2\, \mathcal{D} = 3$: it is well known that the Desargues configuration can always be viewed as a configuration of points and lines in projective 3-space.

Example 9.2. The well-known *Pappus configuration*, a configuration 9_3, is used to characterize those projective geometries $PG(n,K)$ defined over a skewfield K for which K has commutative multiplication (and is therefore a field). Again, the Pappus configuration $\mathcal{D}$ is always embedded in $PG(2,q)$, so that $\mathrm{gdim}_q\, \mathcal{D} = 2$ for every prime power $q \geq 3$. As in the case of the Desargues configuration, $\mathcal{D}$ cannot live in $PG(2,2)$, as it has too many points. Again, $\mathrm{gdim}_2\, \mathcal{D} = 3$; see [35] for an explicit embedding of the Pappus configuration into $\Pi = PG(3,2)$.

Example 9.3. The smallest interesting configuration is the *Fano configuration* $\mathcal{D}$, often simply denoted as 7_3 – that is, in other notation, the projective plane $PG(2,2)$ of order 2; it is used to characterize those projective geometries $PG(n,K)$ defined over a field K for which K has characteristic 2. In particular, $PG(2,q)$ contains $\mathcal{D}$ as a subplane of order 2 if and only if q is even, and we conclude $\mathrm{gdim}_q\, \mathcal{D} = 2$ if and only if q is a power of 2.

In all other cases, one has $\mathrm{gdim}_q\, \mathcal{D} = 3$. For this, we present an embedding of the Fano configuration into $\Pi = PG(3,q)$ which is taken from [12]. Let $\mathbf{a}, \mathbf{b}, \mathbf{c}, \mathbf{d}$ be a basis for the vector space $GF(q)^4$. We use the seven points with coordinates

$$\mathbf{a},\ \mathbf{b},\ \mathbf{c},\ \mathbf{d},\ \mathbf{a}-\mathbf{b},\ \mathbf{b}-\mathbf{c},\ \mathbf{c}-\mathbf{a}$$

as the point set V for the desired embedding. Then four of the seven lines of 7_3 are induced by lines of Π in the obvious way, and the remaining three lines are induced by the three planes spanned by the sets

$$\{\mathbf{a}, \mathbf{b}-\mathbf{c}, \mathbf{d}\},\ \{\mathbf{b}, \mathbf{c}-\mathbf{a}, \mathbf{d}\} \text{ and } \{\mathbf{c}, \mathbf{a}-\mathbf{b}, \mathbf{d}\}\,,$$

as is easily checked. According to Proposition 5.5, $\mathcal{D}^*$ has to be supported by a four-dimensional code C over $GF(q^2)$, since we used subspaces of codimensions 1 and 2 to induce the lines of $\mathcal{D}$ on V.

We mention in passing that the same setup gives an embedding into $PG(2,q)$ for q a power of 2, if we choose $\mathbf{d} = \mathbf{a} + \mathbf{b} + \mathbf{c}$ in this case.

Our final example is similar to Example 9.3, but probably even more interesting:

Example 9.4. Let $\mathcal{D} = AG(2,3)$ be the unique affine plane of order 3, and let $\mathcal{D}_p$ be the configuration 8_3 which arises from $\mathcal{D}$ by omitting a point p together with all lines through p; this configuration is sometimes called the *biaffine plane* of order 3. Then any embedding of $\mathcal{D}_p$ into a projective plane $PG(2,q)$ can be extended to an embedding of $\mathcal{D}$, and such an embedding is possible if and only if q is a power of 3 or a prime power congruent to 1 modulo 3; see [1]. Thus $\operatorname{gdim}_q \mathcal{D} = \operatorname{gdim}_q \mathcal{D}_p = 2$ if and only if $q = 3^b$ or $q \equiv 1 \pmod 3$. In all other cases, one has $\operatorname{gdim}_q \mathcal{D} = \operatorname{gdim}_q \mathcal{D}_p = 3$; see [12].

Remark 9.5. It should be emphasized that Example 9.4 is quite exceptional: by a result of Rigby [41], the affine plane $AG(2,q')$ with $q' \geq 4$ can be embedded into $PG(2,q)$ if and only if q is a power of q'. Therefore, $\operatorname{gdim}_q AG(2,q') \geq 3$, whenever $q' \geq 4$ and q is not a power of q'.

Let us remark that the geometric q-dimension is known in just one such cross-characteristic case: $AG(2,4)$ admits an embedding into $PG(3,3)$, which shows

$$\operatorname{gdim}_{3^m} AG(2,4) = 3\,.$$

Trivially,

$$\operatorname{gdim}_{4^m} AG(2,4) = 2\,.$$

Using Rigby's result and Proposition 5.3, one also has

$$\operatorname{gdim}_q AG(2,4) \in \{3,4\}$$

for every prime power $q \geq 17$ which is not a power of 4; see [12].

Further results on the geometric dimension of small configurations can also be found in [12].

10 Conclusion and Open Problems

We hope that we have been able to convince the reader that the new invariants introduced in [35] connecting low dimensional embeddings of simple incidence structures and trace codes of the complementary structures are, if not of practical, at least of great theoretical interest. Let us conclude with a selection of open problems connected to the material we have covered.

Research Problem 10.1. *Let $\mathcal{D}$ be a design with the parameters of $AG_d(n,q)$, where $2 \leq d \leq (n-2)/2$, and assume that $\mathcal{D}$ is embedded in $PG(n,q)$. Settle Conjectures* 6.5 *and* 6.6 *by proving that $\mathcal{D}$ has to be classical.*

Research Problem 10.2. *Determine the geometric q-dimension for further interesting classes of incidence structures. In particular, find $\operatorname{gdim}_q \mathcal{D}$ for $q = 2^a$, $a \geq 2$ if $\mathcal{D}$ is one of the two large Witt designs $S(4,8,23)$ and $S(5,8,24)$. Also, what is $\operatorname{gdim}_q \mathcal{D}$ for the two small Witt designs $S(4,5,11)$ and $S(5,6,12)$ in even characteristic, in particular, for $q = 2$?*

Research Problem 10.3. *Determine the geometric q-dimension for affine or dual affine subplanes in some further cross characteristic cases, cf. Remark 9.5.*

Research Problem 10.4. *Determine the geometric q-dimension for some interesting classes of non-Desarguesian projective planes, for example, the Figueroa planes* [17].

Research Problem 10.5. *Settle the conjecture concerning "natural embeddings" of distorted designs which was made in* [18]. *More specifically, determine the geometric q-dimension of the polarity designs constructed in* [32].

Research Problem 10.6. *Find infinite families of counterexamples to the strong version of Hamada's conjecture when q is not a prime. In particular, the case $q = 4$ should be studied.*

Research Problem 10.7. *Are there counterexamples to the weak version of Hamada's conjecture?*

Research Problem 10.8. *Settle the status of Hamada's conjecture for $n = 2$ (projective and affine planes) and, more generally, for $d = n - 1$ (symmetric and affine designs).*

References

[1] M. S. Abdul-Elah, M. W. Al-Dhahir, and D. Jungnickel: 8_3 in $PG(2, q)$. *Arch. Math.* **49** (1987), 141–150.

[2] E. F. Assmus Jr. and J. D. Key: *Designs and their Codes*. Cambridge University Press, Cambridge, 1992.

[3] T. Beth, D. Jungnickel, and H. Lenz: *Design Theory*, 2nd edition (2 Volumes). Cambridge University Press, 1999.

[4] A. Beutelspacher and U. Rosenbaum: *Projective Geometry*, 2nd edition. Cambridge University Press, Cambridge, 2004.

[5] J. Bierbrauer: *Introduction to Coding Theory*. Chapman & Hall/CRC, 2005.

[6] R. Calderbank and W. M. Kantor: The geometry of two-weight codes. *Bull. London Math. Soc.* **18** (1986), 97–122.

[7] D. Clark, D. Jungnickel, and V. D. Tonchev: Affine geometry designs, polarities, and Hamada's conjecture. *J. Combin. Theory Ser. A.* **118** (2011), 231–239.

[8] D. Clark, D. Jungnickel, and V. D. Tonchev: Correction to: "Exponential bounds on the number of designs with affine parameters". *J. Combin. Des.* **19** (2011), 156–166.

[9] C. J. Colbourn and J. H. Dinitz: *Handbook of Combinatorial Designs (2nd edition)*. CRC Press, Boca Raton, 2007.

[10] H. S. M. Coxeter: Twelve points in *PG*(5, 3) with 95040 self-transformations. *Phil. Trans. Roy. Soc. London (A)* **247** (1958), 279–293.

[11] F. De Clerck and N. Durante: Constructions and characterizations of classical sets in $PG(n, q)$, in: *Current Research Topics in Galois Geometry*, pp. 1–33. Nova Science Publishers, New York, 2012.

[12] S. De Winter and D. Jungnickel: The geometric dimension of some small configurations. *J. Geom.* **103** (2012), 417–430.

[13] P. Delsarte: Weights of linear codes and strongly regular normed spaces. *Discrete Math.* **3** (1972), 47–64.

[14] P. Delsarte: On subfield subcodes of modified Reed–Solomon codes. *IEEE Trans. Inf. Theory* **21** (1975), 575–576.

[15] P. Dembowski: *Finite Geometries*, Springer, Berlin, 1968.

[16] J. Doyen, X. Hubaut, and M. Vandensavel: Ranks of incidence matrices of Steiner triple systems. *Math. Z.* **163** (1978), 251–259.

[17] R. Figueroa: A family of not (V, ℓ)-transitive projective planes of order q^3, $q \not\equiv 1 \pmod 3$ and $q > 2$. *Math. Z.* **181** (1982), 471–479.

[18] D. Ghinelli, D. Jungnickel, and K. Metsch: Remarks on polarity designs. *Des. Codes Cryptogr.* (2012), DOI 10.1007/s10623-012-9748-5.

[19] M. Giorgetti and A. Previtali: Galois invariance, trace codes and subfield subcodes. *Finite Fields Appl.* **16** (2010), 96–99.

[20] J. M. Goethals and P. Delsarte: On a class of majority-logic decodable cyclic codes. *IEEE Trans. Inform. Theory* **14** (1968), 182–188.

[21] B. Grünbaum: *Configurations of points and lines.* Graduate Studies in Mathematics 103. American Mathematical Society, Providence, RI, 2009.

[22] N. Hamada: On the p-rank of the incidence matrix of a balanced or partially balanced incomplete block design and its application to error correcting codes. *Hiroshima Math. J.* **3** (1973), 154–226 .

[23] N. Hamada and H. Ohmori: On the BIB-design having the minimum p-rank. *J. Combin. Theory Ser. A* **18** (1975), 131–140.

[24] M. Harada, C. W. H. Lam, and V. D. Tonchev: Symmetric (4,4)-nets and generalized Hadamard matrices over groups of order 4. *Des. Codes Cryptogr.* **34** (2005), 71–87.

[25] T. Helleseth, T. Kløve, and J. Mykkeltveit: The weight distribution of the coset leaders of some classes of codes with related parity-check matrices. *Discrete Math.* **28** (1979), 161–171.

[26] J. W. P. Hirschfeld: Projective spaces of square size. *Simon Stevin* **65** (1991), 319–329.

[27] J. W. P. Hirschfeld: *Projective Geometries over Finite Fields (2nd edition).* Oxford University Press, 1998.

[28] J. W. P. Hirschfeld: *Finite Projective Spaces of Three Dimensions.* Oxford University Press, 1985.

[29] J. W. P. Hirschfeld and J. A. Thas: *General Galois Geometries*. Oxford University Press, 1991.

[30] W. C. Huffman and V. Pless: *Fundamentals of Error-Correcting Codes*. Cambridge University Press, 2003.

[31] D. Jungnickel: Recent results on designs with classical parameters. *J. Geom.* **101** (2011), 137–155.

[32] D. Jungnickel and V. D. Tonchev: Polarities, quasi-symmetric designs, and Hamada's conjecture. *Des. Codes Cryptogr.* **51** (2009), 131–140.

[33] D. Jungnickel and V. D. Tonchev: The number of designs with geometric parameters grows exponentially. *Des. Codes Cryptogr.* **55** (2010), 131–140.

[34] D. Jungnickel and V. D. Tonchev: A Hamada type characterization of the classical geometric designs. *Des. Codes Cryptogr.* **65** (2012), 15–28.

[35] D. Jungnickel and V. D. Tonchev: New Invariants for Incidence Structures. *Des. Codes Cryptogr.* (2012), DOI: 10.1007/s10623-012-9636-z.

[36] R. Jurrius: Weight enumeration of codes from finite spaces. *Des. Codes Cryptogr.* **63** (2012), 321–330.

[37] R. Jurrius and G. R. Pellikaan: Codes, arrangements and matroids, in: *Algebraic geometry modeling in information theory*. Series on Coding Theory and Cryptology. World Scientific Publishing, Hackensack, NJ, 2012.

[38] T. Kløve: The weight distribution of linear codes over $GF(q^l)$ having generator matrix over $GF(q)$. *Discrete Math.* **23** (1978), 159–168.

[39] I. Landjev and L. Storme: Galois geometries and Coding Theory, in: *Current Research Topics in Galois Geometry*, pp. 187–214. Nova Science Publishers, New York, 2012.

[40] F. J. MacWilliams and N. J. A. Sloane: *The Theory of Error-Correcting Codes*. North-Holland, Amsterdam, 1977.

[41] J. F. Rigby: Affine subplanes of finite projective planes. *Can. J. Math.* **17** (1965), 977–1009.

[42] H. Stichtenoth: On the dimension of subfield subcodes. *IEEE Trans. Inform. Theory* **36** (1990), 90–93.

[43] G. Tallini: On caps of kind s in a Galois r-dimensional space. *Acta Arith.* **7** (1961), 19–28.

[44] L. Teirlinck: On projective and affine hyperplanes. *J. Combin. Theory Ser. A* **28** (1980), 290–306.

[45] V. D. Tonchev: Quasi-symmetric 2-(31, 7, 7)-designs and a revision of Hamada's conjecture. *J. Combin. Theory Ser. A* **42** (1986), 104–110.

[46] V. D. Tonchev: Linear perfect codes and a characterization of the classical designs, *Des. Codes Cryptogr.* **17** (1999), 121–128.

[47] V. D. Tonchev: A note on MDS codes, n-arcs and complete designs. *Des. Codes Cryptogr.* **29** (2003), 247–250.

[48] V. D. Tonchev: Finite geometry, designs, codes, and Hamada's conjecture, in: *Information Security, Coding Theory and Related Combinatorics*, D. Crnković and V. Tonchev eds., IOS Press, Amsterdam, 2011, pp. 437–448.

[49] M. A. Tsfasman and S. G. Vlădut: Geometric approach to higher weights. *IEEE Trans. Inf. Theory* **41** (1995), 1564–1588.

Gohar M. Kyureghyan

Special Mappings of Finite Fields

Abstract: Mappings of finite fields play an important role in many applications like coding theory, combinatorics, cryptology or finite geometry. In this article we survey recent progress on classification and explicit constructions of almost perfect nonlinear, bent, crooked mappings and those having a linear structure. We present the switching method, which proved itself as a powerful tool for constructing mappings satisfying additive properties. We describe main open challenges in this research area.

Keywords: Perfect Non-Linear Mapping, Bent Mapping, Crooked Mapping, Planar Mapping, Linear Structure, Linearized Polynomial, Permutation Polynomial, Switching Method, Sparse Polynomial

2010 Mathematics Subject Classifications: 11T06, 11T71, 12K10, 12Y05

Gohar M. Kyureghyan: Department of Mathematics, Otto-von-Guericke University of Magdeburg, Magdeburg, Germany; Project SECRET, INRIA Paris – Rocquencourt, Le Chesnay, France, e-mail: gohar.kyureghyan@ovgu.de

1 Introduction

Let q be a prime power, $\mathbb{F}_q$ be the finite field with q elements and $\mathbb{F}_q^* = \mathbb{F}_q \setminus \{0\}$. Given a univariate polynomial $F(X) \in \mathbb{F}_q[X]$, the *associated mapping* F of it is defined by

$$F\colon \mathbb{F}_q \to \mathbb{F}_q,\ x \mapsto F(x)\,.$$

The associated mappings of polynomials $F(X)$ and $G(X)$ are equal on $\mathbb{F}_q$ if and only if $F(X) \equiv G(X) \pmod{X^q - X}$. In particular, the associated mappings of two different polynomials of degree less than q are different. This shows that any mapping of $\mathbb{F}_q$ into itself is the associated mapping of a unique polynomial over $\mathbb{F}_q$ of degree less than q.

Indeed, the number of different mappings of $\mathbb{F}_q$ into itself is q^q, which is also the number of different polynomials of degree less than q in $\mathbb{F}_q[X]$. If the mapping F is the associated mapping of the polynomial $F(X)$ and the degree of $F(X)$ is less than q, then $F(X)$ is called the *reduced polynomial* describing the mapping F or briefly the reduced polynomial of F. The *degree of the mapping* F is the degree of its reduced poly-

The author thanks Pascale Charpin for many stimulating discussions especially on crooked mappings and Reed–Muller codes. She thanks also Yves Edel, Alex Pott, Valentin Suder and Arne Winterhof for their comments on the preliminary version of this survey.

nomial. A mapping F is called *monomial*, resp. *binomial*, if its reduced polynomial is a monomial αX^d, $\alpha \in \mathbb{F}_q^*$, or a binomial $\alpha X^d + \beta X^t$, $\alpha, \beta \in \mathbb{F}_q^*$. A polynomial over $\mathbb{F}_q$ is called a *permutation polynomial* of $\mathbb{F}_q$ if it induces a permutation on $\mathbb{F}_q$.

Let $q = s^n$ with $n \geq 1$. Then $\mathbb{F}_q$ is an n-dimensional vector space over its subfield $\mathbb{F}_s$. Let $(x_1, \ldots, x_n)$ be the coordinate vector of $x \in \mathbb{F}_q$ with respect to a fixed basis $\mathcal{B}$ of $\mathbb{F}_q$ over $\mathbb{F}_s$. Then any multivariate polynomial $F(X_1, \ldots, X_n) \in \mathbb{F}_q[X_1, \ldots, X_n]$ defines a mapping of $\mathbb{F}_q$ via $(x_1, \ldots, x_n) \mapsto F(x_1, \ldots, x_n)$. Two polynomials $F(X_1, \ldots, X_n)$ and $G(X_1, \ldots, X_n)$ define the same mapping on $\mathbb{F}_q$ if and only if $F(X_1, \ldots, X_n) \equiv G(X_1, \ldots, X_n) \pmod{(X_1^s - X), \ldots, (X_n^s - X)}$. Again a simple counting argument shows that for any mapping F there is a unique polynomial $F(X_1, \ldots, X_n) \in \mathbb{F}_q[X_1, \ldots, X_n]$ describing this mapping such that its degree in every variable X_i is less than s. We call this polynomial the *$\mathbb{F}_s$-reduced multivariate representation* of the mapping F with respect to the basis $\mathcal{B}$. The *algebraic $\mathbb{F}_s$-degree* of the mapping F is the total degree of a reduced multivariate polynomial of it. Note that the $\mathbb{F}_s$-reduced multivariate representation of F is basis dependent, while its algebraic $\mathbb{F}_s$-degree is independent of the choice of basis.

The algebraic $\mathbb{F}_s$-degree of a mapping can be computed from its reduced (univariate) polynomial. Recall that the *s-weight* of a non-negative integer d is the sum of the digits in its s-ary representation, i.e. if $d = \sum_{i=0}^{l} d_i s^i$ with $0 \leq d_i \leq s-1$, then the s-weight of d is $\mathrm{wt}_s(d) = \sum_{i=0}^{l} d_i \in \mathbb{Z}$.

Lemma 1.1. *The algebraic $\mathbb{F}_s$-degree of the mapping $F(x) = \sum_{k=0}^{q-1} \alpha_k x^k$ on $\mathbb{F}_q$ is equal to $\max_{k, \alpha_k \neq 0}\{\mathrm{wt}_s(k)\}$.*

Proof. Let $\mathcal{B} = (\beta_1, \ldots, \beta_n)$ be a basis of $\mathbb{F}_q$ over $\mathbb{F}_s$. Any $x \in \mathbb{F}_q$ can be represented as

$$x = \beta_1 x_1 + \cdots + \beta_n x_n$$

with $x_i \in \mathbb{F}_s$ and thus

$$F(x) = F(\beta_1 x_1 + \cdots + \beta_n x_n) = \sum_{k=0}^{q-1} \alpha_k \Big(\sum_{j=1}^{n} \beta_j x_j \Big)^k$$

$$= \sum_{k=0}^{q-1} \alpha_k \Big(\sum_{j=1}^{n} \beta_j x_j \Big)^{\sum_{i=0}^{n-1} k_i s^i} = \sum_{k=0}^{q-1} \alpha_k \prod_{i=0}^{n-1} \Big(\sum_{j=1}^{n} \beta_j^{s^i} x_j \Big)^{k_i},$$

which shows that the algebraic $\mathbb{F}_s$-degree of F does not exceed $\max_{k,\alpha_k \neq 0}\{\mathrm{wt}_s(k)\}$. Further, let $m = \sum_{i=0}^{n-1} m_i s^i$ be such that $\alpha_m \neq 0$ and $\mathrm{wt}_s(m) = \max_{k,\alpha_k \neq 0}\{\mathrm{wt}_s(k)\}$. Then the coefficient of the monomial $x_1^{m_0} \ldots x_n^{m_{n-1}}$ is $\alpha_m \beta_1^{m_0} \beta_2^{m_1 s} \ldots \beta_n^{m_{n-1} s^{n-1}} \neq 0$, completing the proof. □

Non-zero $\mathbb{F}_s$-linear mappings L of $\mathbb{F}_q$ satisfy $L(0) = 0$ and have algebraic $\mathbb{F}_s$-degree 1. Hence the reduced polynomial description of an $\mathbb{F}_s$-linear mapping is given by

$$L(X) = \alpha_{n-1}X^{s^{n-1}} + \alpha_{n-2}X^{s^{n-2}} + \cdots \alpha_0 X \in \mathbb{F}_q[X]\,. \tag{1.1}$$

The polynomials of shape (1.1) are called *s-polynomials* over $\mathbb{F}_q$. In the case when s is fixed, such polynomials are also referred to as *linearized polynomials*.

The $\mathbb{F}_s$*-affine* mappings of $\mathbb{F}_q$ are those of algebraic $\mathbb{F}_s$-degree at most 1. Similarly, the $\mathbb{F}_s$*-quadratic*, $\mathbb{F}_s$*-cubic* mappings of $\mathbb{F}_q$ are those of algebraic $\mathbb{F}_s$-degree 2, 3, respectively. In general, if $\mathbb{F}_s$ is the prime subfield of $\mathbb{F}_q$, then the abbreviations algebraic degree, affine, quadratic, cubic are used instead of algebraic $\mathbb{F}_s$-degree, $\mathbb{F}_s$-affine, $\mathbb{F}_s$-quadratic, $\mathbb{F}_s$-cubic.

Two mappings $F, G\colon \mathbb{F}_q \to \mathbb{F}_q$ are called *affine equivalent* if $G = A_1 \circ F \circ A_2$, and *extended affine equivalent* (EA-equivalent) if $G = A_1 \circ F \circ A_2 + A$ for some affine permutations $A_1, A_2\colon \mathbb{F}_q \to \mathbb{F}_q$ and an affine mapping $A\colon \mathbb{F}_q \to \mathbb{F}_q$. EA-equivalent non-affine mappings have the same algebraic degree.

Two mappings F, $G\colon \mathbb{F}_q \to \mathbb{F}_q$ are called *Carlet–Charpin–Zinoviev equivalent* (CCZ-equivalent) if the set $\{(x, G(x)) \mid x \in \mathbb{F}_q\} \subseteq \mathbb{F}_q^2$ is the image of the set $\{(x, F(x)) \mid x \in \mathbb{F}_q\} \subseteq \mathbb{F}_q^2$ under an affine permutation of $\mathbb{F}_q^2$. In other words, two mappings of $\mathbb{F}_q$ are CCZ-equivalent if their graphs in $\mathbb{F}_q^2$ are affine equivalent. It is observed in [29] that EA-equivalence is a special case of CCZ-equivalence, and that the inverse mapping of a permutation is CCZ-equivalent to it. The latter results are stated in [29] only for CCZ-equivalence over fields of even order, but they can be easily extended to the fields of odd order.

The *trace* of an element α of $\mathbb{F}_q$ over $\mathbb{F}_s$ is the sum of its conjugate elements over $\mathbb{F}_s$ and it is denoted here by $\mathrm{tr}_{q/s}(\alpha)$, i.e.

$$\mathrm{tr}_{q/s}(\alpha) = \alpha + \alpha^s + \cdots + \alpha^{s^{n-1}}\,.$$

The trace mapping $\mathrm{tr}_{q/s}$ of $\mathbb{F}_q$ is given by $x \mapsto \mathrm{tr}_{q/s}(x)$. The mappings T_α, $\alpha \in \mathbb{F}_q$, with $T_\alpha(x) = \mathrm{tr}_{q/s}(\alpha x)$ are all the $\mathbb{F}_s$-linear mappings from $\mathbb{F}_q$ into $\mathbb{F}_s$. If $\alpha \neq \beta$, then $T_\alpha \neq T_\beta$, and $T_\alpha\colon \mathbb{F}_q \to \mathbb{F}_s$ is surjective for every $\alpha \neq 0$. For any $\mathbb{F}_s$-hyperplane $\mathcal{H}$ of $\mathbb{F}_q$, i.e. an $(n-1)$-dimensional $\mathbb{F}_s$-vector subspace of $\mathbb{F}_q$, there is a unique nonzero $\alpha \in \mathbb{F}_q$ such that

$$\mathcal{H} = \{x \in \mathbb{F}_q \mid \mathrm{tr}_{q/s}(\alpha x) = 0\}\,.$$

We use the term *function* for the mappings $f\colon \mathbb{F}_q \to \mathbb{F}_s$ in order to emphasize that the image set of f is definitely contained in the subfield $\mathbb{F}_s$. Any function $f\colon \mathbb{F}_q \to \mathbb{F}_s$ can be represented as a composition of a suitable mapping $F\colon \mathbb{F}_q \to \mathbb{F}_q$ and the trace function $\mathrm{tr}_{q/s}$, i.e. $f = \mathrm{tr}_{q/s} \circ F$. Indeed, for $x \in \mathbb{F}_q$ the value $F(x)$ must be just chosen to satisfy $\mathrm{tr}_{q/s}(F(x)) = f(x)$. A function $f\colon \mathbb{F}_q \to \mathbb{F}_s$ is called *monomial* if it can be represented by $\mathrm{tr}_{q/s}(\alpha x^d)$ for some $\alpha \in \mathbb{F}_q^*$ and an integer d.

Let $F\colon \mathbb{F}_q \to \mathbb{F}_q$ and $(\gamma_1, \ldots, \gamma_n)$ be an $\mathbb{F}_s$-basis of $\mathbb{F}_q$. The uniquely determined functions $f_i\colon \mathbb{F}_q \to \mathbb{F}_s$, $1 \le i \le n$, such that

$$F(x) = f_1(x) \cdot \gamma_1 + \cdots + f_n(x) \cdot \gamma_n\,,$$

are called the *coordinate functions* of F with respect to the basis $(y_1, \dots, y_n)$. The algebraic degree of F is equal to the maximum of algebraic degrees of its coordinate functions. The *component functions* of F over the subfield $\mathbb{F}_s$ are the functions $\mathrm{tr}_{q/s}(\alpha F(x))$ with $\alpha \in \mathbb{F}_q^*$. The set of component functions of a mapping coincides with the one of its coordinate functions:

Proposition 1.2. *Any component function over $\mathbb{F}_s$ of a mapping $F\colon \mathbb{F}_q \to \mathbb{F}_q$ is a coordinate function with respect to some $\mathbb{F}_s$-basis, and vice versa.*

Proof. Recall that any basis $(y_1, \dots, y_n)$ has a unique dual basis $(\bar{y}_1, \dots, \bar{y}_n)$ defined by

$$\mathrm{tr}_{q/s}(y_i \bar{y}_j) = \begin{cases} 1 & \text{if } i = j \\ 0 & \text{if } i \neq j \end{cases}$$

for all $1 \le i, j \le n$. In particular for any $a \in \mathbb{F}_q$ the coefficients a_i in $a = \sum_{i=1}^{n} a_i y_i$ are given by

$$a_i = \mathrm{tr}_{q/s}(\bar{y}_i a)\,.$$

Consequently, the coordinate function $f_i(x)$ of $F(x)$ with respect to $(y_1, \dots, y_n)$ is the component function $\mathrm{tr}_{q/s}(\bar{y}_i F(x))$. On the other hand, a given component function $\mathrm{tr}_{q/s}(\alpha F(x))$ is a coordinate function of $F(x)$ with respect to the dual basis of any basis containing α. □

Given $F\colon \mathbb{F}_q \to \mathbb{F}_q$ and $\alpha \in \mathbb{F}_q^*$, the mapping

$$D_{F,\alpha} : \mathbb{F}_q \to \mathbb{F}_q, \quad x \mapsto F(x + \alpha) - F(x)$$

is called the *difference mapping* of F defined by α (or the derivative of F in direction α). The *differential spectrum* of F is defined as the (multi)set collecting the integers

$$|\{x \in \mathbb{F}_q \mid D_{F,\alpha}(x) = y\}|$$

for all $\alpha \neq 0, y \in \mathbb{F}_q$.

2 Different Notions for Optimal Non-linearity

Mappings which are used in some cryptological applications must be as far as possible from being linear. There are several criteria which measure the non-linearity of a mapping:

(1) *Algebraic degree:* A linear mapping has algebraic degree 1, so a mapping with the largest possible algebraic degree can be seen as an optimal non-linear mapping.
(2) *Differential properties:* Given a linear mapping $L\colon \mathbb{F}_q \to \mathbb{F}_q$, for any fixed $a \in \mathbb{F}_q^*$ the set $\{L(x + a) - L(x) \mid x \in \mathbb{F}_q\}$ contains only one element, namely $L(a)$. Hence a mapping $F\colon \mathbb{F}_q \to \mathbb{F}_q$ for which all the sets $\{F(x + a) - F(x) \mid x \in \mathbb{F}_q\}$ are as large as possible can be considered as perfectly non-linear.

(3) *Linear approximation:* The idea here is that a non-linear mapping F does not allow a good affine approximation. More precisely, the sets $\{x \in \mathbb{F}_q \mid F(x) = L(x) + c\}$ must be as small as possible for all linear mappings $L\colon \mathbb{F}_q \to \mathbb{F}_q$ and all constant $c \in \mathbb{F}_q$.

The mappings having the best non-linearity with respect to the differential properties are called perfect non-linear or planar when q is odd, and almost perfect non-linear when q is even. The bent and almost bent mappings have the worst possible affine approximations. Further non-linearity criteria are discussed in [84].

In the rest of this section it is assumed that $\mathbb{F}_q$ has characteristic 2. The analogous concepts for finite fields of odd characteristic are introduced in Section 5.

2.1 Almost Perfect Nonlinear (APN) Mappings

Let $F\colon \mathbb{F}_{2^n} \to \mathbb{F}_{2^n}$ and $\alpha \in \mathbb{F}_{2^n}^*$. The image set of a difference mapping $D_{F,\alpha} : x \mapsto F(x+\alpha) + F(x)$ contains at most 2^{n-1} elements, since $D_{F,\alpha}(x) = D_{F,\alpha}(x+\alpha)$ for any $x \in \mathbb{F}_{2^n}$. Clearly, the image set of a difference mapping $D_{F,\alpha}$ is of maximal size if and only if $D_{F,\alpha}$ is 2-to-1. A mapping is called *almost perfect non-linear*, abbreviated APN, if all its difference mappings are 2-to-1. Note that the APN mappings can be defined also as those having differential spectrum containing only the integers 0 and 2. APN mappings provide the optimal resistance against the differential cryptanalysis when they are used as an S-box [87]. The APN property is invariant under CCZ-equivalence [29].

The following theorem is essentially from [7]:

Theorem 2.1. *Let $\mathcal{H}$ be a hyperplane in $\mathbb{F}_{2^n}$, that is $\mathcal{H}$ is an $(n-1)$-dimensional $\mathbb{F}_2$-subspace of $\mathbb{F}_{2^n}$. A mapping $F\colon \mathbb{F}_{2^n} \to \mathbb{F}_{2^n}$ is APN if and only if the difference mappings $D_{F,\beta}$ are 2-to-1 for all non-zero $\beta \in \mathcal{H}$.*

Proof. Necessity of the condition follows clearly from definition of APN mappings. To prove that it is also sufficient, suppose that $\alpha \in \mathbb{F}_{2^n} \setminus \mathcal{H}$ and $D_{F,\alpha}$ is not 2-to-1. Then there are two distinct $x, y \in \mathbb{F}_{2^n}$ such that $x + y \neq \alpha$ and

$$D_{F,\alpha}(x) = F(x+\alpha) + F(x) = F(y+\alpha) + F(y) = D_{F,\alpha}(y)\,.$$

The above equality implies

$$\begin{aligned} D_{F,x+y}(x) &= F(x) + F(x + (x+y)) \\ &= F(y+\alpha) + F(y + \alpha + (x+y)) \\ &= D_{F,x+y}(y+\alpha) \end{aligned}$$

and

$$\begin{aligned} D_{F,x+y+\alpha}(x) &= F(x) + F(x + (x + y + \alpha)) \\ &= F(x + \alpha) + F(x + \alpha + (x + y + \alpha)) \\ &= D_{F,x+y+\alpha}(x + \alpha)\,. \end{aligned}$$

Consequently, both $D_{F,x+y}$ and $D_{F,x+y+\alpha}$ are not 2-to-1, which is a contradiction since either $x + y$ or $x + y + \alpha$ belongs to $\mathcal{H}$. □

Theorem 2.1 reduces the computational costs for verifying the APN property significantly.

To show that a monomial mapping $x \mapsto x^d$ is APN, it is enough to check that its difference mapping $D_{d,1}$ defined by $\alpha = 1$ is 2-to-1. Indeed, for an arbitrary element $\alpha \in \mathbb{F}_{2^n}^*$

$$D_{d,\alpha}(x) = (x+\alpha)^d + x^d = \alpha^d \left(\left(\frac{x}{\alpha} + 1\right)^d + \left(\frac{x}{\alpha}\right)^d \right) = \alpha^d \cdot D_{d,1}\left(\frac{x}{\alpha}\right)$$

holds. This property simplifies the study of APN monomial mappings, since only one difference mapping, instead of $2^{n-1} - 1$ in general case, must be proved to be 2-to-1. In spite of that, the complete characterization of APN monomial mappings is a difficult open problem.

An integer d defining an APN monomial mapping is called an APN exponent. The known APN exponents are:

$$\begin{aligned} &2^k + 1\,, \ \gcd(k,n) = 1 \text{ and } 1 \le k < \tfrac{n}{2} \quad \text{(Gold's exponent [59])}, \\ &2^{2k} - 2^k + 1\,, \ \gcd(k,n) = 1 \text{ and } 1 \le k < \tfrac{n}{2} \quad \text{(Kasami's exponent [70])}, \\ &2^{4k} + 2^{3k} + 2^{2k} + 2^k - 1\,, \ \text{where } n = 5k \quad \text{(Dobbertin's exponent [52])}, \\ &2^m + 3\,, \ \text{where } n = 2m+1 \quad \text{(Welch's exponent [51])}, \\ &2^m + 2^{\frac{m}{2}} - 1\,, \ \text{where } n = 2m+1 \text{ and } m \text{ is even, and} \\ &2^m + 2^{\frac{3m+1}{2}} - 1\,, \ \text{where } n = 2m+1 \text{ and } m \text{ is odd} \quad \text{(Niho's exponent [50])}, \\ &2^n - 2\,, \ \text{where n is odd} \quad \text{(field inverse [88])}. \end{aligned}$$

The above list contains only one representative from the cyclotomic coset $\{d, \dots, 2^{n-1}d\}$ and only one from the inverse elements d, d^{-1} modulo $2^n - 1$ in the case when the inverse for d exists. Hans Dobbertin conjectured that no other APN exponents do exist. Exhaustive computations by Yves Edel show that the conjecture is true for all $n \le 34$ and $n \in \{36, 38, 40, 42\}$. The explicit representations for inverses of the APN exponents listed above are given in [76, 88, 89].

In the list of APN exponents, the Gold and Kasami exponents are *exceptional* in the sense that they define APN monomial mappings for infinitely many choices of n. It is known that Gold and Kasami exponents are the only such ones [64] (see also

Chapter 6 of this book). More generally, in [4], it is conjectured that any exceptional APN polynomial, that is a polynomial defining an APN mapping on infinitely many finite fields, must be CCZ-equivalent to a monomial mapping with Gold or Kasami exponent. Results supporting this conjecture are obtained in [4, 43, 91].

The APN monomial mappings are bijective if n is odd and are 3-to-1 otherwise [28]. The "biggest open problem" on APN mappings asking whether APN permutations do exist for an even n was answered affirmative in [17] by presenting such an example for $n = 6$. Currently, no example of an APN permutation for even $n \geq 8$ is known. There are no APN permutations on $\mathbb{F}_{2^n}$ which is defined by a polynomial over its subfield $\mathbb{F}_{2^{n/2}}$ as shown in [67].

An easy way to construct new APN mappings is the application of the EA-equivalence: Take $A_1, A_2 \colon \mathbb{F}_{2^n} \to \mathbb{F}_{2^n}$ to be affine permutations and $A \colon \mathbb{F}_{2^n} \to \mathbb{F}_{2^n}$ an affine mapping and construct $A_1 \circ F \circ A_2 + A$ from the given APN mapping F. Further possibility is the application of more sophisticated CCZ-equivalence [21]. However, recognizing whether two given APN mappings are CCZ- or even EA-equivalent is in general difficult. In some cases the (in)equivalence of mappings is indicated only by computational results about coding/design theoretical invariants for fields of small sizes [15, 17, 54, 100].

Besides the monomials, quadratic polynomials deliver another class of mappings, for which checking the APN property is easier compared with the general case. This is explained by the fact that the difference mappings of a quadratic one $Q(x) = \sum_{i,j} a_{i,j} x^{2^i+2^j}$ are $\mathbb{F}_2$-affine. Indeed, for $\alpha \in \mathbb{F}_{2^n}^*$ the difference mapping

$$D_{Q,\alpha}(x) = \sum_{i,j} a_{i,j} \left(\alpha^{2^i} x^{2^j} + \alpha^{2^j} x^{2^i} + \alpha^{2^i+2^j} \right)$$

is given by sum of a 2-polynomial and the constant $Q(\alpha)$. Hence $D_{Q,\alpha}(x)$ is 2-to-1 if and only if the linear mapping $D_{Q,\alpha}(x) - Q(\alpha)$ has one dimensional kernel.

It was widely believed that every APN quadratic mapping is EA-equivalent to a Gold monomial mapping. In [54] this conjecture was disproved by showing that for suitable chosen u the quadratic binomials

$$x^3 + ux^{36} \text{ in } \mathbb{F}_{2^{10}} \text{ and } x^3 + ux^{528} \text{ in } \mathbb{F}_{2^{12}}, \tag{2.1}$$

define APN mappings which are CCZ-inequivalent to the monomial APN mappings. Several families of quadratic APN mappings have been constructed ever since. The list of currently known pairwise CCZ-inequivalent infinite families of non-monomial APN mappings is given in [15]. Yves Edel conjectured that two CCZ-equivalent APN quadratic mappings must be EA-equivalent [53]. This was proved to be true for the quadratic mappings CCZ-equivalent to quadratic APN monomials in [16]. The general case was settled in [102]. Even a stronger fact was established in [45]: A mapping obtained from a quadratic APN one by applying CCZ-equivalence, which is not EA-equivalence, is not quadratic.

Besides of applications in cryptology, APN mappings yield optimal objects in finite geometry, combinatorics and coding theory, cf. [29, 58, 61, 95, 100, 101]. An interesting connection between APN mappings and reversed Dickson polynomials is given in [68].

2.2 Bent and Almost Bent (AB) Mappings

Bent functions are the Boolean functions $f\colon \mathbb{F}_{2^n} \to \mathbb{F}_2$ having the worst possible affine approximations. Recall that any linear function $l\colon \mathbb{F}_{2^n} \to \mathbb{F}_2$ is given by the trace function $T_\alpha(x) = \mathrm{tr}_{2^n/2}(\alpha x)$ with a suitable $\alpha \in \mathbb{F}_{2^n}$. The Hamming distance $d(f,g)$ between two Boolean functions f and g is defined as the cardinality of the set $\{x \in \mathbb{F}_{2^n} \mid f(x) \neq g(x)\}$. The non-linearity $\mathrm{nl}(f)$ of a Boolean function f is its Hamming distance to the set of affine Boolean functions, i.e.

$$\mathrm{nl}(f) := \min_{\alpha\in\mathbb{F}_{2^n}, c\in\mathbb{F}_2} \{d(f, T_\alpha + c)\}\,.$$

Hence if a function f has high non-linearity then it has no good affine approximations. The Walsh coefficient of $f\colon \mathbb{F}_{2^n} \to \mathbb{F}_2$ at $\alpha \in \mathbb{F}_{2^n}$ is the integer

$$\mathcal{W}_f(\alpha) = \sum_{x\in\mathbb{F}_{2^n}} (-1)^{f(x)+T_\alpha(x)} = \sum_{x\in\mathbb{F}_{2^n}} (-1)^{f(x)+\mathrm{tr}_{2^n/2}(\alpha x)}\,.$$

It is easy to see that

$$\mathcal{W}_f(\alpha) = 2^n - 2\cdot d(f, T_\alpha)\,,$$

and thus

$$\mathrm{nl}(f) = 2^{n-1} - \tfrac{1}{2}\max_{\alpha\in\mathbb{F}_{2^n}} \{|\mathcal{W}_f(\alpha)|\}\,.$$

The Walsh coefficients satisfy the Parseval equation

$$\sum_{\alpha\in\mathbb{F}_{2^n}} \mathcal{W}_f(\alpha)^2 = 2^{2n}\,,$$

implying

$$\max_{\alpha\in\mathbb{F}_{2^n}} \{|\mathcal{W}_f(\alpha)|\} \geq 2^{n/2}\,. \tag{2.2}$$

It is easy to see that in (2.2) equality holds if and only if

$$|\mathcal{W}_f(\alpha)| = 2^{n/2} \text{ for every } \alpha \in \mathbb{F}_{2^n}\,.$$

Hence the worst affine approximation will be achieved by the Boolean functions for which all Walsh coefficients have absolute value $2^{n/2}$. Such functions may exist only for even n, since Walsh coefficients are integers.

Definition 2.2. Let n be even. A Boolean function $f\colon \mathbb{F}_{2^n} \to \mathbb{F}_2$ is called *bent* if $\mathrm{nl}(f) = 2^{n-1} - 2^{n/2-1}$, or equivalently if $\mathcal{W}_f(\alpha) = \pm 2^{n/2}$ for every $\alpha \in \mathbb{F}_{2^n}$.

When n is odd, the highest possible non-linearity for Boolean functions on $\mathbb{F}_{2^n}$ is currently unknown, see [27, Chapter 4.1.2] for more details.

The algebraic degree of a bent function on $\mathbb{F}_{2^n}$ does not exceed $n/2$ for $n \geq 4$, and there are bent functions with algebraic degree m for any $2 \leq m \leq n/2$ as shown in [46, 93].

Finding explicit families of sparse univariate polynomials describing bent functions is a challenging problem [85]. An integer d is called a *bent exponent* of $\mathbb{F}_{2^n}$ if there is an element $\lambda \in \mathbb{F}_{2^n}$ such that the monomial function $x \mapsto \mathrm{tr}_{2^n/2}(\lambda x^d)$ is bent. All presently known monomial bent functions belong to an infinite family. These families are defined by:

- $d = 2^k + 1$, $\gcd(k, n) = 1$ and $\lambda \notin \{y^d | y \in \mathbb{F}_{2^n}\}$ (folklore)
- $d = (2^{n/2} - 1)\ell$, $\gcd(\ell, 2^{n/2} + 1) = 1$ and λ corresponds to a zero of the Kloosterman sum (see [46, 78])
- $d = 2^{2k} - 2^k + 1$, $\gcd(k, n) = 1$ and $\lambda \notin \{y^3 | y \in \mathbb{F}_{2^n}\}$ (see [48, 81])
- If $n = 4r$ and r is odd, then $d = 2^{2r} + 2^{r+1} + 1$ and $\lambda = \lambda' a^{2^{2r}+2^{r+1}+1}$, where $\lambda' \in \omega\mathbb{F}_{2^r}$, $\omega \in \mathbb{F}_4 \setminus \mathbb{F}_2$ and $a \in \mathbb{F}_{2^{4r}}$ (see [32, 82]).
- If $n = 6r$, then $d = 2^{2r} + 2^r + 1$ and $\lambda = \lambda' a^{2^{2r}+2^r+1}$, where $a \in \mathbb{F}_{2^{6r}}$ and $\lambda' \in \mathbb{F}_{2^{3r}}$ such that $\mathrm{tr}_{2^{3r}/2^r}(\lambda') = 0$ (see [24]).

The non-linearity of a mapping $F\colon \mathbb{F}_{2^n} \to \mathbb{F}_{2^n}$ is defined as the minimal non-linearity of its component functions $\mathrm{tr}_{2^n/2}(\alpha F(x))$. If n is odd, the maximal achievable non-linearity is $2^{n-1} - 2^{(n-1)/2}$ as shown in [30]. In the case of n even, the best possible non-linearity is not known.

Definition 2.3. Let n be odd. A mapping $F\colon \mathbb{F}_{2^n} \to \mathbb{F}_{2^n}$ is called *almost bent*, abbreviated AB, if

$$\sum_{x \in \mathbb{F}_{2^n}} (-1)^{\mathrm{tr}_{2^n/2}(\alpha F(x)+\beta x)} \in \left\{0, \pm 2^{(n+1)/2}\right\}$$

for all $\alpha, \beta \in \mathbb{F}_{2^n}$ and $\alpha \neq 0$.

AB mappings are APN [30], the converse is not true in general. The algebraic degree of an AB mapping does not exceed $(n+1)/2$ for $n \geq 3$, and CCZ-equivalence preserves AB property [29]. For an odd n, the Gold, Kasami, Welch and Niho exponents define AB mappings [23, 66], while the power mappings defined by Dobbertin and the inverse exponents are not AB. If d defines an AB monomial mapping on the field $\mathbb{F}_{2^n}$, then d defines an AB mapping on every subfield of $\mathbb{F}_{2^n}$ as well [60].

3 Functions with a Linear Structure

Let f be a function from $\mathbb{F}_q$ into its subfield $\mathbb{F}_s$. Given $a \in \mathbb{F}_s$, an element $\alpha \in \mathbb{F}_q^*$ is called an *a-linear translator* (or a-linear structure) for f if for every $u \in \mathbb{F}_s$

$$f(x + u\alpha) - f(x) = ua \tag{3.1}$$

holds for all $x \in \mathbb{F}_q$. Note that (3.1) with $x = 0$ and $u = 1$ implies $a = f(\alpha) - f(0)$. The concept of a linear translator was introduced in cryptography for Boolean functions. The functions with linear translators are considered to be weak for some cryptographic applications because of the cryptanalysis suggested in [56].

Denote by $\Lambda^*(f)$ the set of all linear translators of f and let $\Lambda(f) = \Lambda^*(f) \cup \{0\}$. The set $\Lambda(f)$ is called the *linear space* of f. We say that the function f has a *linear structure* if $\Lambda^*(f)$ is not empty. The next result follows directly from the definition of a linear translator.

Proposition 3.1 ([79]). *Let $\alpha, \beta \in \mathbb{F}_q^*$, $\alpha + \beta \neq 0$ and $a, b, c \in \mathbb{F}_s$, $c \neq 0$. If α is an a-linear translator and β is a b-linear translator of $f\colon \mathbb{F}_q \to \mathbb{F}_s$, then $\alpha + \beta$ is an $(a + b)$-linear translator of f and $c \cdot \alpha$ is a $(c \cdot a)$-linear translator of f. In particular, $\Lambda(f)$ is an $\mathbb{F}_s$-linear subspace of $\mathbb{F}_q$.*

Proposition 3.1 shows that the restriction of the function $f(x) - f(0)$ on the subspace $\Lambda(f)$ is an $\mathbb{F}_s$-linear function. In particular, $\Lambda(f) = \mathbb{F}_q$ if and only if f is an $\mathbb{F}_s$-affine function, or equivalently if $f(x) = \mathrm{tr}_{q/s}(\beta x) + b$ for some $\beta \in \mathbb{F}_q$ and $b \in \mathbb{F}_s$. More generally, the following theorem holds:

Theorem 3.2 ([34, 79]). *A function $f\colon \mathbb{F}_q \to \mathbb{F}_s$ has a linear structure if and only if there is a non-bijective $\mathbb{F}_s$-linear mapping $L\colon \mathbb{F}_q \to \mathbb{F}_q$ such that*

$$f(x) = \mathrm{tr}_{q/s}(H \circ L(x) + \beta x) \tag{3.2}$$

for some $H\colon \mathbb{F}_q \to \mathbb{F}_q$ and $\beta \in \mathbb{F}_q$. In this case, the kernel of L is contained in the subspace $\Lambda(f)$.

The following examples describe functions for which a given element $\gamma \in \mathbb{F}_q^*$ is a linear translator.

Example 3.3.

(a) Let $H\colon \mathbb{F}_q \to \mathbb{F}_q$ be an arbitrary function, $\gamma, \beta \in \mathbb{F}_q$, $\gamma \neq 0$ and $c = \mathrm{tr}_{q/s}(\beta\gamma)$. Then γ is a c-linear translator of $f(x) = \mathrm{tr}_{q/s}(G(x))$ where

$$G(x) = H(x^s - \gamma^{s-1}x) + \beta x\,.$$

(b) Let $g\colon \mathbb{F}_q \to \mathbb{F}_s$ and $\gamma \in \mathbb{F}_q^*$. Then for any $c \in \mathbb{F}_s^*$ the element $c\gamma$ is a 0-linear translator of

$$f(x) = \sum_{u \in \mathbb{F}_s} g(x + u\gamma)\,.$$

In general, for a given polynomial $F(X) \in \mathbb{F}_q[X]$ it is difficult to determine whether the function $\mathrm{tr}_{q/s}(F(x))$ has a linear structure. The complete characterization of the monomial functions from a finite field into its prime field having a linear structure is obtained in [33] for fields of even order and in [35] for fields of odd order.

Theorem 3.4 ([33]). *Let $0 \leq s \leq 2^n - 2$, $\delta \in \mathbb{F}_q^*$ be such that the Boolean function $\mathrm{tr}_{2^n/2}(\delta x^s)$ is a non-zero function. Then $\alpha \in \mathbb{F}_{2^n}^*$ is a linear translator of the Boolean function $\mathrm{tr}_{2^n/2}(\delta x^s)$ if and only if*
(a) *$s = 2^i$ and α is arbitrary*
(b) *$s = 2^i + 2^j$ $(i \neq j)$ and $(\delta \alpha^{2^i+2^j})^{2^{n-i}} + (\delta \alpha^{2^i+2^j})^{2^{n-j}} = 0$.*

Theorem 3.5 ([35]). *Let p be an odd prime number, $\delta \in \mathbb{F}_{p^n}$ and $1 \leq d \leq p^n - 2$ be such that $f(x) = \mathrm{tr}_{p^n/p}(\delta x^d)$ is not the zero function. Then f has a linear structure if and only if one of the following cases holds:*
(i) *$d = p^j$, $0 \leq j \leq n - 1$, and $\delta \in \mathbb{F}_{p^n}^*$. In this case f is $\mathbb{F}_p$-linear and hence $\Lambda(f) = \mathbb{F}_{p^n}$.*
(ii) *$d = p^j(p^i + 1)$ where $0 \leq i, j \leq n - 1$ and $i \neq 0, n/2$. Moreover, n/t is even, where $t = \gcd(n, i)$, and*
- *if $n/2t$ is even, then δ is a $(p^t + 1)$-th power in $\mathbb{F}_{p^n}$;*
- *if $n/2t$ is odd, then δ is a $(p^t + 1)/2$-th power but not a $(p^t + 1)$-th power in $\mathbb{F}_{p^n}$;*

In this case $\Lambda(f) = \epsilon \mathbb{F}_{p^{2t}}$, where ϵ is a fixed element satisfying $\epsilon^{p^{2i}-1} = -\delta^{-p^{n-j}(p^i-1)}$. Moreover, every $\alpha \in \Lambda^*(f)$ is a 0-linear translator of f.

The concept of functions with a linear structure can be naturally extended to mappings: An element $\alpha \in \mathbb{F}_q^*$ is called an *a-linear translator* for $F\colon \mathbb{F}_q \to \mathbb{F}_q$ if for every $u \in \mathbb{F}_s$

$$F(x + u\alpha) - F(x) = ua \tag{3.3}$$

holds for all $x \in \mathbb{F}_q$ and some fixed $a \in \mathbb{F}_q$. In fact, again $a = F(\alpha) - F(0)$. Observe that this definition coincides with the original one for the functions, i.e. when the image set of F is contained in $\mathbb{F}_s$. A mapping is said to have a linear structure if the set of its linear translators is not empty. Most of characterization results for functions with a linear structure can be directly adopted for mappings with a linear structure as shown in [36].

A further generalization of the concept of mappings with a linear structure is suggested in [1]:

Definition 3.6 ([1, Definition 1.8]). Let $S \subseteq \mathbb{F}_q$ and $\gamma, b \in \mathbb{F}_q$. We say that γ is a b-linear translator with respect to S for the mapping $F\colon \mathbb{F}_q \to \mathbb{F}_q$, if

$$F(x + u\gamma) - F(x) = ub$$

for all $x \in \mathbb{F}_q$ and for all $u \in S$.

In contrast to condition (3.3), Definition 3.6 does not require that S is a subfield of $\mathbb{F}_q$. The next proposition shows that the subset S in Definition 3.6 can without loss of generality be taken to be a subspace of $\mathbb{F}_q$:

Proposition 3.7. *Let $\gamma \in \mathbb{F}_q$ be a b-linear translator with respect to S for $F\colon \mathbb{F}_q \to \mathbb{F}_q$, where $b \in \mathbb{F}_q$ and $S \subseteq \mathbb{F}_q$. Then γ is a b-linear translator with respect to the $\mathbb{F}_p$-linear span of S, where $\mathbb{F}_p$ is the prime subfield of $\mathbb{F}_q$.*

Proof. Let $u, v \in S$. Then from Definition 3.6 it follows that

$$F(x + (u+v)\gamma) = F(x + u\gamma) + vb = F(x) + (u+v)b\,,$$

implying the statement. □

Proposition 3.7 shows in particular that if F has a linear translator with respect to a subset S, then it has a linear translator with respect to the subspace $\sigma \cdot \mathbb{F}_p$, where σ is an arbitrary element of S and $\mathbb{F}_p$ is the prime field of $\mathbb{F}_q$. More precisely, let $\gamma \in \mathbb{F}_q$ be a b-linear translator with respect to S for $F\colon \mathbb{F}_q \to \mathbb{F}_q$. Then for any $a \in \mathbb{F}_p$ and any $\sigma \in S$

$$F(x + a\sigma\gamma) = F(x) + a\sigma b$$

holds, implying that $\sigma\gamma$ is a σb-linear translator with respect to $\mathbb{F}_p$ for F. Hence the set of mappings admitting a linear translator with respect to a subset is a subset of those with a linear structure with respect to the prime subfield, and therefore the classification results of [33–35, 79] apply to them too.

4 Crooked Mappings

A mapping $F\colon \mathbb{F}_{2^n} \to \mathbb{F}_{2^n}$ is called *crooked* if the image sets of all its difference mappings are affine hyperplanes. Given $\gamma \in \mathbb{F}_{2^n}^*$ and $c \in \mathbb{F}_2$, let $\mathcal{H}_\gamma(c)$ be the affine hyperplane defined by

$$\mathcal{H}_\gamma(c) := \{y \in \mathbb{F}_{2^n} \mid \mathrm{tr}_{2^n/2}(\gamma y) = c\}\,.$$

Hence F is crooked whenever for every $\alpha \in \mathbb{F}_{2^n}^*$ there are uniquely defined $\gamma \in \mathbb{F}_{2^n}^*$ and $c \in \mathbb{F}_2$ such that

$$\{F(x+\alpha) + F(x) \mid x \in \mathbb{F}_{2^n}\} = \mathcal{H}_\gamma(c)\,,$$

implying that α is a c-linear translator of the component function $\mathrm{tr}_{2^n/2}(\gamma F(x))$. Crookedness is preserved by EA-equivalence, but not by CCZ-equivalence.

Clearly, a crooked mapping is APN. If n is odd, then a crooked mapping is AB [72]. Any quadratic APN mapping is crooked, since all its difference mappings are affine and 2-to-1. The central problem in the research on crooked mappings is the question whether non-quadratic crooked mappings do exist. The answer is negative for monomial and binomial mappings:

Theorem 4.1 ([72]). *The only crooked monomial mappings in $\mathbb{F}_{2^n}$ are the ones with exponents $2^i + 2^j$, where $\gcd(i-j, n) = 1$.*

Theorem 4.2 ([11]). *If $x^d + ux^t$ is a crooked mapping of $\mathbb{F}_{2^n}$, then the exponents d and t are of binary weight at most 2.*

Theorems 4.1 and 4.2 imply that an APN monomial or binomial mapping is crooked if and only if it is quadratic. In [72] it is conjectured that this is true for all APN mappings:

Conjecture 4.3. *Every crooked mapping is quadratic, i.e. it has algebraic degree 2.*

The combinatorial techniques used in [11, 72] to prove Theorems 4.1 and 4.2 become very involved in the case of polynomials with many terms, therefore a novel approach is needed to prove or disprove the above conjecture.

It is possible to determine the algebraic degree of a mapping without having a polynomial representation of it. Lemma 4.4 below follows from well-known properties of Reed–Muller codes and their links with Boolean functions, cf. [3, 95]: For an integer $0 \le r \le n$, the *Reed–Muller code of order r* is the subspace of Boolean functions on $\mathbb{F}_{2^n}$ that consists of all functions of algebraic degree at most r. The Reed–Muller code of order $n - r$ is generated by the characteristic functions of the r-dimensional subspaces of $\mathbb{F}_{2^n}$, or, indeed, by the affine r-dimensional subspaces containing any fixed point of $\mathbb{F}_{2^n}$. The orthogonal code of the Reed–Muller code of order r is the Reed–Muller code of order $n - r - 1$. Thus a Boolean function f is contained in the Reed–Muller code of order r if and only if $\sum_{u \in \mathcal{U}} f(u) = 0$ for all affine $(r + 1)$-dimensional spaces of $\mathbb{F}_{2^n}$ through a fixed point, yielding Lemma 4.4.

Lemma 4.4. *The algebraic degree of $F\colon \mathbb{F}_{2^n} \to \mathbb{F}_{2^n}$ is equal to the maximum dimension k for which there is an affine k-dimensional subspace $\mathcal{U}$ of $\mathbb{F}_{2^n}$ such that $\sum_{u \in \mathcal{U}} F(u) \neq 0$.*

Any affine $(s + 1)$-dimensional subspace is a union of two s-dimensional ones: If W is an affine $(s + 1)$-dimensional subspace generated by $\langle w_1, \ldots, w_{s+1} \rangle$, i.e. $W = \langle w_1, \ldots, w_{s+1} \rangle + y$ for some $y \in \mathbb{F}_{2^n}$, then $W = (\langle w_1, \ldots, w_s \rangle + y) \cup (\langle w_1, \ldots, w_s \rangle + (w_{s+1} + y))$. Hence Lemma 4.4 can be stated also in the following form:

Lemma 4.5. *The algebraic degree of $F\colon \mathbb{F}_{2^n} \to \mathbb{F}_{2^n}$ is equal to the minimum dimension k such that $\sum_{w \in \mathcal{W}} F(w) = 0$ holds for all affine $(k+1)$-dimensional subspaces $\mathcal{W}$ of $\mathbb{F}_{2^n}$.*

Lemma 4.5 implies that Conjecture 4.3 is equivalent to:

Conjecture 4.6. *Every crooked mapping $F\colon \mathbb{F}_{2^n} \to \mathbb{F}_{2^n}$ satisfies $\sum_{u \in \mathcal{U}} F(u) = 0$ for all affine 3-dimensional subspaces $\mathcal{U}$ of $\mathbb{F}_{2^n}$.*

The best currently known upper bound on the algebraic degree of crooked mappings can be obtained by considering their component functions [31]. A Boolean function is called *balanced* if it takes the values 0 and 1 equally often. Let $F\colon \mathbb{F}_{2^n} \to \mathbb{F}_{2^n}$ be crooked and $\gamma \in \mathbb{F}_{2^n}^*$. Then for any $\alpha \in \mathbb{F}_{2^n}$, the difference function $\mathrm{tr}_{2^n/2}(\gamma(F(x+\alpha)+F(x)))$ is either constant or balanced. Boolean functions having the latter property are called *partially bent*, cf. [27, Chapter 6.8]. The algebraic degree of a partially bent Boolean function does not exceed $\lfloor n/2 \rfloor$ as remarked in [26]. Thus the algebraic degree of all component functions of a crooked mapping does not exceed $\lfloor n/2 \rfloor$, and therefore:

Proposition 4.7 ([31]). *Let $F\colon \mathbb{F}_{2^n} \to \mathbb{F}_{2^n}$ be crooked. Then the algebraic degree of F is at most $\lfloor n/2 \rfloor$.*

The crooked mappings were introduced in [5]. The original definition of a crooked mapping from [5] is more restrictive, since it requires that the image sets of all difference mappings are complements of hyperplanes. Such a crooked mapping is necessarily bijective, since $f(x) \neq f(x+a)$ for all $a \neq 0$. Crooked permutations do not exist if n is even as shown in [72]. In the case of crooked permutations, the upper bound of Proposition 4.7 was proved in [95] using the Lemma 4.4. In [5] it is shown that bijective crooked mappings can be characterized as follows:

Proposition 4.8. *Let $F\colon \mathbb{F}_{2^n} \to \mathbb{F}_{2^n}$ be bijective. Then F is crooked if and only if the following two conditions are satisfied:*
(1) *$F(x)+F(y)+F(z)+F(x+y+z) \neq 0$ for any distinct $x, y, z \in \mathbb{F}_{2^n}$;*
(2) *$F(x)+F(y)+F(z)+F(x+a)+F(y+a)+F(z+a) \neq 0$ for any $a \in \mathbb{F}_{2^n}^*$ and $x, y, z \in \mathbb{F}_{2^n}$.*

In [58] the crooked permutations are characterized in terms of the minimum distance of a Preparata-like code, and via distance-regularity of a certain graph.

The notion of crooked mappings is generalized in [25]: A mapping $F\colon \mathbb{F}_{2^n} \to \mathbb{F}_{2^n}$ is called *crooked of codimension d*, if for any $\alpha \in \mathbb{F}_{2^n}^*$, the image set of the difference mapping $D_F(\alpha)$ is an affine subspace of codimension d.

5 Planar Mappings

In this section $q = p^n$, where p is an odd prime number. A mapping $P\colon \mathbb{F}_q \to \mathbb{F}_q$ is called *planar* if all its difference mappings $D_{P,\alpha} : x \mapsto P(x+\alpha) - P(x)$, $\alpha \neq 0$, are permutations of $\mathbb{F}_q$. Planar mappings were introduced in [44]. In cryptology such mappings are called *perfect non-linear* [87].

On a prime field $\mathbb{F}_p$, a mapping is planar if and only if it is given by a quadratic polynomial $X^2 + aX + b \in \mathbb{F}_p[X]$. This was shown independently in papers [57, 65, 92]. In the AMS reviews for [92], the reviewer W. M. Kantor says: "This evidently was a theorem whose time had come. Although it concerns a question in effect raised 20 years ago, it was proved using ingenuity together with methods available at that time independently in Detroit, Osaka and Budapest within a period of a few months." The time for complete characterization of planar mappings on extension fields seems to be far away in future.

Similar to the case of APN mappings, CCZ-equivalence preserves the planarity of mappings. However, CCZ-equivalence coincides with EA-equivalence for the planar mappings [75]. The known planar monomial mappings on $\mathbb{F}_q$ have exponents from the cyclotomic cosets modulo $q-1$ of

$$p^k + 1\,, \quad \text{where } k = 0 \text{ or } n/\gcd(n,k) \text{ is odd} \quad \text{(see [41, 44])},$$

$$\frac{3^k+1}{2}\,, \quad \text{where } p = 3 \text{ and } k \geq 3 \text{ is odd, } \gcd(k,n) = 1 \quad \text{(see [41, 63])}.$$

In fields of order p^2 and $p^4 > 81$, the only planar exponents are those from the cyclotomic coset of 2 as shown in [38, 40]. The classification of planar exponents for the other cases is open.

Except of the planar mappings EA-equivalent to $x \mapsto x^{(3^k+1)/2}$ on $\mathbb{F}_{3^n}$, all known planar mappings are quadratic, that is they are given by polynomials of shape

$$\sum_{i,j=0}^{n-1} a_{i,j} X^{p^i+p^j} + L(X) + c \in \mathbb{F}_q[X]\,,$$

where $L(X)$ is a p-polynomial and c is a constant. The polynomials $\sum_{i,j=0}^{n-1} a_{i,j} X^{p^i+p^j}$ are called *Dembowski–Ostrom polynomials* in [41].

Quadratic planar mappings define commutative semifields and vice versa. A finite *presemifield* is a finite set S with two binary operations $+$ and $*$ satisfying:

- $(S,+)$ is an Abelian group with identity 0.
- $a*(b+c) = a*b + a*c$ and $(a+b)*c = a*c + b*c$ for all $a,b,c \in S$.
- If $a*b = 0$, then a or b is 0.

If, in addition to this, there exists an element $1 \neq 0$ such that $1*a = a = a*1$ for all $a \in S$, then the presemifield is called a *semifield*. Presemifields are commutative if $a*b = b*a$ for all $a,b \in S$. The additive group of a finite presemifield is necessarily elementary Abelian. Hence any finite presemifield can be represented by $(\mathbf{F},+,*)$, where $\mathbf{F}$ is the underlying set of a finite field $\mathbb{F}_q$ and the addition of the semifield coincides with the one of the finite field. In the rest of this section we always use $\mathbf{F}$ to represent the underlying set of that $\mathbb{F}_q$. Two finite presemifields $(\mathbf{F},+,*)$ and $(\mathbf{F},+,\star)$ are called *isotopic* if there exist linearized permutation polynomials L, M, N over $\mathbb{F}_q$ such that

$$M(x) \star N(y) = L(x*y) \quad \text{for any } x,y \in \mathbf{F}\,.$$

Any presemifield $\mathbf{S} = (\mathbf{F},+,*)$ is isotopic to a semifield.

Given a planar Dembowski–Ostrom polynomial P over $\mathbb{F}_q$, the multiplication $*$ defined by

$$x*y = \tfrac{1}{2}(P(x+y) - P(x) - P(y)) \tag{5.1}$$

yields a commutative presemifield, which we denote by $\mathbf{S}_P = (\mathbf{F}, +, *)$. Conversely, any commutative presemifield $\mathbf{S} = (\mathbf{F}, +, *)$ of an odd order induces a planar mapping $P_\mathbf{S} : \mathbb{F}_q \to \mathbb{F}_q$ by $P_\mathbf{S} : x \mapsto x * x$. Moreover, the mapping $P_\mathbf{S}$ has a polynomial representation given by a sum of a planar Dembowski–Ostrom polynomial and a linearized polynomial [39].

EA-equivalent planar Dembowski–Ostrom polynomials define isotopic commutative presemifields. However isotopic commutative semifields may yield two EA-inequivalent Dembowski–Ostrom planar polynomials. Such an example is given in [106].

In [104] the APN binomials from [8, 19] are used to define planar mappings over fields of odd characteristic.

Theorem 5.1 ([104]). *Let p be prime, $n = 3k$, $\gcd(3, k) = 1$ and u be a primitive element of $\mathbb{F}_{p^n}$. Choose a positive integer s such that $k - s \equiv 0 \pmod 3$ and let $t := \gcd(s, n)$. Then the mapping*

$$P(x) = x^{p^s+1} - u^{p^k-1} x^{p^k+p^{2k+s}}$$

- *is planar if $p \geq 3$ and n/t is odd,*
- *is APN if $p = 2$ and $t = 1$.*

The planar mappings of Theorem 5.1 describe commutative semifields of order p^{3k}. These semifields along with the finite fields and so-called Albert's twisted fields ([2]) are the only currently known commutative semifields of order p^n with $p \geq 5$ and odd n.

In [22, 105] it is shown that the pattern of APN mappings given in [14, 18] may be used to define planar mappings as well:

Theorem 5.2. *Let p be an odd prime number and $q = p^m, m \geq 1$. Then the mapping $M \colon \mathbb{F}_{q^2} \to \mathbb{F}_{q^2}$ given by*

$$M(x) = x^{q+1} + \omega\, \mathrm{tr}_{q^2/q}\left(\alpha x^{p^i+p^j}\right), \quad i \geq j \geq 0,$$

is planar if and only if all the following conditions are fulfilled:
- *$i = j$ or $v(i - j) \neq v(m)$,*
- *$\omega \in \mathbb{F}_{q^2} \setminus \mathbb{F}_q$,*
- *α is a non-square in $\mathbb{F}_{q^2}$,*

where $v_2(e)$ is the highest power of 2 which divides the integer e.

The above theorem is stated in the form given in [12]. Further constructions of planar Dembowski–Ostrom polynomials leading new commutative semifields can be found in [9, 10, 49]. Moreover, in [9], a new beautiful method for proving planarity and APN property is introduced. In [86], planar polynomials corresponding to the known commutative semifields are listed.

A planar mapping $P \colon \mathbb{F}_q \to \mathbb{F}_q$ is never bijective, since $P(x + a) - P(x) = 0$ has a solution for every fixed non-zero a. In [75, 98] it is shown that the image set of a planar mapping cannot be too small. For an odd q, a mapping $F\colon \mathbb{F}_q \to \mathbb{F}_q$ is called 2-to-1 if all but one elements in the image set of F have two preimages and the exceptional element has one preimage.

Theorem 5.3. *Let $P\colon \mathbb{F}_q \to \mathbb{F}_q$ be a planar mapping and I be its image set. Then $|I| \geq \frac{q+1}{2}$. Moreover, $|I| = \frac{q+1}{2}$ if and only if P is 2-to-1.*

Observe that Theorem 5.3 and the observation that a planar mapping is never bijective imply that a planar exponent s on $\mathbb{F}_q$ must satisfy $\gcd(s, q-1) = 2$.

Surprisingly, for Dembowski–Ostrom polynomials, also the inverse of Theorem 5.3 holds as independently shown in [37, 42, 99]:

Theorem 5.4 ([37]). *Let $P\colon \mathbb{F}_q \to \mathbb{F}_q$ be given by a Dembowski–Ostrom polynomial. The following statements are equivalent:*
(a) *The mapping P is planar.*
(b) *The mapping P is 2-to-1.*
(c) *There is a permutation polynomial G over $\mathbb{F}_q$ such that $P(x) = G(x^2)$ for all $x \in \mathbb{F}_q$.*

Remark 5.5. The authors of [37] assume that the Dembowski–Ostrom polynomials $O(x) = \sum_{i,j=0}^{n-1} a_{i,j}X^{p^i+p^j}$ are exactly those for which the difference mappings $O(x+y) - O(x) - O(y)$ are additive. This is false, since obviously the latter property is fulfilled for any sum $O + L$ of a Dembowski–Ostrom polynomial O with a linearized one L. The statement of Theorem 5.4 does not hold in general for such sums $O + L$ as shown in [75], answering negative Questions 2.5 and 2.6 from [37].

The presently known main (up to the addition of an affine mapping) examples of planar mappings are given by either monomial or Dembowski–Ostrom polynomials, which define 2-to-1 mappings. Questions 2.5 and 2.6 from [37] can be reformulated as follows:

Open Question 5.6. *Find upper bounds on the image size of a planar mapping. Is it true that for any planar mapping P there is a linear mapping L such that $P + L$ is 2-to-1?*

6 Switching Construction

Suppose a mapping $F\colon \mathbb{F}_{q^n} \to \mathbb{F}_{q^n}$ has an additively defined property, like being APN or planar. Then usually this property can be easily stated also in terms of the coordinate functions $(f_1, \dots, f_n)$ of F with respect to a basis of $\mathbb{F}_{q^n}$ over $\mathbb{F}_q$. For instance, $F\colon \mathbb{F}_{q^n} \to \mathbb{F}_{q^n}$ with $F = (f_1, \dots, f_n)$ is planar (resp. APN) if and only if for every non-zero $a \in \mathbb{F}_{q^n}$ and $(b_1, \dots, b_n) \in \mathbb{F}_q^n$ the system of equations

$$\begin{cases} f_1(x+a) - f_1(x) &= b_1 \\ &\vdots \\ f_n(x+a) - f_n(x) &= b_n \end{cases}$$

has exactly 1 (resp. at most 2) solutions in $\mathbb{F}_{q^n}$. Thus constructing a mapping with a desired property is equivalent to finding n coordinate functions fulfilling the corresponding conditions. The idea of the *switching (over $\mathbb{F}_q$)*, introduced by John Dillon and extensively studied in [55], is to obtain a mapping on $\mathbb{F}_{q^n}$ with the required property by replacing one or small number of coordinate functions of a given mapping, which more or less satisfies this property. The switching appeared to be very successful for generating APN mappings which are CCZ-inequivalent to the monomial ones [20, 47, 55] as well as permutations [33–35].

We say that the mapping $F\colon \mathbb{F}_{q^n} \to \mathbb{F}_{q^n}$ is a switching of $G\colon \mathbb{F}_{q^n} \to \mathbb{F}_{q^n}$ over $\mathbb{F}_q$ if there is a $\mathbb{F}_q$-basis $(\beta_1, \dots, \beta_n)$ of $\mathbb{F}_{q^n}$ such that

$$F(x) = f_1(x) \cdot \beta_1 + f_2(x) \cdot \beta_2 + \cdots + f_n(x) \cdot \beta_n$$

and

$$G(x) = g_1(x) \cdot \beta_1 + f_2(x) \cdot \beta_2 + \cdots + f_n(x) \cdot \beta_n \,.$$

The first and nicest example of APN mapping obtained by switching is:

Theorem 6.1 ([20]). *For any $n \geq 1$, the mapping $x^3 + \mathrm{tr}_{2^n/2}(x^9)$ is APN on $\mathbb{F}_{2^n}$.*

The mapping $F(x) = x^3 + \mathrm{tr}_{2^n/2}(x^9)$ is indeed a switching over $\mathbb{F}_2$ of the Gold monomial mapping $G(x) = x^3$: Let $\mathcal{B} = (1, \beta_1, \dots, \beta_{n-1})$ be a basis of $\mathbb{F}_{2^n}$ over $\mathbb{F}_2$ and $(g_0(x), g_1(x), \dots, g_{n-1}(x))$ be the coordinate functions of $G(x)$ with respect to $\mathcal{B}$. Then

$$G(x) = g_0(x) \cdot 1 + g_1(x) \cdot \beta_1 + \cdots + g_{n-1}(x) \cdot \beta_{n-1}$$

and consequently

$$F(x) = (g_0(x) + \mathrm{tr}_{2^n/2}(x^9)) \cdot 1 + g_1(x) \cdot \beta_1 + \cdots + g_{n-1}(x) \cdot \beta_{n-1} \,,$$

showing that $F(x)$ is obtained from $G(x)$ by switching over $\mathbb{F}_2$.

Switching was used to find the first, and up to EA-equivalence currently the only known example of a non-quadratic APN mapping, which is not CCZ-equivalent to a monomial APN mapping [55].

Switching for planar mappings must be yet better understood [90]. The following result produces a planar mapping on $\mathbb{F}_{q^3}$ by switching the mapping x^2 over $\mathbb{F}_q$:

Theorem 6.2 ([71]). *Let $\gamma \in \mathbb{F}_{q^3}$, $\gamma \neq 0$ and $\beta = \eta(\gamma)\gamma^{(q+1)/2}$, where*

$$\eta(\gamma) = \begin{cases} 1 & \text{if } \gamma \text{ is a square in } \mathbb{F}_{q^3} \\ -1 & \text{otherwise.} \end{cases}$$

Then the mapping $f: \mathbb{F}_{q^3} \to \mathbb{F}_{q^3}$ *given by* $f(x) = \mathrm{tr}_{q^3/q}(\beta x^{q+1}) + \gamma x^2$ *is planar.*

Switching results large explicit families of permutation polynomials, which additionally can be designed to satisfy several properties like being sparse, having certain (algebraic) degree, differential spectrum.

Let $G: \mathbb{F}_{q^n} \to \mathbb{F}_{q^n}$, $f: \mathbb{F}_{q^n} \to \mathbb{F}_q$ and $\gamma \in \mathbb{F}_q^*$ be such that $F(x) = G(x) + \gamma f(x)$ is a permutation on $\mathbb{F}_{q^n}$, i.e. a switching of G over $\mathbb{F}_q$ yields a permutation. Then it is easy to see that any element in the image set of G has at most q preimages. Switchings of permutations are best studied when the mapping G is a permutation itself or it is a $\mathbb{F}_q$-linear mapping with 1-dimensional kernel [34].

Theorem 6.3 ([73]). *Let* $n \geq 2$ *and* $L: \mathbb{F}_{q^n} \to \mathbb{F}_{q^n}$ *be an* $\mathbb{F}_q$*-linear permutation of* $\mathbb{F}_{q^n}$. *Let* $b \in \mathbb{F}_q$, $h: \mathbb{F}_q \to \mathbb{F}_q$ *and* $\gamma \in \mathbb{F}_{q^n}$ *be a b-linear translator of* $f: \mathbb{F}_{q^n} \to \mathbb{F}_q$. *Then the mapping*

$$F(x) = L(x) + L(\gamma)\, h(f(x))$$

permutes $\mathbb{F}_{q^n}$ *if and only if* $g(u) = u + bh(u)$ *permutes* $\mathbb{F}_q$.

Special cases of Theorem 6.3 were originally proved in [34] and [83] where h is the identity mapping of $\mathbb{F}_{q^n}$ and f is the trace mapping from $\mathbb{F}_{q^n}$ onto $\mathbb{F}_q$, respectively.

Besides giving a sufficient condition for obtaining permutations via switching, Theorem 6.3 describes also a method to lift a single permutation on $\mathbb{F}_q$ to a variety of permutations on its extension $\mathbb{F}_{q^n}$: Given a permutation g on $\mathbb{F}_q$ take $b \in \mathbb{F}_q^*$ and set $h(u) = b^{-1}(g(u) - u)$. Further choose $\gamma \in \mathbb{F}_{q^n}$ and $f: \mathbb{F}_{q^n} \to \mathbb{F}_q$ such that γ is a b-linear translator of f using Example 3.3. Then by Theorem 6.3 the mapping $L(x) + L(\gamma)\, h(f(x))$ permutes $\mathbb{F}_{q^n}$ for any linear permutation L. Moreover, the functions h, f and the mapping L can be chosen such that the resulting permutation F fulfills certain restrictions on the degree, algebraic degree, number of terms or differential spectrum [35].

The next theorem describes permutations obtained via switching of $\mathbb{F}_q$-linear mappings of $\mathbb{F}_{q^n}$ with 1-dimensional kernels:

Theorem 6.4 ([73]). *Let* $n \geq 2$ *and* $L: \mathbb{F}_{q^n} \to \mathbb{F}_{q^n}$ *be an* $\mathbb{F}_q$*-linear mapping of* $\mathbb{F}_{q^n}$ *with kernel* $\alpha\mathbb{F}_q$, $\alpha \neq 0$. *Suppose* α *is a b-linear translator of* $f: \mathbb{F}_{q^n} \to \mathbb{F}_q$ *and* $h: \mathbb{F}_q \to \mathbb{F}_q$ *is a permutation on* $\mathbb{F}_q$. *Then the mapping*

$$F(x) = L(x) + \gamma\, h(f(x))$$

permutes $\mathbb{F}_{q^n}$ *if and only if* $b \neq 0$ *and* γ *does not belong to the image set of* L.

The study of switchings for a permutation $G(x)$ can be reduced to the study of switchings of the identity mapping: Recall that any function $f: \mathbb{F}_{q^n} \to \mathbb{F}_q$ can be represented as $\mathrm{tr}_{q^n/q}(R(x))$ with an appropriate mapping $R: \mathbb{F}_{q^n} \to \mathbb{F}_{q^n}$. Let G^{-1} be the inverse mapping of G. Then

$$G(x) + \gamma f(x) = G(x) + \gamma\, \mathrm{tr}_{q^n/q}(R(x)) = \left(x + \gamma\, \mathrm{tr}_{q^n/q}(R \circ G^{-1}(x))\right) \circ G(x)\,.$$

Theorem 6.3 can be used to result permutations of shape $x + \gamma h \circ f(x)$. Note that in general the function $h \circ f$ need not have a linear structure. The situation is more special for switchings over $\mathbb{F}_2$. A switching $L(x) + \gamma g(x)$ of an $\mathbb{F}_2$-linear mapping $L(x)$ over $\mathbb{F}_2$ gives rise to a permutation, only if the involved function $g\colon \mathbb{F}_{2^n} \to \mathbb{F}_2$ has a linear structure. The next theorem summarizes results on switchings of permutations over $\mathbb{F}_2$ from [33] and [73]:

Theorem 6.5. *Let $n \ge 2, \gamma \in \mathbb{F}_{2^n}^*, L\colon \mathbb{F}_{2^n} \to \mathbb{F}_{2^n}$ be an $\mathbb{F}_2$-linear mapping of $\mathbb{F}_{2^n}$ and $g\colon \mathbb{F}_{2^n} \to \mathbb{F}_2$. Then the mapping*

$$F(x) := L(x) + \gamma g(x)$$

is a permutation on $\mathbb{F}_{2^n}$ if and only if
- *$L(x)$ is a permutation and δ is a 0-linear translator for g, where $L(\delta) = \gamma$.*
- *$L(x)$ has a one-dimensional kernel $\{0, \alpha\}$, γ does not belong to the image set of $L(x)$ and α is a 1-linear translator of g.*

Proof. If $F(x)$ is a permutation then $L(x)$ has at most one-dimensional kernel. Let $L(x)$ be a permutation. Then

$$F(x) = L(x + \delta \cdot g(x)) .$$

Hence $F(x)$ is a permutation if and only if $x + \delta \cdot g(x)$ is a permutation. Further note that if there are distinct $x, y \in \mathbb{F}_{2^n}$ such that

$$x + \delta \cdot g(x) = y + \delta \cdot g(y) ,$$

then $y = x + \delta$. It remains to note that

$$x + \delta \cdot g(x) \ne x + \delta + \delta \cdot g(x + \delta)$$

if and only if $g(x + \delta) = g(x)$ for all $x \in \mathbb{F}_{2^n}$, proving the first case.

Suppose L has the kernel $\{0, \alpha\}$. Then clearly $F(x)$ is a permutation only if γ does not belong to the image set of $L(x)$. Observe that if for distinct $x, y \in \mathbb{F}_{2^n}$ the equality

$$L(x) + \gamma g(x) = L(y) + \gamma g(y)$$

holds, then $\gamma(g(x) + g(y)) = L(x + y)$. This forces $g(x) + g(y) = 0$ and thus $y = x + \alpha$. To complete the proof it remains to note that

$$L(x + \alpha) + \gamma g(x + \alpha) = L(x) + \gamma g(x + \alpha) \ne L(x) + \gamma g(x)$$

if and only if α is a 1-linear translator of g. □

Theorem 6.5 combined with Theorem 3.5 allows the complete characterization of the following classes of sparse permutation polynomials:

Corollary 6.6 ([33]). *Let $n \geq 2$.*
(a) *Let $1 \leq d, t \leq 2^n - 2$. Then*

$$X^d + \mathrm{tr}_{2^n/2}(X^t) \in \mathbb{F}_2[X]$$

is a permutation polynomial over $\mathbb{F}_{2^n}$ if and only if the following conditions are satisfied:

- *n is even*
- $\gcd(d, 2^n - 1) = 1$
- *$t = d \cdot s \pmod{2^n - 1}$ for some s such that $1 \leq s \leq 2^n - 2$ and has binary weight 1 or 2.*

(b) *Let $1 \leq k \leq n - 1$ and $1 \leq s \leq 2^n - 2$. Then*

$$X^{2^k} + X + \mathrm{tr}_{2^n/2}(X^s) \in \mathbb{F}_2[X]$$

is a permutation polynomial over $\mathbb{F}_{2^n}$ if and only if the following conditions are satisfied:

- *n is odd*
- $\gcd(k, n) = 1$
- *s has binary weight 1 or 2.*

Further explicit families of permutation polynomials which can be explained using the switching method can be found in [1, 103]. Permutations produced by switching of two or three coordinate functions are presented in [73, 77].

Open Problem 6.7. *Construct (non-trivial) examples of permutation polynomials of shape $G(X) + \gamma\, \mathrm{tr}_{q^n/q}(R(x))$ with $G(X), R(X) \in \mathbb{F}_{q^n}[X]$ such that G is neither a permutation on $\mathbb{F}_{q^n}$ nor $\mathbb{F}_q$-linear. (Trivial examples are those with $G(X) = G'(X) + \gamma\, \mathrm{tr}_{q^n/q}(R'(X))$ where G' is either a permutation on $\mathbb{F}_{q^n}$ or $\mathbb{F}_q$-linear.)*

7 Products of Linearized Polynomials

Given a polynomial $F(X) \in \mathbb{F}_q[X]$ of degree $1 < d < q$, what can be said about the induced mapping $F\colon \mathbb{F}_q \to \mathbb{F}_q$. A natural question is for example what is the size of the image set of F. An easy observation is that every $\alpha \in \mathbb{F}_q$ has at most d preimages under F, since the equation $F(X) = \alpha$ has at most d solutions. In particular, the image set of F has at least $\lceil q/d \rceil$ elements. In the case when F is non-bijective, its image set contains at most $\lfloor q - (q-1)/d \rfloor$ elements by Wan's bound [94, 96]. Better bounds are given in [97], however for generic polynomials the parameters involved in these bounds are difficult to compute.

For some classes of polynomials the associated mappings are easier to study. For instance, a monomial mapping $x \mapsto x^k$ induces a homomorphism on the multiplicative group of $\mathbb{F}_q$, and consequently its image set is of size $1 + (q-1)/\gcd(q-1,k)$. Linearized polynomials define linear mappings, this allows to have more information on their image sets. What are the properties of the mappings defined by products of two linearized polynomials? This question was firstly studied in [13] and later in [6, 74, 80], where bijective, APN and planar products of two linearized polynomials are considered. In this section we use $L_i(X)$, $i \in \mathbb{N}$, to denote linearized polynomials.

Lemma 7.1. *Let $L_1(X), L_2(X)$ be linearized polynomials over $\mathbb{F}_q$.*

(a) *If $L_1(X) \cdot L_2(X)$ is a permutation polynomial, then q is even and both L_1 and L_2 are permutation polynomials as well.*
(b) *Let q be even. If $L_1(X) \cdot L_2(X)$ is APN (equivalently crooked), then the kernels of L_1 and L_2 are at most one dimensional and intersect trivially.*
(c) *Let q be odd. If $L_1(X) \cdot L_2(X)$ is planar, then both L_1 and L_2 are permutation polynomials.*
(d) *Let q be odd. Then the image set of $L_1(X) \cdot L_2(X)$ has at most $(q+1)/2$ elements. Moreover, the image set of $L_1(X) \cdot L_2(X)$ has $(q+1)/2$ elements if and only if $L_1(X) \cdot L_2(X)$ is planar.*

Proof. The statement in (a) follows from (d) and the observation that if one of L_1 or L_2 is not bijective, then the product $L_1 \cdot L_2$ has more than one zero. To prove (b) note that the difference mapping of $L_1 \cdot L_2$ is

$$D_{L_1 \cdot L_2, a}(x) = L_1(a)L_2(x) + L_2(a)L_1(x) + L_1(a)L_2(a)\,,$$

which is 2-to-1 for any non-zero a. Suppose the kernel of L_1 has dimension ≥ 2, and let $b \neq 0$ and $c \neq 0$ be two distinct elements with $L_1(b) = L_1(c) = 0$. Then the kernel of the difference mapping $D_{L_1 \cdot L_2, b}(x) = L_2(b)L_1(x)$ contains both b and c, and thus the product $L_1 \cdot L_2$ cannot be APN. It remains to note that if b belongs the intersection of kernels of L_1 and L_2, then the difference mapping $D_{L_1 \cdot L_2, b}(x)$ is constantly zero, implying (b). The statements of (c) and (d) follow from Theorem 5.4 and the fact $L_1(x) \cdot L_2(x) = L_1(-x) \cdot L_2(-x)$ for every $x \in \mathbb{F}_q$. □

When studying properties, like the size of the image set, planarity or differential properties, of $L_1(X) \cdot L_2(X)$ with $L_1(X)$ a permutation polynomial, it is enough to consider $X \cdot L(X)$. Indeed, composition the mapping $L_1 \cdot L_2$ with the inverse L_1^{-1} of L_1 is

$$L_1 \cdot L_2 \circ L_1^{-1}(x) = x \cdot L_2 \circ L_1^{-1}(x)\,,$$

where $L_2 \circ L_1^{-1}$ is linear as well.

Several families of permutation polynomials of shape $X \cdot L(X)$ are listed in [13]:

Theorem 7.2 ([13]). *The following linearized polynomials $L(X) \in \mathbb{F}_{2^n}[X]$ yield permutation polynomials of shape $X \cdot L(X)$ on $\mathbb{F}_{2^n}$:*

(a) $L(X) = X^{2^k}$ *with* $n/\gcd(n,k)$ *odd.*
(b) $L(X) = X^{2^k} + aX^{2^{n-k}}$ *with* $n/\gcd(n,k)$ *odd and* $a^{(2^n-1)/(2^{\gcd(n,k)}-1)} \neq 1$.
(c) $L(X) = X^{2^{2k}} + a^{2^k+1}X^{2^k} + aX$ *with* $n = 3k$ *and* $a^{(2^n-1)/(2^k-1)} \neq 1$.
(d) $L(X) = \mathrm{tr}_{2^n/2^l}(X) + aX$ *with* n/l *odd and* $a \in \mathbb{F}_{2^l} \setminus \mathbb{F}_2$.

In [80] a recursive method is presented, which generates linearized polynomials $L(X)$ such that $X \cdot L(X)$ are permutation polynomials:

Theorem 7.3 ([80]). *Let $l(X) \in \mathbb{F}_{2^t}[X]$ be such that $X \cdot l(X)$ is a permutation polynomial on $\mathbb{F}_{2^t}$. Then for any odd k and $a \in \mathbb{F}_{2^t}^*$ the linearized polynomial*

$$L(X) = l(\mathrm{tr}_{2^{tk}/2^t}(X)) + a\ \mathrm{tr}_{2^{tk}/2^t}(X) + aX$$

yields a permutation polynomial $X \cdot L(X)$ on $\mathbb{F}_{2^{tk}}$.

Also planar polynomials given by $X \cdot L(X)$ seem to be rare. Currently known examples are:

Theorem 7.4 ([41, 62, 69, 74]). *Let p be an odd prime number and $q = p^n$. The following linearized polynomials $L(X) \in \mathbb{F}_q[X]$ yield planar polynomials of shape $X \cdot L(X)$ on $\mathbb{F}_{2^n}$:*
(a) $L(X) = X^{p^k}$ *with* $n/\gcd(n,k)$ *odd.*
(b) $L(X) = X^{p^k} - aX^{p^{n-k}}$ *with* $n/\gcd(n,k)$ *odd and* $a^{(p^n-1)/(p^{\gcd(n,k)}-1)} \neq 1$.
(c) $L(X) = X + aX^{p^k}$ *with* $n = 2k$ *and* $(1 - a^{p^k+1})$ *a non-square in the subfield* $\mathbb{F}_{p^k}$.
(d) $L(X) = X^{p^{2k}} + X^{p^k}$ *with* $n = 3k$.
(e) $L(X) = X^{p^{2k}} + X^{p^k} - X$ *with* $n = 3k$.

Open Problem 7.5. *Are there further families of linearized polynomials $L(X)$ implying permutation (resp. planar) polynomials of shape $X \cdot L(X)$?*

In [6] all APN mappings of shape $X \cdot L(X)$ are classified:

Theorem 7.6 ([6]). *Let $L(X)$ be a linearized polynomial over $\mathbb{F}_{2^n}$. Then $X \cdot L(X)$ is APN on $\mathbb{F}_{2^n}$ if and only if $L(X) = c \cdot X^{2^i}$ with a non-zero $c \in \mathbb{F}_{2^n}$ and $\gcd(i,n) = 1$.*

Theorem 7.6 implies the complete characterization of APN polynomials $L_1(X) \cdot L_2(X)$ where at least one of factors $L_1(X)$ or $L_2(X)$ is a permutation polynomial:

Corollary 7.7. *Let $L_1, L_2 \in \mathbb{F}_{2^n}[X]$ be linearized polynomials and additionally L_2 be a permutation polynomial. Then the mapping $L_1 \cdot L_2 \colon \mathbb{F}_{2^n} \to \mathbb{F}_{2^n}$ is APN if and only if $L_1(X) = c \cdot L_2(X)^{2^i}$ with a non-zero $c \in \mathbb{F}_{2^n}$ and $\gcd(i,n) = 1$.*

There are examples of APN polynomials $L_1(X) \cdot L_2(X)$ such that the factors define non-bijective mappings. For instance, the product $(X^2 + aX)(X^2 + bX)$ with $a, b \in \mathbb{F}_{2^n}, a + b \neq 0$, is APN.

References

[1] A. Akbary, D. Ghioca, and Q. Wang, On constructing permutations of finite fields, *Finite Fields Appl.* **17** (2011), 51–67.

[2] A. A. Albert, On nonassociative division algebras, *Trans. Amer. Math. Soc.* **72** (1952), 296–309.

[3] E. F. Assmus, Jr. and J. D. Key, *Polynomial codes and finite geometries*, Handbook of coding theory, Vol. I, II, pp. 1269–1343, North-Holland, Amsterdam, 1998.

[4] Y. Aubry, G. McGuire, and F. Rodier, *A few more functions that are not APN infinitely often*, Finite fields: theory and applications, Contemp. Math. 518, pp. 23–31, Amer. Math. Soc., Providence, RI, 2010.

[5] T. D. Bending and D. Fon-Der-Flaass, Crooked functions, bent functions, and distance regular graphs, *Electron. J. Combin.* **5** (1998), Research Paper 34, 14 (electronic).

[6] T. P. Berger, A. Canteaut, P. Charpin, and Y. Laigle-Chapuy, On almost perfect nonlinear functions over F_2^n, *IEEE Trans. Inform. Theory* **52** (2006), 4160–4170.

[7] T. Beth and C. Ding, *On almost perfect nonlinear permutations*, Advances in cryptology – EUROCRYPT '93 (Lofthus, 1993), Lecture Notes in Comput. Sci. 765, pp. 65–76, Springer, Berlin, 1994.

[8] J. Bierbrauer, A family of crooked functions, *Des. Codes Cryptogr.* **50** (2009), 235–241.

[9] J. Bierbrauer, New semifields, PN and APN functions, *Des. Codes Cryptogr.* **54** (2010), 189–200.

[10] J. Bierbrauer, Commutative semifields from projection mappings, *Des. Codes Cryptogr.* **61** (2011), 187–196.

[11] J. Bierbrauer and G. M. Kyureghyan, Crooked binomials, *Des. Codes Cryptogr.* **46** (2008), 269–301.

[12] J. Bierbrauer and G. M. Kyureghyan, *On the projection construction of APN and planar mappings*, Koninklijke Vlaamse Academie van België voor Wetenschappen en Kunsten, Brussel, 2012, p. to appear.

[13] A. Blokhuis, R. S. Coulter, M. Henderson, and C. M. O'Keefe, *Permutations amongst the Dembowski-Ostrom polynomials*, Finite fields and applications (Augsburg, 1999), pp. 37–42, Springer, Berlin, 2001.

[14] C. Bracken, E. Byrne, N. Markin, and G. McGuire, New families of quadratic almost perfect nonlinear trinomials and multinomials, *Finite Fields Appl.* **14** (2008), 703–714.

[15] C. Bracken, E. Byrne, N. Markin, and G. McGuire, A few more quadratic APN functions, *Cryptogr. Commun.* **3** (2011), 43–53.

[16] C. Bracken, E. Byrne, G. McGuire, and G. Nebe, On the equivalence of quadratic APN functions, *Des. Codes Cryptogr.* **61** (2011), 261–272.

[17] K. A. Browning, J. F. Dillon, M. T. McQuistan, and A. J. Wolfe, *An APN permutation in dimension six*, Finite fields: theory and applications, Contemp. Math. 518, pp. 33–42, Amer. Math. Soc., Providence, RI, 2010.

[18] L. Budaghyan and C. Carlet, Classes of quadratic APN trinomials and hexanomials and related structures, *IEEE Trans. Inform. Theory* **54** (2008), 2354–2357.

[19] L. Budaghyan, C. Carlet, and G. Leander, Two classes of quadratic APN binomials inequivalent to power functions, *IEEE Trans. Inform. Theory* **54** (2008), 4218–4229.

[20] L. Budaghyan, C. Carlet, and G. Leander, Constructing new APN functions from known ones, *Finite Fields Appl.* **15** (2009), 150–159.

[21] L. Budaghyan, C. Carlet, and A. Pott, New classes of almost bent and almost perfect nonlinear polynomials, *IEEE Trans. Inform. Theory* **52** (2006), 1141–1152.

[22] L. Budaghyan and T. Helleseth, *New perfect nonlinear multinomials over $F_{p^{2k}}$ for any odd prime p*, Sequences and their applications – SETA 2008, Lecture Notes in Comput. Sci. 5203, pp. 403–414, Springer, Berlin, 2008.

[23] A. Canteaut, P. Charpin, and H. Dobbertin, Binary m-sequences with three-valued crosscorrelation: a proof of Welch's conjecture, *IEEE Trans. Inform. Theory* **46** (2000), 4–8.
[24] A. Canteaut, P. Charpin, and G. M. Kyureghyan, A new class of monomial bent functions, *Finite Fields Appl.* **14** (2008), 221–241.
[25] A. Canteaut and M. Naya-Plasencia, *Structural weaknesses of permutations with a low differential uniformity and generalized crooked functions*, Finite fields: theory and applications, Contemp. Math. 518, pp. 55–71, Amer. Math. Soc., Providence, RI, 2010.
[26] C. Carlet, Partially-bent functions, *Des. Codes Cryptogr.* **3** (1993), 135–145.
[27] C. Carlet, *Boolean functions for cryptography and error correcting codes*, Boolean Models and Methods in Mathematics, Computer Science and Engineering, pp. 257–397, Cambridge University Press, London, 2010.
[28] C. Carlet, *Vectorial Boolean Functions for Cryptography*, Boolean Models and Methods in Mathematics, Computer Science and Engineering, pp. 398–469, Cambridge University Press, London, 2010.
[29] C. Carlet, P. Charpin, and V. Zinoviev, Codes, bent functions and permutations suitable for DES-like cryptosystems, *Des. Codes Cryptogr.* **15** (1998), 125–156.
[30] F. Chabaud and S. Vaudenay, *Links between differential and linear cryptanalysis*, Advances in cryptology – EUROCRYPT '94 (Perugia), Lecture Notes in Comput. Sci. 950, pp. 356–365, Springer, Berlin, 1995.
[31] P. Charpin, Private communication, 2012.
[32] P. Charpin and G. M. Kyureghyan, Cubic monomial bent functions: a subclass of $\mathcal{M}$, *SIAM J. Discrete Math.* **22** (2008), 650–665.
[33] P. Charpin and G. M. Kyureghyan, *On a class of permutation polynomials over* $\mathbb{F}_{2^n}$, Sequences and their applications – SETA 2008, Lecture Notes in Comput. Sci. 5203, pp. 368–376, Springer, Berlin, 2008.
[34] P. Charpin and G. M. Kyureghyan, When does $G(x) + \gamma\mathrm{Tr}(H(x))$ permute $\mathbb{F}_{p^n}$?, *Finite Fields Appl.* **15** (2009), 615–632.
[35] P. Charpin and G. M. Kyureghyan, *Monomial functions with linear structure and permutation polynomials*, Finite fields: theory and applications, Contemp. Math. 518, pp. 99–111, Amer. Math. Soc., Providence, RI, 2010.
[36] P. Charpin and S. Sarkar, Polynomials with linear structure and Maiorana–McFarland construction, *IEEE Trans. Inform. Theory* **57** (2011), 3796–3804.
[37] Y. Q. Chen and J. Polhill, Paley type group schemes and planar Dembowski–Ostrom polynomials, *Discrete Math.* **311** (2011), 1349–1364.
[38] R. S. Coulter, The classification of planar monomials over fields of prime square order, *Proc. Amer. Math. Soc.* **134** (2006), 3373–3378 (electronic).
[39] R. S. Coulter and M. Henderson, Commutative presemifields and semifields, *Adv. Math.* **217** (2008), 282–304.
[40] R. S. Coulter and F. Lazebnik, On the classification of planar monomials over fields of square order, *Finite Fields Appl.* **18** (2012), 316–336.
[41] R. S. Coulter and R. W. Matthews, Planar functions and planes of Lenz–Barlotti class II, *Des. Codes Cryptogr.* **10** (1997), 167–184.
[42] R. S. Coulter and R. W. Matthews, On the number of distinct values of a class of functions over a finite field, *Finite Fields Appl.* **17** (2011), 220–224.
[43] M. Delgado and H. Janwa, On the Conjecture on APN Functions, *arXiv:1207.5528*, 2012.
[44] P. Dembowski and T. G. Ostrom, Planes of order n with collineation groups of order n^2, *Math. Z.* **103** (1968), 239–258.
[45] U. Dempwolff and Y. Edel, Dimensional Dual Hyperovals and APN Functions with Translation Groups, *submitted*.

[46] J. F. Dillon, *Elementary Hadamard difference sets*, PhD thesis, University of Maryland, 1974.
[47] J. F. Dillon, *APN Polynomials and Related Codes, Polynomials over Finite Fields and Applications*, Banff International Research Station, Nov. 2006.
[48] J. F. Dillon and H. Dobbertin, New cyclic difference sets with Singer parameters, *Finite Fields Appl.* **10** (2004), 342–389.
[49] C. Ding and J. Yuan, A family of skew Hadamard difference sets, *J. Combin. Theory Ser. A* **113** (2006), 1526–1535.
[50] H. Dobbertin, Almost perfect nonlinear power functions on $GF(2^n)$: the Niho case, *Inform. and Comput.* **151** (1999), 57–72.
[51] H. Dobbertin, Almost perfect nonlinear power functions on $GF(2^n)$: the Welch case, *IEEE Trans. Inform. Theory* **45** (1999), 1271–1275.
[52] H. Dobbertin, *Almost perfect nonlinear power functions on $GF(2^n)$: a new case for n divisible by 5*, Finite fields and applications (Augsburg, 1999), pp. 113–121, Springer, Berlin, 2001.
[53] Y. Edel, *Enhancing Cryptographic Primitives with Techniques from Error Correcting Codes*, Veliko Tarnovo, Bulgaria, October 2008.
[54] Y. Edel, G. Kyureghyan, and A. Pott, A new APN function which is not equivalent to a power mapping, *IEEE Trans. Inform. Theory* **52** (2006), 744–747.
[55] Y. Edel and A. Pott, A new almost perfect nonlinear function which is not quadratic, *Adv. Math. Commun.* **3** (2009), 59–81.
[56] J.-H. Evertse, *Linear structures in block ciphers*, Advances in Cryptology – EUROCRYPT '87, Lecture Notes in Comput. Sci. 304, pp. 249–266, Springer, Berlin, 1988.
[57] D. Gluck, A note on permutation polynomials and finite geometries, *Discrete Math.* **80** (1990), 97–100.
[58] C. Godsil and A. Roy, Two characterizations of crooked functions, *IEEE Trans. Inform. Theory* **54** (2008), 864–866.
[59] R. Gold, Maximal recursive sequences with 3-valued recursive cross-correlation functions, *IEEE Trans. Inform. Theory* **14** (1968), 154–156.
[60] F. Göloglu, *Almost bent and almost perfect nonlinear functions Exponential sums, geometries and sequences*, PhD thesis, Otto-von-Guericke University of Magdeburg, 2009.
[61] F. Göloglu and A. Pott, *Almost perfect nonlinear functions: a possible geometric approach*, S. Nikova, B. Preneel, L. Strorme, J. Thas (eds.) Coding Theory and Cryptography II, Koninklijke Vlaamse Academie van België voor Wetenschappen en Kunsten, pp. 75–100, Brussel, 2007.
[62] T. Helleseth, G. Kyureghyan, G. J. Ness, and A. Pott, *On a family of perfect nonlinear binomials*, Boolean functions in cryptology and information security, NATO Sci. Peace Secur. Ser. D Inf. Commun. Secur. 18, pp. 126–138, IOS, Amsterdam, 2008.
[63] T. Helleseth and D. Sandberg, Some power mappings with low differential uniformity, *Appl. Algebra Engrg. Comm. Comput.* **8** (1997), 363–370.
[64] F. Hernando and G. McGuire, Proof of a conjecture on the sequence of exceptional numbers, classifying cyclic codes and APN functions, *J. Algebra* **343** (2011), 78–92.
[65] Y. Hiramine, A conjecture on affine planes of prime order, *J. Combin. Theory Ser. A* **52** (1989), 44–50.
[66] H. D. L. Hollmann and Q. Xiang, A proof of the Welch and Niho conjectures on cross-correlations of binary m-sequences, *Finite Fields Appl.* **7** (2001), 253–286.
[67] X.-D. Hou, Affinity of permutations of $\mathbb{F}_2^n$, *Discrete Appl. Math.* **154** (2006), 313–325.
[68] X.-D. Hou, G. L. Mullen, J. A. Sellers, and J. L. Yucas, Reversed Dickson polynomials over finite fields, *Finite Fields Appl.* **15** (2009), 748–773.
[69] X.-D. Hou and C. Sze, On certain diagonal equations over finite fields, *Finite Fields Appl.* **15** (2009), 633–643.

[70] T. Kasami, The weight enumerators for several classes of subcodes of the 2nd order binary Reed–Muller codes, *Information and Control* **18** (1971), 369–394.
[71] G. Kyureghyan and Y. Tan, *On a family of planar mappings*, Enhancing cryptographic primitives with techniques from error correcting codes, NATO Sci. Peace Secur. Ser. D Inf. Commun. Secur. 23, pp. 175–178, IOS, Amsterdam, 2009.
[72] G. M. Kyureghyan, Crooked maps in $\mathbb{F}_{2^n}$, *Finite Fields Appl.* **13** (2007), 713–726.
[73] G. M. Kyureghyan, Constructing permutations of finite fields via linear translators, *J. Combin. Theory Ser. A* **118** (2011), 1052–1061.
[74] G. M. Kyureghyan and F. Özbudak, Planarity of products of two linearized polynomials, *Finite Fields Appl.* in press (2012).
[75] G. M. Kyureghyan and A. Pott, *Some theorems on planar mappings*, Arithmetic of finite fields, Lecture Notes in Comput. Sci. 5130, pp. 117–122, Springer, Berlin, 2008.
[76] G. M. Kyureghyan and V. Suder, *On inverses of APN exponents*, Proceedings of ISIT 2012, Cambridge, MA, USA, July 1–6.
[77] M. Kyureghyan and S. Abrahamyan, A method of constructing permutation polynomials over finite fields, *Int. J. Information Theories and Applications* **17** (2010).
[78] G. Lachaud and J. Wolfmann, The weights of the orthogonals of the extended quadratic binary Goppa codes, *IEEE Trans. Inform. Theory* **36** (1990), 686–692.
[79] X. Lai, *Additive and linear structures of cryptographic functions*, FSE 94, Lecture Notes in Comput. Sci. 1008, pp. 75–85, Springer, Berlin, 1995.
[80] Y. Laigle-Chapuy, *A note on a class of quadratic permutations over* $\mathbb{F}_{2^n}$, Applied algebra, algebraic algorithms and error-correcting codes, Lecture Notes in Comput. Sci. 4851, pp. 130–137, Springer, Berlin, 2007.
[81] P. Langevin and G. Leander, Monomial bent functions and Stickelberger's theorem, *Finite Fields Appl.* **14** (2008), 727–742.
[82] G. Leander, Monomial bent functions, *IEEE Trans. Inform. Theory* **52** (2006), 738–743.
[83] J. E. Marcos, Specific permutation polynomials over finite fields, *Finite Fields Appl.* **17** (2011), 105–112.
[84] W. Meier and O. Staffelbach, *Nonlinearity criteria for cryptographic functions*, Advances in cryptology – EUROCRYPT '89 (Houthalen, 1989), Lecture Notes in Comput. Sci. 434, pp. 549–562, Springer, Berlin, 1990.
[85] S. Mesnager, Bent and hyper-bent functions in polynomial form and their link with some exponential sums and Dickson polynomials, *IEEE Trans. Inform. Theory* **57** (2011), 5996–6009.
[86] N. Nakagawa, *On functions of finite fields*, `http://www.math.is.tohoku.ac.jp/ taya/sendaiNC/2006/report/nakagawa.pdf`, 2006.
[87] K. Nyberg, *Perfect nonlinear S-boxes*, Advances in cryptology – EUROCRYPT '91 (Brighton, 1991), Lecture Notes in Comput. Sci. 547, pp. 378–386, Springer, Berlin, 1991.
[88] K. Nyberg, *Differentially uniform mappings for cryptography*, Advances in cryptology – EUROCRYPT '93 (Lofthus, 1993), Lecture Notes in Comput. Sci. 765, pp. 55–64, Springer, Berlin, 1994.
[89] M. Portmann and M. Rennhard, *Almost Perfect Nonlinear Permutations*, Semester Project, Swiss Federal Institute of Technology Zurich, Zurich, 1997.
[90] A. Pott and Y. Zhou, *Switching construction of planar functions on finite fields*, Arithmetic of finite fields, Lecture Notes in Comput. Sci. 6087, pp. 135–150, Springer, Berlin, 2010.
[91] F. Rodier, Functions of degree $4e$ that are not APN infinitely often, *Cryptogr. Commun.* **3** (2011), 227–240.
[92] L. Rónyai and T. Szőnyi, Planar functions over finite fields, *Combinatorica* **9** (1989), 315–320.
[93] O. S. Rothaus, On "bent" functions, *J. Combinatorial Theory Ser. A* **20** (1976), 300–305.

[94] G. Turnwald, A new criterion for permutation polynomials, *Finite Fields Appl.* **1** (1995), 64–82.

[95] E. R. van Dam and D. Fon-Der-Flaass, Codes, graphs, and schemes from nonlinear functions, *European J. Combin.* **24** (2003), 85–98.

[96] D. Q. Wan, *A p-adic lifting lemma and its applications to permutation polynomials*, Finite fields, coding theory, and advances in communications and computing (Las Vegas, NV, 1991), Lecture Notes in Pure and Appl. Math. 141, pp. 209–216, Dekker, New York, 1993.

[97] D. Q. Wan, P. Jau-Shyong Shiue, and C. S. Chen, Value sets of polynomials over finite fields, *Proc. Amer. Math. Soc.* **119** (1993), 711–717.

[98] G. Weng, W. Qiu, Z. Wang, and Q. Xiang, Pseudo-Paley graphs and skew Hadamard difference sets from presemifields, *Des. Codes Cryptogr.* **44** (2007), 49–62.

[99] G. Weng and X. Zeng, Further results on planar DO functions and commutative semifields, *Des. Codes Cryptogr.* **63** (2012), 413–423.

[100] S. Yoshiara, Dimensional dual hyperovals associated with quadratic APN functions, *Innov. Incidence Geom.* **8** (2008), 147–169.

[101] S. Yoshiara, Notes on APN functions, semibiplanes and dimensional dual hyperovals, *Des. Codes Cryptogr.* **56** (2010), 197–218.

[102] S. Yoshiara, Equivalences of quadratic APN functions, *Journal of Algebraic Combinatorics* **35** (2012), 461–475.

[103] P. Yuan and C. Ding, Permutation polynomials over finite fields from a powerful lemma, *Finite Fields Appl.* **17** (2011), 560–574.

[104] Z. Zha, G. M. Kyureghyan, and X. Wang, Perfect nonlinear binomials and their semifields, *Finite Fields Appl.* **15** (2009), 125–133.

[105] Z. Zha and X. Wang, New families of perfect nonlinear polynomial functions, *J. Algebra* **322** (2009), 3912–3918.

[106] Y. Zhou, A note on the isotopism of commutative semifields, http://arxiv.org/abs/1006.1529.

Fernando Hernando and Gary McGuire

On The Classification of Perfect Nonlinear (PN) and Almost Perfect Nonlinear (APN) Monomial Functions

Abstract: We will present some results towards a classification of exceptional PN and APN functions on finite fields. Regarding APN functions, we outline the proof by Jedlicka and Hernando–McGuire that completes the classification. Our results on PN functions are partial, and represent work in progress. The same techniques are used in both cases,since we use the Weil bound and Bezout's theorem.

Keywords: Absolutely Irreducible Polynomial, Coding Theory, Planar Function

2010 Mathematics Subject Classifications: 11T06

Fernando Hernando: Department of Mathematics, Universidad Jaume I, Spain,
e-mail: carrillf@uji.es
Gary McGuire: School of Mathematical Sciences, University College Dublin, Ireland,
e-mail: gary.mcguire@ucd.ie

1 Introduction

We present an outline of the classification of almost perfect nonlinear (APN) functions, and some new results on the classification of perfect nonlinear (PN) functions. These have connections to finite geometry, coding theory and cryptography.

This paper is arranged as follows. In Section 2 we give the definitions and some background on the two problems under consideration here. These two problems are different, and yet similar in flavour and can, therefore, be attacked with the same techniques. In Section 3 we outline the techniques and proof of the classification of exceptional APN monomial functions. Then in Sections 4, 5 and 6 we apply the same techniques to the classification of exceptional PN functions. We give some partial results; the complete conjecture is still open.

Research of the first author supported by MEC MTM2007-64704 (Spain). Research of the second author supported by the Claude Shannon Institute, Science Foundation Ireland Grant 06/MI/006.

2 Background and Motivation

In this section we present background to the problems under consideration in this paper.

2.1 PN and Planar Functions

Let p be a prime number and let $q = p^n$. Recall that any function $\mathbb{F}_q \longrightarrow \mathbb{F}_q$ can be expressed uniquely as a polynomial function (with coefficients in $\mathbb{F}_q$) of degree less than q. A polynomial function is called a permutation polynomial (PP) if it is a bijective function $\mathbb{F}_q \longrightarrow \mathbb{F}_q$.

Definition 2.1. A function $f : \mathbb{F}_q \longrightarrow \mathbb{F}_q$ is said to be *planar* if the functions $f(x + a) - f(x)$ are PPs for all nonzero $a \in \mathbb{F}_q$.

Planar functions are used to construct finite projective planes, and have been studied by finite geometers since at least 1968 (Dembowski and Ostrom [4]). Note that planar functions cannot exist in characteristic 2, because, if $D_a(x) := f(x + a) - f(x)$ and $D_a(x) = b$, then $D_a(x + a) = b$ also.

Definition 2.2. A function $f : \mathbb{F}_q \longrightarrow \mathbb{F}_q$ is said to be PN (Perfect Nonlinear) if for every $a, b \in \mathbb{F}_q$ with $a \neq 0$ we have

$$\sharp\{x \in \mathbb{F}_q \mid f(x + a) - f(x) = b\} \leq 1 .$$

PN functions were first defined in 1992 by Nyberg and Knudsen [16], in a cryptography paper. Note that PN functions cannot exist in characteristic 2, because if x is a solution to $f(x + a) - f(x) = b$ then $x + a$ is another solution.

It is clear that PN functions and planar functions are the same thing! They have different origins; PN functions come from cryptography whereas planar functions come from finite geometry.

We consider monomial functions in this article. The known planar monomials $f(x) = x^t$ are in Table 2.1.

It is conjectured that this list is complete:

Conjecture 2.3. *All planar functions of the form x^t are listed in Table* 2.1.

Table 2.1: Known PN exponents t

Characteristic	Exponents t	Conditions	Proved by
odd	2	None	Classical
odd	$p^i + 1$	$n/(i, n)$ odd	Dembowski–Ostrom
3	$(3^i + 1)/2$	$(i, n) = 1$, i odd	Coulter–Matthews

In this article we present some partial results towards this conjecture. We consider the classification of functions x^t that are planar/PN on $\mathbb{F}_{p^n}$ for infinitely many n. The known examples in Table 2.1 all have this property. Therefore, a weaker conjecture than Conjecture 2.3 is the following:

Conjecture 2.4. *If x^t is a planar function on $\mathbb{F}_{p^n}$ for infinitely many n, then t is of the values listed in the table.*

For monomial functions $f(x) = x^t$, it was shown in [2] that x^t is planar over $\mathbb{F}_q$ if and only if $(x+1)^t - x^t$ is a PP over $\mathbb{F}_q$, i.e., for monomial functions we only need consider the $a = 1$ case of Definition 1.

Definition 2.5. A PP $f(x) \in \mathbb{F}_q[x]$ is called *exceptional* if f is a PP on infinitely many extension fields of $\mathbb{F}_q$.

Therefore, to prove Conjecture 2.4, we consider the function $f(x) = x^t$ on the base field $\mathbb{F}_p$, and we would like to prove that $(x+1)^t - x^t$ is not an exceptional PP on $\mathbb{F}_p$ when t is not one of the values listed.

Observe that $(x+1)^t - x^t$ is not a PP over $\mathbb{F}_{p^n}$ if there exist $\mathbb{F}_{p^n}$-rational points (x, y) on the curve

$$A_t(x, y) = (x+1)^t - x^t - (y+1)^t + y^t$$

with $x \neq y$. It is obvious that $A_t(x, y)$ has $x - y$ as a factor. Therefore, the problem is to check when

$$B_t(x, y) = \frac{(x+1)^t - x^t - (y+1)^t + y^t}{x - y}$$

has rational points over $\mathbb{F}_{p^n}$. Note that $B_t(x, y)$ is defined over $\mathbb{F}_p$.

Conjecture 2.6. *Suppose $t > 2$. The polynomial $B_t(x, y)$ has an absolutely irreducible factor defined over $\mathbb{F}_p$ for all t not of the form $p^i + 1$ or $(3^i + 1)/2$ in characteristic 3.*

The following is easily proved using the Weil bound.

Theorem 2.7. *If $B_t(x, y)$ has an absolutely irreducible factor defined over $\mathbb{F}_p$, then $B_t(x, y)$ has rational points $(\alpha, \beta) \in (\mathbb{F}_{p^n})^2$ with distinct coordinates for all n sufficiently large.*

Corollary 2.8. *Conjecture 2.6 implies Conjecture 2.4.*

Therefore, the topic of this paper is proving Conjecture 2.6. We give some partial results, the full conjecture is still open. We note that a preprint has been posted on the arxiv by Elodie Leducq [13] with similar results to ours, and similar results have also been achieved by Robert Coulter [3].

2.2 APN Functions

The origin of the names PN and APN comes from cryptography. One of the desired properties for an S-box used in a block cipher is to have the best possible resistance against differential attacks, i.e., any given plaintext difference $a = y - x$ provides a ciphertext difference $f(y) - f(x) = b$ with small probability. Over a field of characteristic 2, PN functions do not exist, and this motivates the following definition.

Definition 2.9. A function $f : \mathbb{F}_{p^n} \longrightarrow \mathbb{F}_{p^n}$ is said to be APN (Almost Perfect Nonlinear) if for every $a, b \in \mathbb{F}_{p^n}$ with $a \neq 0$ we have

$$\sharp\{x \in \mathbb{F}_{p^n} \mid f(x+a) - f(x) = b\} \leq 2\,.$$

APN functions provide optimal resistance to differential cryptanalysis.

Monomial functions $f(x) = x^t$ from $\mathbb{F}_{2^n} \longrightarrow \mathbb{F}_{2^n}$ are often considered for use in applications. Table 2.2 contains all known APN monomial functions.

Conjecture 2.10. *All APN functions of the form x^t are listed in Table* 2.2.

Definition 2.11. The exponent t is called *exceptional* if $f(x) = x^t$ is APN on $\mathbb{F}_{2^n}$ for infinitely many n.

The conjecture stated by Dillon [5] is

Conjecture 2.12. *The only exceptional APN exponents are the Gold and Kasami–Welch numbers.*

Conjecture 2.12 says that for a fixed odd $t \geq 3$, $t \neq 2^i + 1$ or $t \neq 4^i - 2^i + 1$, the function $f(x) = x^t$ is APN on, at most, a finite number of fields $\mathbb{F}_{2^n}$. In [7] we proved Conjecture 2.12.

Table 2.2: Known APN exponents

	Exponents t	**Conditions**	**Proved by**
Gold	$2^i + 1$	$\gcd(i, n) = 1$	Gold 1968
Kasami–Welch	$2^{2i} - 2^i + 1$	$\gcd(i, n) = 1$	Kasami, Welch 1970
			Janwa–Wilson 1993
Welch	$2^t + 3$	$n = 2t + 1$	Dobbertin 1999
Niho	$2^t + 2^{\frac{t}{2}} - 1$, t even	$n = 2t + 1$	Dobbertin 1999
	$2^t + 2^{\frac{3t+1}{2}} - 1$, t odd		
Inverse	$2^{2t} - 1$	$n = 2t + 1$	Nyberg 1993
Dobbertin	$2^{4t} + 2^{3t} + 2^{2t} + 2^t - 1$	$n = 5t$	Dobbertin 1999

It is well known and easy to see that

$$h_t(x,y) = \frac{(x+1)^t + x^t + (y+1)^t + y^t}{(x+y)(x+y+1)}.$$

has no rational points over $\mathbb{F}_{2^n}$ besides those with $x = y$ and $x = y+1$ if and only if x^t is APN over $\mathbb{F}_{2^n}$.

Analogous to Theorem 2.7, Jedlicka [10] showed that as a consequence of the Weil bound we have the following result.

Theorem 2.13. *If $h_t(x,y)$ has an absolutely irreducible factor over $\mathbb{F}_2$ then $h_t(x,y)$ has rational points over $\mathbb{F}_{2^n}$ besides those with $x = y$ and $x = y+1$ for all n sufficiently large.*

The following conjecture is essentially stated in [10].

Conjecture 2.14. *The polynomial $h_t(x,y)$ has an absolutely irreducible factor defined over $\mathbb{F}_2$ for all t not of the form 2^i+1 or 4^i-2^i+1.*

By Theorem 2.13 and the discussion above, it is clear that

$$\text{Conjecture 2.14} \Rightarrow \text{Conjecture 2.12.}$$

In [7] we completed the proof of Conjecture 2.14. In this article we outline the proof.

It is easy to show (see [7]) that the absolute irreducibility of $h_t(x,y)$ is equivalent to the absolute irreducibility of

$$g_t(x,y,z) = \frac{f_t(x,y,z)}{(x+y)(x+z)(y+z)}. \tag{2.1}$$

where

$$f_t(x,y,z) = x^t + y^t + z^t + (x+y+z)^t. \tag{2.2}$$

Our arguments use the polynomials $f_t(x,y,z)$ and $g_t(x,y,z)$. It is shown in [8] that we may work with the affine versions $f_t(x,y,1)$ and $g_t(x,y,1)$.

There is a connection to cyclic codes, fully outlined in [1], see also [7] or [8]. Let C_n^t be the binary cyclic code of length 2^n-1 with two zeros ω, ω^t, where ω is a primitive element in $GF(2^n)$. Then the monomial x^t is an APN function over $\mathbb{F}_{2^n}$ if and only if the code C_n^t has minimum distance 5.

3 Outline of APN Functions Classification Proof

From the previous section, we wish to prove Conjecture 2.14. Writing $t = 2^i\ell + 1$, we divide the proof into two cases. The case $\gcd(\ell, 2^i-1) < \ell$ was proved by Jedlicka, and the case $\gcd(\ell, 2^i-1) = \ell$ was proved by Hernando–McGuire. Both proofs follow the line of argument introduced in Janwa–McGuire–Wilson [8] where the $i = 1$

case was proved. The idea is to show that Bezout's theorem cannot possibly hold, when applied to two (or more) putative factors of the polynomial g_t. This proof depends heavily on analyzing the singular points of the curve g_t. We outline the strategy of the proof here.

Bezout's theorem is a classical result in algebraic geometry and appears frequently in the literature (see for example Chapter 5 of [6]).

Theorem 3.1 (Bezout's Theorem). *Let r and s be two projective plane curves of degrees D_1 and D_2 over an algebraically closed field k having no components in common. Then,*

$$\sum_P I(P, r, s) = D_1 D_2 \,. \tag{3.1}$$

The sum runs over all points P in the projective plane $\mathbb{P}^2(k)$, and by $I(P, r, s)$ we mean the intersection multiplicity of the curves r and s at the point P. Notice that if r or s does not go through P, then $I(P, r, s) = 0$. Therefore, the sum in (3.1) runs over the singular points of the product rs. In our case, the sum will run over the singular points of g_t.

Using properties $I(P, r_1 r_2, s) = I(P, r_1, s) + I(P, r_2, s)$ and $\deg(r_1 r_2) = \deg(r_1) + \deg(r_2)$, one can generalize Bezout's Theorem to several curves f_1, f_2, ..., f_r as follows:

$$\sum_P \sum_{1\le i<j\le r} I(P, f_i, f_j) = \sum_{1\le i<j\le r} \deg(f_i)\deg(f_j) \,. \tag{3.2}$$

Consider a curve f and let $P = (\alpha, \beta)$, be a point in the plane. Write

$$f(x+\alpha, y+\beta) = F_0 + F_1 + F_2 + F_3 + \cdots$$

where F_m is homogeneous of degree m.

Definition 3.2. The multiplicity of f at P is the smallest m with $F_m \neq 0$, and is denoted by $m_P(f)$. In this case, F_m is called the tangent cone.

We refer the reader to Chapter 5 of [6] for the definition of the intersection multiplicity $I(P, r, s)$ of two curves r, s at a point P. The following property of the intersection multiplicity will be useful for us. It is part of the definition of intersection multiplicity in [6]. We state it as a Corollary.

Corollary 3.3.

$$I(P, r, s) \ge m_P(r) m_P(s) \,, \tag{3.3}$$

and the equality holds if and only if the tangent cones of r and s do not share any linear factor.

We note that the degree of g_t is $t - 3 = 2^i \ell - 2$. Therefore, if $g_t = uv$, then our strategy is to show that $\sum_P I(P, u, v) < (\deg u)(\deg v)$ by analyzing the singular points P. We usually lower bound the product of the degrees, and upper bound the sum of intersection multiplicities, and show that the upper bound is strictly less than the lower bound, to obtain our contradiction.

3.1 Singularities in APN case

We summarize the classification of singular points here. For the full details see [7]. Recall that F_m is the degree m part of the Taylor expansion.

A point $P = (\alpha, \beta)$ is a singular point of $f_t(x, y)$ if and only if $F_0 = F_1 = 0$, which happens if and only if α, β and $\lambda := \alpha + \beta + 1$ are ℓ-th roots of unity (see [9]). We distinguish three types of singular points.

(i) $\alpha = \beta = \lambda = 1$.

(ii) Either $\alpha = 1$ and $\beta \neq 1$, or $\beta = 1$ and $\alpha \neq 1$, or $\alpha = \beta \neq 1$ and $\lambda = 1$.
We divide these singular points into two cases:
(ii.a) Where (ii) holds and $\alpha, \beta \in \mathbb{F}_{2^i}$
(ii.b) Where (ii) holds and α, β not both in $\mathbb{F}_{2^i}$.

(iii) $\alpha \neq 1$, $\beta \neq 1$ and $\alpha \neq \beta$.
We divide these singular points into two cases:
(iii.a) Where (iii) holds and $\alpha, \beta \in \mathbb{F}_{2^i}$
(iii.b) Where (iii) holds and α, β not both in $\mathbb{F}_{2^i}$.

Now we summarize some properties already known, for more details see [8].

Lemma 3.4. *If $F_{2^i} \neq 0$, then $F_{2^i} = (Ax + By)^{2^i}$ where $A^{2^i} = \alpha^{1-2^i} + \lambda^{1-2^i}$ and $B^{2^i} = \beta^{1-2^i} + \lambda^{1-2^i}$.*

The proof is obvious, because we are in characteristic 2. The importance of this lemma is that there is only one distinct linear factor in F_{2^i}. Another useful fact is that the opposite is true for F_{2^i+1}, as shown in [8]:

Lemma 3.5. *F_{2^i+1} has $2^i + 1$ distinct linear factors.*

We now list the classification in a table. We let $w(x, y) = (x+1)(y+1)(x+y)$ so that $f_t = wg_t$ and $m_P(f_t) = m_P(g_t) + m_P(w)$. The values of $m_P(w)$ are easy to work out for the various singular points P.

$\gcd(\ell, 2^i - 1) = 1$:

Type	Number of Points	$m_P(f_t)$	$m_P(g_t)$
(i)	1	$2^i + 1$	$2^i - 2$
(ii)	$3(\ell - 1)$	2^i	$2^i - 1$
(iii)	$\leq (\ell - 1)(\ell - 3)$	2^i	2^i

In this case, the type (ii) points are all of type (ii.b), and the type (iii) points are all of type (iii.b).

$\gcd(\ell, 2^i - 1) = \ell$:

Type	Number of Points	$m_P(f_t)$	$m_P(g_t)$
(i)	1	$2^i + 1$	$2^i - 2$
(ii)	$3(\ell - 1)$	$2^i + 1$	2^i
(iii)	$\leq (\ell - 1)(\ell - 3)$	$2^i + 1$	$2^i + 1$

In this case, the type (ii) points are all of type (ii.a), and the type (iii) points are all of type (iii.a).

The case $1 < \gcd(\ell, 2^i - 1) < \ell$ is a mixture of the previous two cases because $f_t(x, y)$ has points with multiplicity 2^i and points with multiplicity $2^i + 1$. Nevertheless, the upper bounds on the *number* of points still hold.

Janwa–McGuire–Wilson [8] have computed the intersection multiplicity at points of type (ii.b) assuming the curve $g_t(x, y)$ factors:

Lemma 3.6. *If P is a point of type* (ii.b) *and $g_t(x, y) = r(x, y)s(x, y)$ then $I(P, r, s) = 0$.*

One of the ideas involved in the proof is that if $g_t(x, y)$ is irreducible over $\mathbb{F}_2$ and splits in several factors (over an extension field), then all factors have the same degree. The next lemma concerns this sort of phenomenon, and its proof can be found in [12] (although it is surely older).

Lemma 3.7. *Suppose that $p(\underline{x}) \in \mathbb{F}_q[x_1, \dots, x_n]$ is of degree t and is irreducible in $\mathbb{F}_q[x_1, \dots, x_n]$. Then there exists $r \mid t$ and an absolutely irreducible polynomial $h(\underline{x}) \in \mathbb{F}_{q^r}[x_1, \dots, x_n]$ of degree t/r such that*

$$p(\underline{x}) = c \prod_{\sigma \in G} \sigma(h(\underline{x})),$$

where $G = \mathrm{Gal}(\mathbb{F}_{q^r}/\mathbb{F}_q)$ and $c \in \mathbb{F}_q$. Furthermore, if $p(\underline{x})$ is homogeneous, then so is $h(\underline{x})$.

One more technical result is recorded now, whose proof is trivial.

Lemma 3.8. *If $i > 2$ and $\ell \mid 2^i - 1$ but $\ell \neq 2^i - 1$ then the following results hold:*

(1) $2^{i-1} + 1 - \ell > 2$.

(2) $\dfrac{\ell - 3}{2^{i+1}} < \dfrac{1}{4}$.

Proof. Since $\ell \mid 2^i - 1$ but $\ell \neq 2^i - 1$, and both numbers are odd, we certainly have that $\ell < 2^{i-1} - 1$. Then $2^{i-1} - 1 - \ell > 0$ so $2^{i-1} + 1 - \ell > 2$, thus (1) holds.

For (2) we have that $\ell < 2^{i-1} - 1 < 2^{i-1} + 3$ which implies $\frac{\ell-3}{2^{i-1}} < 1$ which certainly implies $\frac{\ell-3}{2^{i+1}} < \frac{1}{4}$. □

3.2 A Warm-Up Case

Here we give the proof of a special case, just to give the reader an idea of the overall proof. For the proof of the main theorem in the APN case, we refer the reader to the paper [7]. The principal starting observation when $\gcd(\ell, 2^i - 1) = \ell$ is to notice that all ℓ-th roots of unity lie in $\mathbb{F}_{2^i}$. Therefore, all type (ii) singularities have type (ii.a), and all type (iii) singularities have type (iii.a). Calculations show that $F_{2^i} = 0$ at all singular points (recall that F_{2^i} is the degree 2^i part of the Taylor expansion). Let $\mathrm{Sing}(g_t)$ denote the set of all singular points of $g_t(x, y)$.

Theorem 3.9. *Suppose that $g_t(x,y)$ is irreducible over $\mathbb{F}_2$ and $\ell \mid 2^i-1$ but $\ell \neq 2^i-1$. Then $g_t(x,y)$ cannot split in two factors h_1 and h_2 with $\deg(h_1) = \deg(h_2)$.*

Proof. We apply Bezout's Theorem, which states

$$\sum_{P \in \mathrm{Sing}(g_t)} I(P, h_1, h_2) = \deg(h_1)\deg(h_2)\,.$$

Since $F_{2^i} = 0$, and since the tangent cones have different lines by Lemma 3.4, Corollary 3.3 tells us that the left hand side is equal to $\sum_{P \in \mathrm{Sing}(g_t)} m_P(h_1) m_P(h_2)$. Using the table of singularities described in Section 3 for $\ell \mid 2^i - 1$ we get

$$\sum_{P \in \mathrm{Sing}(g_t)} m_P(h_1) m_P(h_2) \leq (2^{i-1}-1)^2 + 3(\ell-1)2^{2i-2} + (\ell-1)(\ell-3)2^{i-1}(2^{i-1}+1). \tag{3.4}$$

Since the degrees of both components are the same, the right hand side of Bezout's Theorem is exactly,

$$(2^{i-1}\ell - 1)^2 = 2^{2i-2}\ell^2 - 2\ell 2^{i-1} + 1\,. \tag{3.5}$$

Let us compare (3.5) and (3.4). If the quantity in (3.5) is greater than the quantity in (3.4), we are done, and this happens if and only if,

$$2^{2i-2}(-\ell+1) + 2^{i-1}(\ell^2 - 2\ell + 1) < 0 \tag{3.6}$$

which is equivalent to

$$2^{i-1}(\ell - 1) > (\ell^2 - 2\ell + 1) = (\ell - 1)^2\,. \tag{3.7}$$

So we conclude that the condition for (3.5) > (3.4) is

$$2^{i-1} > (\ell - 1) \tag{3.8}$$

which is true by Lemma 3.8(1). □

Remark 3.10. Notice that this proof fails when $\ell = 2^i - 1$, as it should.

4 PN Functions Classification Proof: Analysis of Singularities

Here we begin the proof of our partial result towards Conjecture 2.6. The proof follows the same lines as the proof in [7], outlined in the previous section. We first analyze the singular points, then prove the result assuming $B_t(x,y)$ is irreducible over $\mathbb{F}_p$, and finally prove the theorem without this assumption.

Let r be the residue of t modulo p, so we may write

$$t = p^i\ell + r \quad \text{with} \quad 0 \le r < p$$

and where ℓ is not divisible by p. We similarly also write

$$\ell = ps + j \quad \text{with} \quad 0 < j < p$$

where this time s could be divisible by p.

We will prove Conjecture 2.6 in the case that $r = 1$ and $j = 1$, i.e., $t \equiv 1 \bmod p$ and $\ell \equiv 1 \bmod p$.

Theorem 4.1. *$x + y + 1$ divides $A_t(x,y)$ if and only if t is odd.*

Proof. We substitute $y = -x - 1$ in $A_t(x,y) = (x+1)^t - x^t - (y+1)^t + y^t$ obtaining $(x+1)^t - x^t - (-1)^t(x)^t + (-1)^t(x+1)^t$ which is identically zero if and only if t is odd.

Therefore, $B_t(x,y) = \frac{A_t(x,y)}{x-y}$ has an absolutely irreducible factor over $\mathbb{F}_p$, so x^t is not a PN function over $\mathbb{F}_{p^n}$ for infinitely many n.

We will assume from now that t is even. This implies that ℓ is odd. Moreover $\ell \ge 3$, because for $\ell = 1$, t is known to be PN (Table 2.1).

Notice that

$$\begin{aligned}(x+a)^t &= (x^{p^i} + a^{p^i})^\ell (x+a)^r \\ &= \left(x^{p^i\ell} + \binom{\ell}{1} x^{p^i(\ell-1)} a^{p^i} + \cdots + \binom{\ell}{\ell-1} x^{p^i} a^{p^i(\ell-1)} + a^{p^i\ell}\right) \\ &\quad \times \left(x^r + \binom{r}{1} x^{r-1} a + \cdots + \binom{r}{r-1} x a^{r-1} + a\right).\end{aligned}$$

To study the singularities, we expand at the point $P = (\alpha, \beta)$, so then we need to study

$$A_t(x+\alpha, y+\beta) = (x+\alpha+1)^t - (x+\alpha)^t - (y+\beta+1)^t + (y+\beta)^t.$$

We write this as a sum of homogeneous parts

$$F_0 + F_1 + F_2 + \cdots$$

and we compute that

$$
\begin{aligned}
F_0 &= (\alpha+1)^{p^i\ell+r} - \alpha^{p^i\ell+r} - (\beta+1)^{p^i\ell+r} + \beta^{p^i\ell+r}\,,\\
F_1(x,y) &= \binom{r}{r-1}\Big[((\alpha+1)^{p^i\ell+r-1} - \alpha^{p^i\ell+r-1})x\\
&\qquad - ((\beta+1)^{p^i\ell+r-1} - \beta^{p^i\ell+r-1})y\Big]\,,\\
F_2(x,y) &= \binom{r}{r-2}\Big[((\alpha+1)^{p^i\ell+r-2} - \alpha^{p^i\ell+r-2})x^2\\
&\qquad - ((\beta+1)^{p^i\ell+r-2} - \beta^{p^i\ell+r-2})y^2\Big]\,,\\
&\vdots\\
F_u(x,y) &= \binom{r}{r-u}\Big[((\alpha+1)^{p^i\ell+r-u} - \alpha^{p^i\ell+r-u})x\\
&\qquad - ((\beta+1)^{p^i\ell+r-u} - \beta^{p^i\ell+r-u})y\Big]\,,\\
&\vdots\\
F_r(x,y) &= ((\alpha+1)^{p^i\ell} - \alpha^{p^i\ell})x - ((\beta+1)^{p^i\ell} - \beta^{p^i\ell})y\,,\\
F_{p^i}(x,y) &= jx^{p^i}((\alpha+1)^{p^i(\ell-1)+r} - \alpha^{p^i(\ell-1)+r})\\
&\qquad - jy^{p^i}((\beta+1)^{p^i(\ell-1)+r} - \beta^{p^i(\ell-1)+r})\,,\\
F_{p^i+1}(x,y) &= \binom{r}{r-1}\Big[jx^{p^i+1}((\alpha+1)^{p^i(\ell-1)+r-1} - \alpha^{p^i(\ell-1)+r-1})\\
&\qquad - jy^{p^i+1}((\beta+1)^{p^i(\ell-1)+r-1} - \beta^{p^i(\ell-1)+r-1})\Big]\,. \qquad \square
\end{aligned}
$$

Lemma 4.2. *If $F_1(x,y) = F_2(x,y) = 0$, then $r = 0$ or 1.*

Proof. $F_1(x,y) = 0$ implies that $(\alpha+1)^{p^i\ell+r-1} - \alpha^{p^i\ell+r-1} = 0 \Leftrightarrow (\frac{\alpha+1}{\alpha})^{p^i\ell+r-1} = 1$.
$F_2(x,y) = 0$ implies that $(\alpha+1)^{p^i\ell+r-2} - \alpha^{p^i\ell+r-2} = 0 \Leftrightarrow (\frac{\alpha+1}{\alpha})^{p^i\ell+r-2} = 1$.
Hence,

$$\left(\frac{\alpha+1}{\alpha}\right)^{p^i\ell+r-1} \Big/ \left(\frac{\alpha+1}{\alpha}\right)^{p^i\ell+r-2} = 1 \Leftrightarrow \left(\frac{\alpha+1}{\alpha}\right) = 1 \Leftrightarrow 1 = 0\,.$$

But this is impossible. So, the only possibilities are:

- $F_1(x,y) \neq 0$ and the coefficient $\binom{r}{r-2} = 0$ in $F_2(x,y)$, i.e. $r = 0$ or 1.
- Both coefficients are zero, $\binom{r}{r-2} = \binom{r}{r-1} = 0 \Leftrightarrow r = 0$.

This completes the proof. □

A point P is singular iff $F_0 = F_1 = 0$ at P. Thus, we need to expand the expression

$$\begin{aligned}
&((\alpha+1)^{p^i\ell+r-1} - \alpha^{p^i\ell+r-1}) \\
&= \alpha^{p^i\ell+r-1} + j\alpha^{p^i(\ell-1)+r-1} + \cdots + j\alpha^{p^i+r-1} + \alpha^{r-1} \\
&\quad + \binom{r-1}{1}\alpha^{p^i\ell+r-2} + j\binom{r-1}{1}\alpha^{p^i(\ell-1)+r-2} + \cdots \\
&\qquad\qquad + j\binom{r-1}{1}\alpha^{p^i+r-2} + \binom{r-1}{1}\alpha^{r-2} \\
&\quad + \cdots \\
&\quad + \binom{r-1}{r-2}\alpha^{p^i\ell+1} + j\binom{r-1}{r-2}\alpha^{p^i(\ell-1)+1} + \cdots \\
&\qquad\qquad + j\binom{r-1}{r-2}\alpha^{p^i+1} + \binom{r-1}{r-2}\alpha \\
&\quad + \alpha^{p^i\ell} + j\alpha^{p^i(\ell-1)} + \cdots + j\alpha^{p^i} + 1 - \alpha^{p^i(\ell-1)+r}.
\end{aligned}$$

This is a complicated expression; therefore, we distinguish different cases:

(a) $r \neq 0, 1$.

(b) $r = 1$, i.e., $t = p^i\ell + 1$, which we divide into two subcases:

  (b.1) $j = 1$, i.e., $\ell = ps + 1$.

  (b.2) $j \neq 1$.

(c) $r = 0$, $t = p^i\ell$. Studying this case is equivalent to studying just $t = \ell$, so we can include this case in the previous cases.

We will prove case (b.1) in this paper.

4.1 Singular Points in Case (b.1)

Lemma 4.3. *In case* (b.1), $P = (\alpha, \beta)$ *is a singular point of* A_t *if and only if all the following hold:*

$$\alpha^{(\ell-1)} + \alpha + 1 = 0 \tag{4.1}$$

$$\beta^{(\ell-1)} + \beta + 1 = 0 \tag{4.2}$$

$$\alpha^{\ell} = \beta^{\ell} \tag{4.3}$$

$$\text{either } \alpha + \beta + 1 = 0 \text{ or } \alpha = \beta. \tag{4.4}$$

Proof. P is a singular point iff $F_0 = F_1 = 0$ iff $(\alpha+1)^{p^i\ell} - \alpha^{p^i\ell} = 0$, $(\beta+1)^{p^i\ell} - \beta^{p^i\ell} = 0$ and $(\alpha+1)^{p^i\ell+1} - \alpha^{p^i\ell+1} - (\beta+1)^{p^i\ell+1} + \beta^{p^i\ell+1} = 0$.

From the first two equations we obtain $\alpha^{(\ell-1)} + \alpha + 1 = \beta^{(\ell-1)} + \beta + 1 = 0$ and using this in the third equation we get $\alpha^{\ell} = \beta^{\ell}$ as desired.

We consider $\alpha(4.1) = \beta(4.2)$, i.e. $\alpha^\ell + \alpha^2 + \alpha = \beta^\ell + \beta^2 + \beta$. Using (4.3) we get $\alpha^2 + \alpha = \beta^2 + \beta \Leftrightarrow (\alpha - \beta)(\alpha + \beta) = (\beta - \alpha) \Leftrightarrow$ either $\alpha + \beta + 1 = 0$ or $\alpha = \beta$. This completes the proof. □

It follows from (4.1) that $A_t(x, y)$ has at most $(\ell - 1)^2$ singular points. From (4.4) we can reduce this number to at most $2(\ell - 1)$ singular (affine) points.

We next compute the multiplicity of the affine singular points. The homogeneous component F_{p^i} is nonzero except when $(\alpha + 1)^{p^i(\ell-1)+1} - \alpha^{p^i(\ell-1)+1} = 0$ and $(\beta + 1)^{p^i(\ell-1)+1} - \beta^{p^i(\ell-1)+1} = 0$. Expanding this equation we get the following equations.

Lemma 4.4. *In case (b.1), let $P = (\alpha, \beta)$ be a singular point of A_t. Then $F_{p^i} = 0$ if and only if*

$$\alpha^{p^i(\ell-1)} + \alpha + 1 = 0 \tag{4.5}$$

$$\beta^{p^i(\ell-1)} + \beta + 1 = 0 \tag{4.6}$$

Finally, if we expand the coefficients of F_{p^i+1}, we get the following.

Lemma 4.5. *In case (b.1), let $P = (\alpha, \beta)$ be a singular point of A_t. Then $F_{p^i+1} = x^{p^i+1} - y^{p^i+1}$.*

So the multiplicity is at most $p^i + 1$, and it is actually p^i unless equations (4.5) and (4.6) hold. But if $P = (\alpha, \beta)$ is a singular point then (4.1), (4.2) and (4.3) hold. Thus, using (4.1) = (4.5) we obtain

$$\alpha^{p^i-1} = 1\,. \tag{4.7}$$

and using (4.2) = (4.6) we obtain

$$\beta^{p^i-1} = 1\,. \tag{4.8}$$

4.2 Singular Points at Infinity

Next we consider singular points at infinity. The projective curve we are working with is $A_t(x, y, z) = [(x + z)^t - x^t - (y + z)^t + y^t]/z$. After cancelling terms, we get, $A_t(x, y, z) = tx^{t-1} - ty^{t-1} + z(\text{lower order terms})$. The next result explains when there are singular points at infinity ($z = 0$).

Lemma 4.6. *There are no singular points at infinity in case* (a). *In case* (b.1) $(\alpha, 1, 0)$ *is a singular point of $A_t(x, y, z)$ if and only if $\alpha^{t-1} = 1$, which is equivalent to $\alpha^\ell = 1$.*

Proof. The dehomogenization of $A_t(x,y,z) = [(x+z)^t - x^t - (y+z)^t + y^t]/z$ relative to y is $A'_t(x,z) = A_t(x,1,z)/z$. We compute

$$\begin{aligned}
A'_t(x+\alpha,z) &= \left[(x+z+\alpha)^t - (x+\alpha)^t - (z+1)^t + 1\right]/z \\
&= \frac{1}{z}\Bigg[(x+z)^t + \binom{t}{t-1}(x+z)^{t-1}\alpha + \cdots + \binom{t}{1}(x+z)\alpha^{t-1} + \alpha^t \\
&\quad - x^t - \binom{t}{t-1}(x)^{t-1}\alpha - \cdots - \binom{t}{1}(x)\alpha^{t-1} - \alpha^t \\
&\quad - z^t - \binom{t}{t-1}(z)^{t-1} - \binom{t}{t-2}(z)^{t-2} - \cdots - \binom{t}{1}(z) - 1 + 1\Bigg] \\
&= \binom{t}{1}(\alpha^{t-1}-1) + \binom{t}{2}\left[(\alpha^{t-2}-1)z + 2\alpha^{t-2}x\right] \\
&\quad + \text{higher order terms.}
\end{aligned}$$

Notice that the linear part cannot be zero, unless the coefficient $\binom{t}{2} = 0$, which implies $r = 0$ or 1. Therefore, there are no singular points in case (a). And if $r = 1$, then P is a singular point iff $\alpha^{t-1} - 1 = 0$. □

Lemma 4.7. *In case* (b.1) *let* $(\alpha, 1, 0)$ *be a singular point of* $A'_t(x,z)$ *at infinity. Then:*

$$\begin{aligned}
F'_{p^i-1} &= z^{p^i-1}(\alpha^{p^i(\ell-1)+1} - 1)\,. \\
F'_{p^i} &= z^{p^i}(\alpha^{p^i(\ell-1)} - 1) + (xz^{p^i-1} + x^{p^i})\alpha^{p^i(\ell-1)}\,.
\end{aligned}$$

Proof.

$$\begin{aligned}
&f'_t(x+\alpha+1,z) = 1/z[(x+z+\alpha)^t - (x+\alpha)^t - (z+1)^t + 1] \\
&= 1/z[((x+z)^{p^i\ell+1} + j(x+z)^{p^i(\ell-1)+1}\alpha^{p^i} + \cdots + j(x+z)^{p^i+1}\alpha^{p^i(\ell-1)} \\
&\quad + (x+z)\alpha^{p^i\ell} + (x+z)^{p^i\ell}\alpha + j(x+z)^{p^i(\ell-1)}\alpha^{p^i+1} + \cdots \\
&\quad + j(x+z)^{p^i}\alpha^{p^i(\ell-1)+1} + \alpha^{p^i\ell+1} - (x)^{p^i\ell+1} - j(x)^{p^i(\ell-1)+1}\alpha^{p^i} - \cdots \\
&\quad - j(x)^{p^i+1}\alpha^{p^i(\ell-1)} - (x)\alpha^{p^i\ell} - (x)^{p^i\ell}\alpha - j(x)^{p^i(\ell-1)}\alpha^{p^i+1} - \cdots \\
&\quad - j(x)^{p^i}\alpha^{p^i(\ell-1)+1} - \alpha^{p^i\ell+1} - (z)^{p^i\ell+1} - j(z)^{p^i(\ell-1)+1} - \cdots \\
&\quad - j(z)^{p^i+1} - (z) - (z)^{p^i\ell} - j(z)^{p^i(\ell-1)} - \cdots - (z)^{p^i} - 1 + 1)]\,.
\end{aligned}$$

Therefore, we get:

$$\begin{aligned}
F'_{p^i-1} &= jz^{p^i-1}(\alpha^{p^i(\ell-1)+1} - 1)\,. \\
F'_{p^i} &= jz^{p^i}(\alpha^{p^i(\ell-1)} - 1) + j(xz^{p^i-1} + x^{p^i})\alpha^{p^i(\ell-1)}\,.
\end{aligned}$$
□

Lemma 4.8. *In case* (b.1) *let* $(\alpha, 1, 0)$ *be a singular point of* $A'_t(x,z)$ *at infinity. Then the multiplicity at P is* p^i *if* $\alpha \in GF(p^i)$*, and is* $p^i - 1$ *otherwise.*

Proof. Since $P = (\alpha, 1, 0)$ is a singular point at the infinity, we know that $\alpha^{\ell} = 1$. Moreover, $F_{p^i} = 0$ iff $\alpha^{p^i(\ell-1)+1} - 1 = 0$. Multiplying the last equation by α^{p^i}, we equivalently obtain $\alpha^{p^i(\ell)+1} - \alpha^{p^i} = 0 \Leftrightarrow \alpha^{p^i-1} = 1 \Leftrightarrow \alpha \in GF(p^i)$. This completes the proof. □

4.3 The Multiplicities

Next we pin down the multiplicities of these singular points $P = (\alpha, \beta)$ on $A_t(x, y)$, and how things change for $B_t(x, y)$. We classify the points into the following types:
(i) $\alpha^{p^i-1} = \beta^{p^i-1} = 1$ $(F_{p^i} = 0)$.
(ii) $\alpha^{p^i-1} \neq 1 \neq \beta^{p^i-1}$ $(F_{p^i} \neq 0)$.
(iii) $(\alpha, 1, 0)$ is a singular point at infinity.
  (iii.a) $\alpha \in GF(p^i)$ $(F'_{p^i-1} = 0)$.
  (iii.a) $\alpha \notin GF(p^i)$ $(F'_{p^i-1} \neq 0)$.
  We note that if $\ell = 1$ and $i > 1$, the only singular point is $(-2, -2)$.

Defining $w(x, y) := (x - y)$, we note the following multiplicities on w: $m_P(w) = 1$ if $P = (\alpha, \alpha)$ (type (ii)), and $m_P(w) = 0$ for all other singular points $P = (\alpha, \beta)$. We now have the multiplicities for $B_t(x, y)$.

Type	Number of Points	$m_P(A_t)$	$m_P(B_t)$
(i)	$N_1 \leq \min\{l-1, p^i-1\}$	$p^i + 1$	$\leq p^i + 1$
(ii)	$\leq 2(\ell - 1) - N_1$	p^i	$\leq p^i$
(iii.a)	$N_2 \leq \ell$	p^i	$\leq p^i$
(iii.b)	$\ell - N_2$	$p^i - 1$	$\leq p^i - 1$

4.4 Further Analysis

Next we need some further analysis of the singularities.

The following result is clear, because we are in characteristic p.

Lemma 4.9. *$F_{p^i} = (\sigma x - \tau y)^{p^i}$ where $\sigma^{p^i} = ((\alpha+1)^{p^i(\ell-1)+1} - \alpha^{p^i(\ell-1)+1})$ and $\tau^{p^i} = ((\beta+1)^{p^i(\ell-1)+1} - \beta^{p^i(\ell-1)+1})$. $F'_{p^i} = (Uz)^{p^i}$ where $U^{p^i} = j(\alpha^{p^i(\ell-1)+1} - 1)$.*

Lemma 4.10. *F_{p^i+1} and F'_{p^i+1} consist of $p^i + 1$ different linear factors.*

Proof. We have seen that $F_{p^i+1} = x^{p^i+1} - y^{p^i+1}$. Consider $h(x) = F_{p^i+1}(x, 1) = x^{p^i+1} - 1$. If $h(x)$ has a repeated root at a, then $h'(a) = 0$. The derivative $h'(x) = x^{p^i}$ is zero only for $x = 0$, which is not a root of h, and; therefore, there are no repeated factors in F_{p^i+1}. The same argument holds for F'_{p^i+1}. □

Next we make a crucial observation for our proofs. For the rest of this paper, we let $L = \sigma x - \tau y$, so that $F_{p^i} = L^{p^i}$. Suppose $B_t(x,y) = u(x,y)v(x,y)$, and suppose that the Taylor expansion at a singular point $P = (\alpha, \beta)$ is

$$u(x+\alpha, y+\beta) = L^{r_1} + u_1,\; v(x+\alpha, y+\beta) = L^{r_2} + v_1\,,$$

where wlog $r_1 \leq r_2$. Then $F_{p^i+1} = L^{r_1}(v_1 + L^{r_2-r_1}u_1)$. From Lemma 4.10, we deduce the following.

Lemma 4.11. *With the notation of the previous paragraph,*
(i) *Either* $r_1 = 1$ *or* $r_1 = 0$.
(ii) *If* $r_1 = 1$ *then* $\gcd(L, v_1 + L^{r_2-r_1}u_1) = 1$.

We next make two quick remarks to aid us in moving between $A_t(x,y)$ and $B_t(x,y)$. Suppose that $P = (\alpha,\beta) \neq (1,1)$ is a singular point of $B_t(x,y)$ such that $F_{p^i}(x,y) \neq 0$ at P (type (ii)). We will need to know the greatest common divisor $(G_m(x,y), G_{m+1}(x,y))$ where $m = m_P(B_t)$. This can be found from $(F_{p^i}(x,y), F_{p^i+1}(x,y))$ as follows.

Again letting $w(x,y) = x - y$, we have

$$A_t(x+\alpha, y+\beta) = w(x+\alpha, y+\beta)B_t(x+\alpha, y+\beta)\,,$$

and so

$$F_{p^i}(x,y) + F_{p^i+1}(x,y) + \cdots = (W_0 + W_1(x,y))(G_m(x,y) + G_{m+1}(x,y) + \cdots).$$

where polynomials with subscript i are 0 or homogeneous of degree i.

Remark 4.12. We get:

If $W_0 \neq 0$, i.e., $\alpha \neq \beta$

$$\begin{aligned} F_{p^i} &= W_0 G_{p^i} = (\sigma x + \tau y)^{2^i} \\ F_{p^i+1} &= W_1 G_{p^i} + W_0 G_{p^i+1} \end{aligned} \tag{4.9}$$

then it follows from these equations that $(F_{2^i}, F_{2^i+1}) = (G_{2^i}, G_{2^i+1})$.

If $W_0 = 0$, i.e., $\alpha = \beta$,

$$\begin{aligned} F_{p^i} &= W_1 G_{p^i-1} = (\sigma x + \tau y)^{2^i} \\ F_{p^i+1} &= W_1 G_{p^i}\,. \end{aligned} \tag{4.10}$$

it is clear that (up to scalars) $W_1 = \sigma x - \tau y$, and so $(F_{p^i}, F_{p^i+1}) = \sigma x - \tau y$ because $F_{p^i+1}(x,y)$ has distinct linear factors (Lemma 4.10). Hence, $(G_{p^i-1}, G_{p^i}) = 1$. The same result is true for the points at infinity of type (iii.b), i.e., $(G'_{p^i-1}, G'_{p^i}) = 1$.

The next result will help us to compute intersection multiplicities.

Lemma 4.13. *Let $h(x, y)$ be an affine curve. Write $h(x + \alpha, y + \beta) = H_m + H_{m+1} + \cdots$ where $P = (\alpha, \beta)$ is a point on $h(x, y)$ of multiplicity m. Suppose that H_m and H_{m+1} are relatively prime, and that there is only one tangent direction at P. If $h = uv$ is reducible, then $I(P, u, v) = 0$.*

Proof. See [8]. □

4.5 Type (i)

We upper bound the intersection multiplicity at the type (i) point.

Lemma 4.14. *If $B_t(x, y) = u(x, y)v(x, y)$ and P is of type* (i), *then $I(P, u, v) \leq (\frac{p^i+1}{2})^2$.*

Proof. Let P be of type (i). We know that $m_P(B_t) = p^i + 1 = m_P(u) + m_P(v)$. From Lemma 4.10 we know that F_{p^i+1} has $p^i + 1$ different linear factors. Thus, $I(P, u, v) = m_p(u)m_p(v)$. This quantity is maximized when $m_P(u) = m_P(v)$ and in this case $m_p(u)m_p(v) = (\frac{p^i+1}{2})^2$. □

4.6 Type (iii)

The next result is equivalent to type (i) with using multiplicity p^i instead.

Lemma 4.15. *If $B_t(x, y) = u(x, y)v(x, y)$ and $P = (\alpha, \beta)$ is a point of type* (iii.a) *then $I(P, u, v) \leq p^{2i}/4$.*

We show that intersection multiplicities at type (iii.b) points are 0, so these points may be disregarded.

Lemma 4.16. *If $B_t(x, y) = u(x, y)v(x, y)$ and $P = (\alpha, \beta)$ is a point of type* (iii.b), *then $I(P, u, v) = 0$.*

Proof. Notice that $\gcd(F'_{p^i-1}, F'_{p^i}) = 1$ and, therefore, $\gcd(G'_{p^i-1}, G'_{p^i}) = 1$ by Remark 4.12. The proof concludes using Proposition 4.13. □

4.7 Type (ii)

We show that there are two possibilities for the intersection multiplicity at a type (ii) point.

Lemma 4.17. *Suppose we are in case* (b.1). *If $B_t(x, y) = u(x, y)v(x, y)$ and $P = (\alpha, \beta)$ is a point of type* (ii) *then either $I(P, u, v) = p^i$ or $I(P, u, v) = 0$.*

Proof. Assume $B_t(x,y) = u(x,y)v(x,y)$. Since P is not on $w(x,y) = x - y$ by Lemma 4.11, we know that $m_P(u)$ is either 1 or 0. If $m_P(u) = 0$, then $I(P,u,v) = 0$. If $m_P(u) = 1$, we proceed as follows.

Let $L(x,y) = \sigma x + \tau y$ and suppose we have the following Taylor expansions at P:

$$u(x+\alpha, y+\beta) = L(x,y) + U_2(x,y) + \cdots$$
$$v(x+\alpha, y+\beta) = L(x,y)^{p^i-1} + V_{p^i}(x,y) + \cdots$$

It follows that

$$u(x+\alpha, y+\beta)L(x,y)^{p^i-2} - v(x+\alpha, y+\beta)$$
$$= L(x,y)^{p^i-2}U_2(x,y) - V_{p^i}(x,y) + \cdots .$$

By definition of intersection multiplicity, we have

$$I(P,u,v) = I(0, u(x+\alpha,y+\beta), u(x+\alpha,y+\beta)L(x,y)^{p^i-2} - v(x+\alpha,y+\beta))$$

so we compute the right-hand side. Notice that $L(x,y) \nmid L(x,y)^{p^i-2}U_2(x,y) - V_{2^i}(x,y)$ because if $L(x,y)$ divides $L(x,y)^{p^i-2}U_2(x,y) - V_{p^i}(x,y)$ then $L(x,y)$ also divides $V_{p^i}(x,y)$. Hence, $L(x,y)^2$ divides $L(x,y)(L(x,y)^{p^i-2}U_2(x,y) + V_{p^i}(x,y)) = G_{p^i+1}(x,y)$ which is a contradiction.

Therefore, $u(x+\alpha,y+\beta)$ and $u(x+\alpha,y+\beta)L^{p^i-2} - v(x+\alpha,y+\beta)$ have different tangent cones. It follows from a property of $I(P,u(x,y),v(x,y))$ that

$$I(0, u(x+\alpha,y+\beta), u(x+\alpha,y+\beta)L^{p^i-2} - v(x+\alpha,y+\beta))$$
$$= m_0(u(x+\alpha,y+\beta))m_0(u(x+\alpha,y+\beta)L^{p^i-2} - v(x+\alpha,y+\beta)) = p^i . \qquad \square$$

5 Case (b.1): Assuming $B_t(x,y)$ Irreducible over $\mathbb{F}_p$

In this section we prove our theorem under the assumption in the title. In the next section we remove this assumption.

Theorem 5.1. *If $B_t(x,y)$ is irreducible over $\mathbb{F}_p$ then $B_t(x,y)$ is absolutely irreducible if either $p \geq 5$, $i \geq 1$, $\ell \geq 3$, or $p = 3$, $i \geq 2$, $\ell \geq 3$.*

Proof. Suppose not, then $B_t(x,y,z) = u(x,y,z)v(x,y,z)$. By irreducibility over $\mathbb{F}_p$, we have that $\deg(u) = \deg(v) = (p^i\ell - 1)/2$. We apply Bezout's Theorem to u and v:

$$\sum_{P \in \mathrm{Sing}(B_t)} I(P,u,v) = \deg(u)\deg(v) = ((p^i\ell - 1)/2)^2 . \tag{5.1}$$

We can bound the left hand side as follows,

$$\sum_{P\in\mathrm{Sing}(B_t)} I(P,u,v) = \sum_{P\in(\mathrm{i})} I(P,u,v) + \sum_{P\in(\mathrm{ii})} I(P,u,v) + \sum_{P\in(\mathrm{iii.a})} I(P,u,v) \tag{5.2}$$

$$= \frac{(p^{2i}+2p^i+1)N_1}{4} + \frac{4p^i(2(\ell-1)-N_1)}{4} + \frac{p^{2i}N_2}{4} \tag{5.3}$$

$$= \frac{(p^{2i}-2p^i+1)N_1}{4} + \frac{8p^i(\ell-1)}{4} + \frac{p^{2i}N_2}{4}. \tag{5.4}$$

Notice that $N_1, \leq \ell - 1$ and $N_2 \leq \ell$, so then (5.4) is $\leq$

$$\frac{(p^{2i}-2p^i+1)(\ell-1)}{4} + \frac{8p^i(\ell-1)}{4} + \frac{p^{2i}\ell}{4}. \tag{5.5}$$

We have that (5.1) $\leq$ (5.5) by Bezout's Theorem, so if we prove that (5.1) $>$ (5.5) we get a contradiction. The inequality (5.5) $<$ (5.1) is

$$\frac{(p^{2i}-2p^i+1)(\ell-1)}{4} + \frac{8p^i(\ell-1)}{4} + \frac{p^{2i}\ell}{4} < \frac{(p^i\ell-1)^2}{4} \tag{5.6}$$

or

$$(p^{2i}+6p^i+1)(\ell-1) + p^{2i}\ell < (\ell p^i - 1)^2 .$$

Thus, the problem is reduced to confirming that

$$p^i(8\ell-6) + (\ell-2) < (\ell-1)^2 p^{2i}$$

or

$$\frac{2(4\ell-3)}{p^i(\ell-1)^2} + \frac{(\ell-2)}{p^{2i}(\ell-1)^2} < 1 .$$

$$\frac{8}{p^i(\ell-1)} + \frac{2}{p^i(\ell-1)^2} + \frac{1}{p^{2i}(\ell-1)} + \frac{1}{p^{2i}(\ell-1)^2} < 1 .$$

If $p \geq 5$, $i \geq 1$ and $\ell \geq 3$, then the lefthand side is at most

$$\frac{8}{10} + \frac{2}{20} + \frac{1}{50} + \frac{1}{100} = \frac{93}{100}$$

which is < 1. And, if $p = 3$, $i \geq 2$ and $\ell \geq 3$, then the lefthand side is at most

$$\frac{8}{3^2 2} + \frac{2}{3^2 2^2} + \frac{1}{3^4 2} + \frac{1}{3^4 2^2} = \frac{165}{3^4 2^2}$$

which again is < 1. □

Remark 5.2. Notice that this proof fails for $\ell = 1$, as it should do because it is already known that $t = p^i + 1$ is an exceptional number.

6 Case (b.1): Assuming $B_t(x, y)$ not Irreducible over $\mathbb{F}_p$

Suppose $B_t = f_1 \cdots f_r$ is the factorization into irreducible factors over $\mathbb{F}_p$. Let $f_j = f_{j,1} \cdots f_{j,n_j}$ be the factorization of f_j into n_j absolutely irreducible factors. Each $f_{j,s}$ has degree $\deg(f_j)/n_j$.

Lemma 6.1. *If P is a point of type* (ii) *then one of the following holds:*
(i) $m_P(f_{j,s}) = 0$ *for all* $j \in \{1, \ldots, r\}$ *and* $s \in \{1, \ldots, n_j\}$ *except for a pair* (j_1, s_1) *with* $m_P(f_{j_1,s_1}) = p^i$.
(ii) $m_P(f_{j,s}) = 0$ *for all* $j \in \{1, \ldots, r\}$ *and* $s \in \{1, \ldots, n_j\}$ *except for two pair* (j_1, s_1) *and* (j_2, s_2) *with* $m_P(f_{j_1,s_1}) = 1$ *and* $m_P(f_{j_2,s_2}) = p^i - 1$.

Proof. This is a consequence of Lemma 4.9 and Lemma 4.11. Consider $u = f_{a,b}$ and $v = \prod_{j \neq a, s \neq b} f_{j,s}$ from Lemma 4.11 we know that $m_P(f_{a,b})$ is either 0 or 1 or $p^i - 1$ or p^i (resp $m_p(v)$ is either p^i or $p^i - 1$ or 1 or 0). But this is true for any pair (a, b).

Clearly no two components $f_{a,b}$ and $f_{a',b'}$ have multiplicity greater than or equal to $p^i - 1$ because the total multiplicity $m_P(B_t) = p^i$. And there are no two components $f_{a,b}$ and $f_{a',b'}$ with multiplicity equal to 1, because then $u = f_{a,b} f_{a',b'}$ has L two times in the tangent cone and $v = g/u$ has $L^{p^{i-2}}$ in the tangent cone which is impossible. Hence, the only possibilities are:
(i) There exists (a, b) with $m_P(f_{a,b}) = p^i$, and $m_P(f_{j,s}) = 0$ for $(j, s) \neq (a, b)$.
(ii) There exist (a, b) and (a', b') with $m_P(f_{a,b}) = 1$ and $m_P(f_{a',b'}) = p^i - 1$, and $m_P(f_{j,s}) = 0$ for $(j, s) \neq (a, b)$, $(j, s) \neq (a', b')$. □

Lemma 6.2. *If P is a point of type* (i) *or* (iii.a)*, then for any two components $f_{a,b}$ and $f_{a',b'}$ we have that $I(P, f_{a,b}, f_{a',b'}) = m_P(f_{a,b}) m_P(f_{a',b'})$.*

Proof. From Lemma 4.10 the tangent cones of $f_{a,b}$ and $f_{a',b'}$ have no common factors. □

Lemma 6.3. *If P is a point of type* (iii.b)*, then for any two components $f_{a,b}$ and $f_{a',b'}$ we have that $I(P, f_{a,b}, f_{a',b'}) = 0$.*

Proof. Consider $u = f_{a,b}$ and $v = g_m/u$. From Lemma 4.15 we know that $I(P, u, v) = 0 = \sum_{(j,s) \neq (a,b)} I(P, u, f_{j,s})$, then $I(P, f_{a,b}, f_{a',b'}) = 0$. □

Lemma 6.4. *Let P as a point of type* (ii) *and $B_t(x, y) = \prod_{j=1}^{r} \prod_{s=1}^{n_j} f_{j,s}$. The intersection multiplicity $I(P, f_{a,b}, f_{a',b'})$ of any two components $f_{a,b}$ and $f_{a',b'}$ is either 0 or p^i.*

Proof. Consider $u = f_{a,b}$ and $v = g_m/u$. From Lemma 4.17 we know that either $I(P, u, v) = 0 = \sum_{(j,s) \neq (a,b)} I(P, u, f_{j,s})$, then $I(P, f_{a,b}, f_{a',b'}) = 0$ or $I(P, u, v) = 2^i = \sum_{(j,s) \neq (a,b)} I(P, u, f_{j,s})$ using Lemma 6.1 we have that there exists (a', b') with $I(P, f_{a,b}, f_{a',b'}) = p^i$. □

We need some more technical results for the main theorem, which give us some upper bounds.

Lemma 6.5.
(i) *If $B_t(x, y)$ does not have an absolutely irreducible factor over $\mathbb{F}_p$, then,*

$$\sum_{j=1}^{r} \deg(f_j)^2/n_j < \deg(B_t)^2/2 . \tag{6.1}$$

(ii) $$\sum_{1\le j<l\le r} \sum_{\substack{1\le i\le n_j\\ 1\le s\le n_l}} \sum_{\substack{P\in \mathrm{Sing}(B_t)\\ P\in \text{(ii)}}} I(P, f_{j,i}, f_{l,s}) \le p^i(2(\ell-1) - N_1)$$

(iii) $$\sum_{1\le j<l\le r} \sum_{\substack{1\le i\le n_j\\ 1\le s\le n_l}} \sum_{P\in \text{(i)}} I(P, f_{j,i}, f_{l,s})) \le (p^i+1)(p^i)/2N_1$$

(iv) $$\sum_{1\le j<l\le r} \sum_{\substack{1\le i\le n_j\\ 1\le s\le n_l}} \sum_{P\in \text{(iii.a)}} I(P, f_{j,i}, f_{l,s})) \le (p^i)(p^i-1)/2N_2 .$$

Proof. (i) $\sum_{j=1}^{r} \deg(f_j)^2/n_j \le \sum_{j=1}^{r} \deg(f_j)^2/2 = 1/2(\deg(f_1)^2 + \cdots + \deg(f_r)^2) \le 1/2\ \deg(B_t)^2$

(ii) From Lemma 6.4 we know that if P is a point of type (ii) then $I(P, f_{j,i}, f_{l,s}) = 0$ for every $j, l \in \{1, \ldots, r\}$ and $1 \le i \le n_j$, $1 \le s \le n_l$. From Lemma 6.4 we now that for each point P of type (ii) there are at most two components $f_{a,b}$ and $f_{a',b'}$ for which $I(P, f_{a,b}, f_{a',b'}) = p^i$. Taking into account that there are $(\ell - 1)$ points of type (ii), we get the result.

(iii) From Lemma 6.2 we have that if P is a point of type (i), then for any two components $f_{a,b}$ and $f_{a',b'}$ we have $I(P, f_{a,b}, f_{a',b'}) = m_P(f_{a,b})m_P(f_{a',b'})$. Hence, we have to prove the following,

$$\sum_{j=1}^{r} \sum_{1\le i<s\le n_j} m_P(f_{j,i})m_P(f_{j,s}) + \sum_{1\le j<l\le r} \sum_{\substack{1\le i\le n_j\\ 1\le s\le n_l}} m_P(f_{j,i})m_P(f_{j,s})$$
$$\le (p^i+1)(p^i)/2 .$$

Notice that the left hand side is a maximum when $m_P(f_{j,s}) = 1$ for every $j \in \{1, \ldots, r\}$, $s \in \{1, \ldots, n_j\}$. The latter equation is

$$\sum_{j=1}^{r} \sum_{1\le i<s\le n_j} m_P(f_{j,i})m_P(f_{j,s}) + \sum_{1\le j<l\le r} \sum_{\substack{1\le i\le n_j\\ 1\le s\le n_l}} m_P(f_{j,i})m_P(f_{j,s})$$
$$\le \sum_{j=1}^{r} \sum_{1\le i<s\le n_j} 1 + \sum_{1\le j<l\le r} \sum_{\substack{1\le i\le n_j\\ 1\le s\le n_l}} 1 = \binom{p^i+1}{2} = (p^i+1)(p^i)/2 .$$

(iv) This is the same proof as (iii) but taking into account that has N_2 singular points of this type with multiplicity p^i. □

Finally, here is our main result.

Theorem 6.6. *In case* (b.1) $B_t(x, y)$ *always has an absolutely irreducible factor over* $\mathbb{F}_p$, *if either* $p \geq 5$, $i \geq 1$, $\ell \geq 5$ *or* $p = 3$, $i \geq 2$, $\ell \geq 5$.

Proof. We apply Bezout's Theorem one more time to the product

$$f_1 f_2 \dots f_r = (f_{1,1} \dots f_{1,n_1})(f_{2,1} \dots f_{2,n_2}) \dots (f_{r,1} \dots f_{r,n_r}) .$$

The sum of the intersection multiplicities can be written

$$\sum_{j=1}^{r} \sum_{1 \leq i < s \leq n_j} \sum_{P \in \mathrm{Sing}(B_t)} I(P, f_{j,i}, f_{j,s}) + \sum_{1 \leq j < l \leq r} \sum_{\substack{1 \leq i \leq n_j \\ 1 \leq s \leq n_l}} \sum_{P \in \mathrm{Sing}(B_t)} I(P, f_{j,i}, f_{l,s})$$

where the first term is for factors within each f_j, and the second term is for cross factors between f_j and f_l. Using Lemma 6.5, part (ii), (iii) and (iv), the previous sums can be bounded by

$$\leq (p^i + 1)(p^i)/2N_1 + p^i(2(l-1) - N_1) + (p^i)(p^i - 1)/2N_2 \tag{6.2}$$

On the other hand, the right-hand side of Bezout's Theorem is

$$\sum_{j=1}^{r} \sum_{1 \leq i < s \leq n_j} \deg(f_{j,i}) \deg(f_{j,s}) + \sum_{1 \leq j < l \leq r} \sum_{\substack{1 \leq i \leq n_j \\ 1 \leq s \leq n_l}} \deg(f_{j,i}) \deg(f_{l,s}) . \tag{6.3}$$

Since each $f_{j,s}$ has the same degree for all s, the first term is equal to

$$\sum_{j=1}^{r} \deg(f_j)^2 \frac{n_j - 1}{2n_j} = \frac{1}{2} \sum_{j=1}^{r} \deg(f_j)^2 - \frac{1}{2} \sum_{j=1}^{r} \frac{\deg(f_j)^2}{n_j} .$$

Note that

$$\begin{aligned}
(\deg(B_t))^2 &= \Big(\sum_{j=1}^{r} \deg(f_j) \Big)^2 \\
&= \sum_{j=1}^{r} \deg(f_j)^2 + 2 \Big(\sum_{1 \leq j < l \leq r} \deg(f_j) \deg(f_l) \Big) \\
&= \sum_{j=1}^{r} \deg(f_j)^2 + 2 \sum_{1 \leq j < l \leq r} \Big(\sum_{s=1}^{n_j} \deg(f_{j,s}) \Big) \Big(\sum_{i=1}^{n_l} \deg(f_{l,i}) \Big) \\
&= \sum_{j=1}^{r} \deg(f_j)^2 + 2 \sum_{1 \leq j < l \leq r} \sum_{\substack{1 \leq i \leq n_j \\ 1 \leq s \leq n_l}} \deg(f_{j,i}) \deg(f_{l,s}) .
\end{aligned}$$

Substituting both of these into (6.3) shows that (6.3) is equal to

$$\frac{1}{2}\left(\deg(B_t)^2 - \sum_{j=1}^{r} \frac{\deg(f_j)^2}{n_j}\right). \tag{6.4}$$

Using (6.1) we get

$$\frac{1}{2}\left(\deg(B_t)^2 - \sum_{j=1}^{r} \frac{\deg(f_j)^2}{n_j}\right) > \frac{1}{2}\left(\deg(B_t)^2 - \deg(B_t)^2/2\right) = \deg(B_t)^2/4\,. \tag{6.5}$$

Comparing (6.5) and (6.2), so far we have shown that Bezout's Theorem implies the following inequality:

$$\deg(B_t)^2/4 \le (p^i+1)(p^i)/2N_1 + p^i(2(l-1)-N_1) + (p^i)(p^i-1)/2N_2$$

Notice that $N_1 \le (\ell - 1)$ and $N_2 \le \ell$. Hence, we have the following inequality

$$(p^i\ell - 1)^2/4 \le (\ell-1)(p^i+1)(p^i)/2 + (\ell-1)p^i + \ell(p^i)(p^i-1)/2$$

Let us now show that the opposite is true, to get a contradiction. The opposite inequality is

$$\frac{\ell^2p^{2i} - 2\ell p^i + 1}{4} > \frac{2(\ell-1)(p^i+1)(p^i)}{4} + \frac{4(\ell-1)p^i}{4} + \frac{2\ell(p^i)(p^i-1)}{4}$$

or

$$\ell^2p^{2i} - 2\ell p^i + 1 > 2(\ell-1)(p^i+1)(p^i) + 4(\ell-1)p^i + 2\ell(p^i)(p^i-1)\,.$$

This simplifies to

$$p^{2i}(\ell^2 - 4\ell + 2) > p^i(6\ell - 6) - 1$$

giving

$$p^i > \frac{6(\ell-1)}{(\ell^2 - 4\ell + 2)}\,.$$

The right-hand side $\frac{6(\ell-1)}{(\ell^2-4\ell+2)}$ is a decreasing function of ℓ which is continuous for $\ell \ge 5$. Thus, substituting ℓ by 5 we get,

$$p^i > \frac{24}{7} = 3 + 3/7\,.$$

This is true for either $p \ge 5$, $i \ge 1$ or $p = 3$, $i \ge 2$. □

Remark 6.7. Notice that this proof fails for $\ell = 1$, as it should, because it is already known that $t = p^i + 1$ is an exceptional number.

We recall that $t = \frac{3^k+1}{2}$ is an exceptional number over $\mathbb{F}_{3^m}$ whenever k is odd and $(m, k) = 1$. The previous theorem does not contradict this either because we are assuming $t = p^i\ell + 1$ (case (b.1)). The number $\frac{3^k+1}{2}$ cannot be written in the form $3^i\ell + 1$, as is easy to check.

References

[1] C. Carlet, P. Charpin, and V. Zinoviev, Codes, bent functions and permutations suitable for DES-like cryptosystems, *Designs, Codes and Cryptography* **15(2)** (1998), 125–156.

[2] R. Coulter and R. Matthews, Planar Functions and Planes of Lenz–Barlotti Class II, *Designs Codes and Cryptography* **10** (1997), 167–184.

[3] R. Coulter, private communication.

[4] P. Dembowski and T. G. Ostrom, Planes of order n with collineation groups of order n^2, *Math. Z.* **103** (1968), 239–258.

[5] J. F. Dillon, *Geometry, codes and difference sets: exceptional connections*, pp. 73–85, Codes and designs (Columbus, OH), 2000.

[6] W. Fulton, *Algebraic curves*, Mathematics Lecture Notes Series. W. A. Benjamin, Inc., New York-Amsterdam, 1969.

[7] F. Hernando and G. McGuire, Proof of a conjecture on the sequence of exceptional numbers, classifying cyclic codes and APN functions, *J. Algebra* **343** (2011), 78–92.

[8] H. Janwa, G. McGuire, and R. M. Wilson, Double-error-correcting cyclic codes and absolutely irreducible polynomials over $GF(2)$, *J. Algebra* **178(2)** (1995), 665–676.

[9] H. Janwa and R. M. Wilson, *Hyperplane sections of Fermat varieties in* P^3 *in* $\mathrm{char.}\,2$ *and some applications to cyclic codes*, Applied algebra, algebraic algorithms and error-correcting codes (San Juan, PR), Lecture Notes in Comput. Sci. 673, pp. 180–194, Springer, Berlin, 1993.

[10] D. Jedlicka, APN monomials over $GF(2^n)$ for infinitely many n, *Finite Fields Appl.* **13(4)** (2007), 1006–1028.

[11] T. Kasami, The weight enumerators for several classes of subcodes of the 2nd order binary Reed–Muller codes, *Information and Control* 18 (1971), 369–394.

[12] S. Kopparty and S. Yekhanin, *Detecting Rational Points on Hypersurfaces over Finite Fields*, Computational Complexity, 2008. CCC '08. 23rd Annual IEEE Conference 2008, pp. 311–320.

[13] E. Leducq, *Functions which are PN on infinitely many extensions of* F_p*,* p *odd*, http://arxiv.org/abs/1006.2610

[14] R. Lidl and H. Niederreiter, *Finite Fields*, Addison-Wesley, 1983.

[15] J. H. van Lint and R. M. Wilson, On the minimum distance of cyclic codes, *IEEE Trans. Inform. Theory* **32(1)** (1986), 23–40.

[16] K. Nyberg and L. R.Knudsen, *Provable security against differential cryptanalysis*, Advances in Cryptology CRYPTO92, LNCS 740, pp. 566–574, Springer-Verlag, 1992.

Harald Niederreiter
Finite Fields and Quasirandom Points

Abstract: Quasi-Monte Carlo methods are deterministic and often more effective analogs of Monte Carlo methods in scientific computing and rely on quasirandom points that form low-discrepancy point sets and sequences. The explicit construction of quasirandom points is a fundamental aspect of quasi-Monte Carlo methods and finite fields play an important role in this context. The article covers all major construction algorithms for quasirandom points that are based on finite fields. The most powerful current constructions employ the theory of digital (t, m, s)-nets and digital (t, s)-sequences, and the overwhelming majority of these types of constructions uses finite fields in one way or another. The tools range from polynomials over finite fields to global function fields. The necessary background on digital (t, m, s)-nets and digital (t, s)-sequences is also provided in the article.

Keywords: Quasi-Monte Carlo Method, Finite Field, Quasirandom Points, (t, m, s)-Net, (t, s)-Sequence, $(\mathbf{T}, s)$-Sequence, Digital Method, Permutation Polynomial, Ordered Orthogonal Array, Niederreiter–Rosenbloom–Tsfasman Weight, Global Function Field, Polynomial Lattice, Hyperplane Net, Cyclic Digital Net, Sobol' Sequence, Faure Sequence, Niederreiter Sequence, Niederreiter–Xing Sequence

2010 Mathematics Subject Classifications: 05B15, 11K38, 11K45, 11R58, 11T06, 14H05, 65C05

Harald Niederreiter: Johann Radon Institute for Computational and Applied Mathematics, Austrian Academy of Sciences, Linz, Austria, e-mail: ghnied@gmail.com

1 Introduction

Since finite fields are basic structures of discrete mathematics, it is to be expected that they have fruitful applications to problems in a discrete setting such as those arising in coding theory and cryptology. It is less obvious that finite fields should have applications to problems in a continuous setting such as numerical integration, global optimization, and related problems in scientific computing. However, it is nowadays a well recognized fact that finite fields can play a significant role in methods for solving the latter types of problems. These methods belong to the family of *quasi-Monte Carlo methods*, that is, deterministic and often more effective analogs of the statistical Monte Carlo methods in scientific computing. For recent general introductions to quasi-Monte Carlo methods, we refer to the book by Dick and Pillichshammer [6] and the survey article by Niederreiter [38]. The book by Lemieux [19] covers quasi-

Monte Carlo methods in the wider context of Monte Carlo methods and discusses also various practical applications of quasi-Monte Carlo methods.

Quasi-Monte Carlo methods rely on special sample points called *quasirandom points* which are generated by a deterministic algorithm and evenly distributed over an underlying domain in Euclidean space $\mathbb{R}^s$. Usually, this domain can be normalized to be the s-dimensional unit cube $[0,1]^s$, and in the sequel we assume that this has been done. Algorithms for the generation of quasirandom points have a long history, and for the early work on this topic we refer to the survey article by Niederreiter [24]. In the last few decades, finite fields have assumed a prominent role in the generation of quasirandom points. It is the aim of the present paper to discuss the connections between finite fields and quasirandom points.

Section 2 assembles basic definitions and facts concerning discrepancy and quasirandom points. Section 3 describes the digital method for the construction of (t,m,s)-nets, (t,s)-sequences, and $(\mathbf{T},s)$-sequences over finite fields. The combinatorial aspects of the theory of (t,m,s)-nets and the resulting necessary conditions for the existence of (t,m,s)-nets and (t,s)-sequences are discussed in Section 4. The duality theory for digital nets and digital sequences is presented in Section 5. Special constructions of digital nets and digital sequences are described in Section 6 and Section 7, respectively.

2 General Background

We mentioned in Section 1 that quasirandom points should be evenly distributed over the s-dimensional unit cube $[0,1]^s$. The deviation from a perfect uniform distribution over $[0,1]^s$ is measured by the notion of discrepancy which is defined as follows. For a given point set P consisting of the N points $\mathbf{x}_0, \mathbf{x}_1, \dots, \mathbf{x}_{N-1} \in [0,1]^s$ and a subinterval J of $[0,1]^s$, let $A(J;P)$ be the number of integers n with $0 \le n \le N-1$ for which $\mathbf{x}_n \in J$. The *discrepancy* of P is given by

$$D_N = D_N(P) = \sup_J \left| \frac{A(J;P)}{N} - \mathrm{Vol}(J) \right| ,$$

where the supremum is extended over all subintervals J of $[0,1]^s$. If S is a sequence (by a "sequence" we mean as usual an infinite sequence) of elements of $[0,1]^s$, then $D_N = D_N(S)$ is defined to be the discrepancy of the point set consisting of the first N terms of S.

A point set P of N quasirandom points should have a small discrepancy $D_N(P)$. A sequence S of quasirandom points should have a small discrepancy $D_N(S)$ for all sufficiently large N. Here "small" means $O(N^{-1})$ up to powers of $\log N$. For instance, it is well known that there are sequences S of points in $[0,1]^s$ for which $D_N(S) = O\left(N^{-1}(\log N)^s\right)$ for all $N \ge 2$ (see [32, Chapter 3]).

Classical constructions of point sets and sequences of quasirandom points (such as the good lattice point sets and Halton sequences described in [32]) use number-theoretic techniques. Finite fields entered the scene via the more recent theory of (t, m, s)-nets and $(\mathbf{T}, s)$-sequences which will be the focus of this article. The following definition is basic.

Definition 2.1. Let $0 \le t \le m$, $s \ge 1$, and $b \ge 2$ be integers. Then a *(t, m, s)-net in base b* is a point set P of b^m points in $[0, 1)^s$ such that $A(J; P) = b^t$ for every interval J of the form

$$J = \prod_{i=1}^{s} [a_i b^{-d_i}, (a_i + 1) b^{-d_i}) \tag{2.1}$$

with integers $d_i \ge 0$ and $0 \le a_i < b^{d_i}$ for $1 \le i \le s$ and with volume $\mathrm{Vol}(J) = b^{t-m}$.

Definition 2.1 means, of course, that every interval J of the form (2.1) has the "right share" of points from P. It is easily seen that any (t, m, s)-net in base b is also a (u, m, s)-net in base b for all integers u with $t \le u \le m$. Any point set consisting of b^m points in $[0, 1)^s$ is a (t, m, s)-net in base b with $t = m$. Smaller values of t signify stronger uniform distribution properties of a (t, m, s)-net in base b. The number t is called the *quality parameter* of a (t, m, s)-net in base b.

Example 2.2. Let $s = 2$ and let $b \ge 2$ and $m \ge 1$ be given integers. For any integer n with $0 \le n < b^m$, let $n = \sum_{j=0}^{m-1} a_j(n) b^j$ with all $a_j(n) \in \{0, 1, \dots, b-1\}$ be the digit expansion of n in base b. Put $\phi_b(n) = \sum_{j=0}^{m-1} a_j(n) b^{-j-1}$. Then the point set consisting of the points

$$\left(\frac{n}{b^m}, \phi_b(n) \right) \in [0, 1)^2, \quad n = 0, 1, \dots, b^m - 1,$$

is easily seen to be a $(0, m, 2)$-net in base b. This point set is called a *Hammersley net* in base b.

Let P be a (t, m, s)-net in base b with $m \ge 1$. Then with $N = b^m$, the discrepancy D_N of P satisfies

$$D_N = O\left(b^t N^{-1} (\log N)^{s-1} \right) \tag{2.2}$$

with an implied constant depending only on b and s. The currently best values of the implied constant can be found in Faure and Lemieux [8] and Kritzer [13]. The bound (2.2) shows again that small values of the quality parameter t are preferable when the aim is to obtain low-discrepancy point sets. It is an important fact that $N^{-1}(\log N)^{s-1}$ is the smallest order of magnitude that is currently known for the discrepancy of N points in $[0, 1]^s$. For general lower bounds on the discrepancy, we refer to [6, Section 3.2].

There is an analog of (t, m, s)-nets for sequences of points in $[0, 1]^s$. This leads to the concepts of (t, s)-sequences and $(\mathbf{T}, s)$-sequences in base b. For instance, the idea of the concept of (t, s)-sequence in base b is that certain finite segments of length b^m of the sequence form (t, m, s)-nets in base b, and this for all sufficiently large m.

Some constructions of (t, s)-sequences yield points in the closed unit cube $[0, 1]^s$ and not just in the half-open unit cube $[0, 1)^s$. This may cause a problem with Definition 2.1 where all points must be in $[0, 1)^s$. For this reason, we use the following truncation operation which takes points from $[0, 1]^s$ into points from $[0, 1)^s$. For an integer $b \geq 2$ and a real number $x \in [0, 1]$, let

$$x = \sum_{j=1}^{\infty} y_j b^{-j} \quad \text{with all } y_j \in \{0, 1, \ldots, b-1\}$$

be a b-adic expansion of x, where the case $y_j = b - 1$ for all sufficiently large j is allowed. For any integer $m \geq 1$, we define the truncation

$$[x]_{b,m} = \sum_{j=1}^{m} y_j b^{-j} \in [0, 1) .$$

Note that this truncation operates on the expansion of x and not on x itself, since it may yield different results depending on which b-adic expansion of x is used. If $\mathbf{x} = (x^{(1)}, \ldots, x^{(s)}) \in [0, 1]^s$ and the $x^{(i)}$, $1 \leq i \leq s$, are given by prescribed b-adic expansions, then we define

$$[\mathbf{x}]_{b,m} = ([x^{(1)}]_{b,m}, \ldots, [x^{(s)}]_{b,m}) \in [0, 1)^s .$$

Definition 2.3. Let $t \geq 0$, $s \geq 1$, and $b \geq 2$ be integers. Then a sequence $\mathbf{x}_0, \mathbf{x}_1, \ldots$ of points in $[0, 1]^s$ is a *(t, s)-sequence in base b* if for all integers $k \geq 0$ and $m > t$ the points $[\mathbf{x}_n]_{b,m}$ with $kb^m \leq n < (k+1)b^m$ form a (t, m, s)-net in base b. Here the coordinates of all points $\mathbf{x}_n$, $n = 0, 1, \ldots$, are given by prescribed b-adic expansions.

It is easily seen that any (t, s)-sequence in base b is also a (u, s)-sequence in base b for all integers $u \geq t$. Smaller values of t signify stronger uniform distribution properties of a (t, s)-sequence in base b. The number t is called the *quality parameter* of a (t, s)-sequence in base b.

Example 2.4. Let $s = 1$ and let $b \geq 2$ be an integer. For $n = 0, 1, \ldots$, let $n = \sum_{j=0}^{\infty} a_j(n) b^j$ with all $a_j(n) \in \{0, 1, \ldots, b-1\}$ and $a_j(n) = 0$ for all sufficiently large j be the digit expansion of n in base b. Put $\phi_b(n) = \sum_{j=0}^{\infty} a_j(n) b^{-j-1}$. Then the sequence $\phi_b(0), \phi_b(1), \ldots$ is easily seen to be a $(0, 1)$-sequence in base b. This sequence is called the *van der Corput sequence* in base b.

For a (t, s)-sequence S in base b, there is an analog of the discrepancy bound (2.2), namely

$$D_N(S) = O\left(b^t N^{-1} (\log N)^s\right) \quad \text{for all } N \geq 2 , \tag{2.3}$$

where the implied constant depends only on b and s. We refer again to [8] and [13] for the current best values of the implied constant. It is worth pointing out that no smaller asymptotic rate than $N^{-1}(\log N)^s$ is currently known for the discrepancy of a sequence of points in $[0, 1]^s$.

A generalization of the concept of (t, s)-sequence was introduced by Larcher and Niederreiter [17]. As usual, $\mathbb{N}$ denotes the set of positive integers and $\mathbb{N}_0$ the set of nonnegative integers.

Definition 2.5. Let $\mathbf{T} : \mathbb{N} \to \mathbb{N}_0$ be a function with $\mathbf{T}(m) \le m$ for all $m \in \mathbb{N}$. Then a sequence $\mathbf{x}_0, \mathbf{x}_1, \ldots$ of points in $[0, 1]^s$ is a $(\mathbf{T}, s)$*-sequence in base* b if for all $k \in \mathbb{N}_0$ and $m \in \mathbb{N}$ the points $[\mathbf{x}_n]_{b,m}$ with $kb^m \le n < (k+1)b^m$ form a $(\mathbf{T}(m), m, s)$-net in base b. Here the coordinates of all points $\mathbf{x}_n$, $n = 0, 1, \ldots$, are given by prescribed b-adic expansions.

Remark 2.6. If the function $\mathbf{T}$ in Definition 2.5 is such that for some integer $t \ge 0$ we have $\mathbf{T}(m) = m$ for $m \le t$ and $\mathbf{T}(m) = t$ for $m > t$, then the definition of a $(\mathbf{T}, s)$-sequence in base b reduces to that of a (t, s)-sequence in base b.

It was shown in [17] that if the function $\mathbf{T}$ satisfies $\sum_{m=1}^{M} b^{\mathbf{T}(m)} = O(M)$ for all $M \in \mathbb{N}$, then for any $(\mathbf{T}, s)$-sequence S in base b we get

$$D_N(S) = O\left(N^{-1}(\log N)^s\right) \quad \text{for all } N \ge 2, \tag{2.4}$$

and thus a discrepancy bound comparable to (2.3). The paper [17] contains also discrepancy bounds for arbitrary functions $\mathbf{T}$.

The theory of (t, m, s)-nets and $(\mathbf{T}, s)$-sequences has become a well developed research area. A detailed account of this theory is available in the recent book by Dick and Pillichshammer [6]. The first expository account of the theory of (t, m, s)-nets and (t, s)-sequences was given in [32, Chapter 4], and in the sequel relevant survey articles have appeared periodically (see [2, 22, 34, 36, 37]).

3 General Construction Principles

We have seen in Section 2 that point sets and sequences of quasirandom points can be obtained via the theory of (t, m, s)-nets and $(\mathbf{T}, s)$-sequences, respectively; compared with the discrepancy bounds (2.2), (2.3), and (2.4). So far however, we have not yet touched upon the actual construction of these mathematical objects. In the present section, we describe general principles for the construction of (t, m, s)-nets and $(\mathbf{T}, s)$-sequences. Explicit constructions based on these general principles will be presented in later sections.

We recall that (t, m, s)-nets and $(\mathbf{T}, s)$-sequences are defined relative to an integer base $b \ge 2$. If b is a prime power q, then it turns out that one can use the finite field $\mathbb{F}_q$ of order q for the construction of (t, m, s)-nets and $(\mathbf{T}, s)$-sequences in base q. In fact, the general construction principles described below employ linear algebra or polynomials over $\mathbb{F}_q$. For an arbitrary base $b \ge 2$, we can write b as a product of prime powers, say $b = \prod_{v=1}^{h} q_v$, and then in these construction principles the finite field $\mathbb{F}_q$ is replaced by the ring direct product $\mathbb{F}_{q_1} \times \ldots \times \mathbb{F}_{q_h}$ of finite fields (see e.g. [28] and [32, Section 4.3]). Since this article is devoted to applications of finite fields, we restrict the attention to prime-power bases from now on.

We start with a technique for the construction of (t, m, s)-nets which, in its general form, goes back to Niederreiter [27]. In some special cases, this method was also used earlier by Sobol' [53] and Faure [7]. This technique is nowadays called the *digital method*, the reason for this terminology being that the coordinates of the s-dimensional quasirandom points are constructed digit by digit relative to a chosen base.

Let q be an arbitrary prime power and let $\mathbb{F}_q$ again denote the finite field of order q. We write $\mathbb{Z}_q = \{0, 1, \ldots, q-1\} \subset \mathbb{Z}$ for the set of digits in base q. Let the integers $m \geq 1$ and $s \geq 1$ be given. We choose $m \times m$ matrices $C^{(1)}, \ldots, C^{(s)}$ over $\mathbb{F}_q$. Next, we define the map $\Psi_m : \mathbb{F}_q^m \to [0, 1)$ by

$$\Psi_m(\mathbf{h}) = \sum_{j=1}^{m} \psi(h_j) q^{-j} \tag{3.1}$$

for any column vector $\mathbf{h} = (h_1, \ldots, h_m)^{\mathrm{T}} \in \mathbb{F}_q^m$, where $\psi : \mathbb{F}_q \to \mathbb{Z}_q$ is a chosen bijection. With a fixed column vector $\mathbf{a} \in \mathbb{F}_q^m$, we associate the point

$$\left(\Psi_m(C^{(1)}\mathbf{a}), \ldots, \Psi_m(C^{(s)}\mathbf{a})\right) \in [0, 1)^s . \tag{3.2}$$

By letting $\mathbf{a}$ range over all q^m possibilities in $\mathbb{F}_q^m$, we arrive at a point set consisting of q^m points in $[0, 1)^s$.

Definition 3.1. The point set P consisting of the q^m points in (3.2) is called a *digital net over* $\mathbb{F}_q$, and if P forms a (t, m, s)-net in base q, then P is called a *digital* (t, m, s)*-net over* $\mathbb{F}_q$. The matrices $C^{(1)}, \ldots, C^{(s)}$ are called the *generating matrices* of P.

Example 3.2. Let $s = 2$, let q be a prime, and let $m \geq 1$ be an integer. Then ψ in (3.1) can be taken as an identity map. Let $C^{(1)} = (c_{ij})_{1 \leq i,j \leq m}$ be the $m \times m$ antidiagonal matrix over $\mathbb{F}_q$ with $c_{ij} = 1$ if $i + j = m + 1$ and $c_{ij} = 0$ otherwise. Let $C^{(2)}$ be the $m \times m$ identity matrix over $\mathbb{F}_q$. Then the Hammersley net in Example 2.2 is a digital $(0, m, 2)$-net over $\mathbb{F}_q$ with generating matrices $C^{(1)}$ and $C^{(2)}$.

A digital net over $\mathbb{F}_q$ as constructed in (3.2) is always a digital (t, m, s)-net over $\mathbb{F}_q$ with $t = m$. However, the quality parameter t may be much smaller. The exact determination of the quality parameter of a digital net proceeds as follows. Like all basic results on (digital) nets, Theorem 3.4 below is due to Niederreiter [27].

Definition 3.3. Let $C^{(1)}, \ldots, C^{(s)}$ be $m \times m$ matrices over $\mathbb{F}_q$ and for $1 \leq i \leq s$ and $1 \leq j \leq m$ let $\mathbf{c}_j^{(i)}$ denote the j-th row of the matrix $C^{(i)}$. Then $\varrho(C^{(1)}, \ldots, C^{(s)})$ is defined to be the largest nonnegative integer d such that, for any integers $0 \leq d_1, \ldots, d_s \leq m$ with $\sum_{i=1}^{s} d_i = d$, the vectors $\mathbf{c}_j^{(i)}$, $1 \leq j \leq d_i$, $1 \leq i \leq s$, are linearly independent over $\mathbb{F}_q$ (this property is assumed to be vacuously satisfied for $d = 0$).

Theorem 3.4. *A digital net over $\mathbb{F}_q$ with $m \times m$ generating matrices $C^{(1)}, \dots, C^{(s)}$ is a digital (t, m, s)-net over $\mathbb{F}_q$ with*

$$t = m - \varrho(C^{(1)}, \dots, C^{(s)}) .$$

Proof. Let

$$J = \prod_{i=1}^{s} [a_i q^{-d_i}, (a_i + 1) q^{-d_i})$$

be an interval of the form (2.1) with $b = q$ and with $\mathrm{Vol}(J) = b^{t-m}$, that is, with

$$\sum_{i=1}^{s} d_i = m - t = \varrho(C^{(1)}, \dots, C^{(s)}) =: r .$$

We can assume that $r \geq 1$. For a given column vector $\mathbf{a} \in \mathbb{F}_q^m$, the corresponding point in (3.2) lies in J if and only if

$$\Psi_m(C^{(i)}\mathbf{a}) \in [a_i q^{-d_i}, (a_i + 1) q^{-d_i}) \quad \text{for } 1 \leq i \leq s .$$

This condition means that for $1 \leq i \leq s$ the first d_i q-adic digits of $\Psi_m(C^{(i)}\mathbf{a})$ and $a_i q^{-d_i}$ agree, and this is equivalent to $C\mathbf{a} = \mathbf{b}$ for some $\mathbf{b} \in \mathbb{F}_q^r$ depending only on J. Here C is an $r \times m$ matrix whose rows are given by the rows $\mathbf{c}_j^{(i)}$, $1 \leq j \leq d_i$, $1 \leq i \leq s$, of the generating matrices. These row vectors are linearly independent over $\mathbb{F}_q$ by the definition of $\varrho(C^{(1)}, \dots, C^{(s)})$, and so the matrix C has rank r. Thus, for any given $\mathbf{b} \in \mathbb{F}_q^r$, the equation $C\mathbf{a} = \mathbf{b}$ has exactly $q^{m-r} = q^t$ solutions $\mathbf{a} \in \mathbb{F}_q^m$. This yields the required property of a (t, m, s)-net in base q. □

Let P be a digital net over $\mathbb{F}_q$ with $m \times m$ generating matrices $C^{(1)}, \dots, C^{(s)}$. With $\varrho = \varrho(C^{(1)}, \dots, C^{(s)})$ and $N = q^m$, it follows from (2.2) and Theorem 3.4 that the discrepancy D_N of P satisfies

$$D_N = O\left(q^{-\varrho} (\log N)^{s-1}\right)$$

with an implied constant depending only on q and s. A lower bound on D_N of the order of magnitude $q^{-\varrho}$ was shown in Niederreiter [30]. These results demonstrate that the discrepancy D_N of a digital net is small if and only if ϱ is large.

A generalization of the digital method for the construction of (t, m, s)-nets was introduced in [37]. In this construction principle, the matrix-vector products $C^{(i)}\mathbf{a}$, $1 \leq i \leq s$, in (3.2) are replaced by operations based on polynomial maps. Let the integers $m \geq 1$ and $s \geq 1$ be given. For each $i = 1, \dots, s$ and $j = 1, \dots, m$, choose a polynomial $f_j^{(i)}$ over $\mathbb{F}_q$ in m variables. For $1 \leq i \leq s$, let the map $\mathbf{f}^{(i)} : \mathbb{F}_q^m \to \mathbb{F}_q^m$ be defined by

$$\mathbf{f}^{(i)}(\mathbf{a}) = \left(f_1^{(i)}(\mathbf{a}), \dots, f_m^{(i)}(\mathbf{a})\right) \quad \text{for } \mathbf{a} \in \mathbb{F}_q^m .$$

With a fixed $\mathbf{a} \in \mathbb{F}_q^m$, we associate the point

$$\left(\Psi_m(\mathbf{f}^{(1)}(\mathbf{a})), \dots, \Psi_m(\mathbf{f}^{(s)}(\mathbf{a}))\right) \in [0, 1)^s , \tag{3.3}$$

where Ψ_m is given by (3.1). By letting **a** range over all q^m elements of $\mathbb{F}_q^m$, we arrive at a point set consisting of q^m points in $[0,1)^s$. This construction technique may still be called a *digital method* since the coordinates of the s-dimensional points are again constructed digit by digit relative to the base q. To single out the special case using (3.2), it is proposed in [37] to call that special case the *linear digital method* and to speak of a *linear digital net over* $\mathbb{F}_q$. However, this terminology is not yet generally accepted.

We return to the construction of (t,m,s)-nets using (3.3) and consider now the question of determining quality parameters for such nets. The following classical concept is needed for this purpose (see [20, Section 7.5]).

Definition 3.5. A polynomial $f \in \mathbb{F}_q[X_1,\dots,X_m]$ in m variables is called a *permutation polynomial* over $\mathbb{F}_q$ if the equation $f(X_1,\dots,X_m) = c$ has exactly q^{m-1} solutions in $\mathbb{F}_q^m$ for each $c \in \mathbb{F}_q$.

The following criterion is now an immediate consequence of [37, Corollary 2 and Proposition 1].

Theorem 3.6. *Let P be the point set consisting of the points in* (3.3) *with* **a** *ranging over* $\mathbb{F}_q^m$. *Then P is a (t,m,s)-net in base q if and only if, for any integers $d_1,\dots,d_s \geq 0$ with $\sum_{i=1}^s d_i = m-t$, the polynomials $f_j^{(i)}$, $1 \leq j \leq d_i$, $1 \leq i \leq s$, over $\mathbb{F}_q$ in m variables have the property that all their nontrivial linear combinations with coefficients from $\mathbb{F}_q$ are permutation polynomials over $\mathbb{F}_q$.*

There is an analog of the (linear) digital method for the construction of sequences (see again Niederreiter [27]). Let the finite field $\mathbb{F}_q$ and the integer $s \geq 1$ be given. We choose $\infty \times \infty$ matrices $C^{(1)},\dots,C^{(s)}$ over $\mathbb{F}_q$, where by an $\infty \times \infty$ matrix we mean a matrix with denumerably many rows and columns. In analogy with (3.1), we define $\Psi_\infty : \mathbb{F}_q^\infty \to [0,1]$ by

$$\Psi_\infty(\mathbf{h}) = \sum_{j=1}^{\infty} \psi(h_j) q^{-j} \tag{3.4}$$

for any column vector $\mathbf{h} = (h_1,h_2,\dots)^{\mathrm{T}} \in \mathbb{F}_q^\infty$, where $\psi : \mathbb{F}_q \to \mathbb{Z}_q$ is a chosen bijection. For $n = 0,1,\dots$, let $n = \sum_{j=1}^\infty a_j(n) q^{j-1}$ with all $a_j(n) \in \mathbb{Z}_q$ and $a_j(n) = 0$ for all sufficiently large j be the digit expansion of n in base q. With n we associate the column vector $\mathbf{n} = (\eta(a_j(n)))_{j=1}^\infty \in \mathbb{F}_q^\infty$, where $\eta : \mathbb{Z}_q \to \mathbb{F}_q$ is a given bijection with $\eta(0) = 0$. Now we define the sequence

$$\mathbf{x}_n = \left(\Psi_\infty(C^{(1)}\mathbf{n}),\dots,\Psi_\infty(C^{(s)}\mathbf{n})\right) \in [0,1]^s \quad \text{for } n = 0,1,\dots. \tag{3.5}$$

Note that the matrix-vector products $C^{(i)}\mathbf{n}$, $1 \leq i \leq s$, are meaningful since $\mathbf{n}$ has only finitely many nonzero coordinates.

Definition 3.7. The sequence S in (3.5) is called a *digital sequence over* $\mathbb{F}_q$. If S forms a (t,s)-sequence in base q and $(\mathbf{T},s)$-sequence in base q, then S is called a *digital*

(t, s)-sequence over $\mathbb{F}_q$ and *digital $(\mathbf{T}, s)$-sequence over $\mathbb{F}_q$*, respectively. The matrices $C^{(1)}, \ldots, C^{(s)}$ are called the *generating matrices* of S.

Example 3.8. Let $s = 1$ and let q be a prime. Then ψ in (3.4) can be taken as an identity map. Let $C^{(1)}$ be the $\infty \times \infty$ identity matrix over $\mathbb{F}_q$. Then the van der Corput sequence in Example 2.4 is a digital $(0, 1)$-sequence over $\mathbb{F}_q$ with generating matrix $C^{(1)}$.

The following analog of Theorem 3.4 was shown in [17].

Theorem 3.9. *Let S be a digital sequence over $\mathbb{F}_q$ with generating matrices $C^{(1)}, \ldots,$ $C^{(s)}$. For $1 \le i \le s$ and $m \in \mathbb{N}$, let $C_m^{(i)}$ denote the left upper $m \times m$ submatrix of $C^{(i)}$. Then S is a digital $(\mathbf{T}, s)$-sequence over $\mathbb{F}_q$ with*

$$\mathbf{T}(m) = m - \varrho(C_m^{(1)}, \ldots, C_m^{(s)}) \quad \textit{for all } m \in \mathbb{N}\,,$$

where the function ϱ is given by Definition 3.3.

The construction of (t, m, s)-nets using the points (3.3) has an analog for sequences. We refer to [37] for the details.

4 The Combinatorics of Nets

There is an interesting combinatorial angle to the theory of (t, m, s)-nets which leads, in particular, to combinatorial obstructions to the existence of (t, m, s)-nets with certain parameters. This combinatorial connection was first observed in Niederreiter [27]. As usual in this article, we focus on the case of a prime-power base q.

A (t, m, s)-net in base q always exists for $m = t$ since, as already noted in Section 2, any point set consisting of q^t points in $[0, 1)^s$ is a (t, t, s)-net in base q. Furthermore, the point set consisting of the points

$$\left(\frac{n}{q}, \ldots, \frac{n}{q}\right) \in [0, 1)^s\,, \quad n = 0, 1, \ldots, q - 1\,,$$

each taken with multiplicity q^t, is a $(t, t + 1, s)$-net in base q. Thus, in investigations concerning the existence of (t, m, s)-nets in base q, we can always assume that $m \ge t + 2$.

For the case $m = t + 2$, it was shown in [31] that the existence of a $(t, t + 2, s)$-net in base q is equivalent to the existence of an orthogonal array with certain parameters. In the general case, one has to use a combinatorial structure called an ordered orthogonal array; see [18] and [23] for Definition 4.1 and Theorem 4.2 below.

Definition 4.1. Let s, k, T, λ be positive integers with $sT \ge k$ and let q be a prime power. An *ordered orthogonal array* $\mathrm{OOA}_q(s, k, T, \lambda)$ is a $(\lambda q^k) \times (sT)$ matrix over $\mathbb{F}_q$ with column labels (i, j) for $1 \le i \le s$ and $1 \le j \le T$ such that, for any integers

$0 \le d_1, \dots, d_s \le T$ with $\sum_{i=1}^{s} d_i = k$, the $(\lambda q^k) \times k$ submatrix obtained by restricting to the columns (i, j), $1 \le j \le d_i$, $1 \le i \le s$, contains among its rows every vector from $\mathbb{F}_q^k$ with the same frequency λ.

Theorem 4.2. *Let $s \ge 2$, $k \ge 2$, and $t \ge 0$ be integers and let q be a prime power. Then there exists a $(t, t+k, s)$-net in base q if and only if there exists an ordered orthogonal array $OOA_q(s, k, k-1, q^t)$.*

Corollary 4.3 (Niederreiter [27]). *For $m \ge 2$, a $(0, m, s)$-net in a prime-power base q exists if and only if $s \le q + 1$.*

The following simple principle from [27] allows a transfer of results between nets and (t, s)-sequences.

Lemma 4.4. *Given a (t, s)-sequence in any base $b \ge 2$, we can construct a $(t, m, s+1)$-net in base b for any integer $m \ge t$.*

Corollary 4.5 (Niederreiter [27]). *A $(0, s)$-sequence in a prime-power base q exists if and only if $s \le q$.*

It was shown in Niederreiter [28] that for any integers $b \ge 2$ and $s \ge 1$ there exists a (t, s)-sequence in base b for some value of t. For any prime-power base q, we can thus define $t_q(s)$ as the least value of t for which there exists a (t, s)-sequence in base q. It follows from Corollary 4.5 that $t_q(s) \ge 1$ for $s > q$. Indeed, combinatorial constraints imply that $t_q(s)$ grows at least linearly as a function of s for $s > q$, according to Niederreiter and Xing [43]. The currently best implied constant is given in the following result of Schürer [52].

Theorem 4.6. *For any prime power q and any integer $s \ge 1$, we have*

$$t_q(s) \ge \frac{s}{q-1} - c_q \log(s+1)$$

with an effective constant $c_q > 0$ depending only on q.

It is another consequence of the paper [28] that for any prime power q and any integer $s \ge 1$, there exists a *digital* (t, s)-sequence over $\mathbb{F}_q$ for some value of t. In analogy with $t_q(s)$ above, we define $d_q(s)$ as the least value of t for which there exists a digital (t, s)-sequence over $\mathbb{F}_q$. We have $d_q(s) = 0$ for $1 \le s \le q$. It is trivial that $t_q(s) \le d_q(s)$, and so Theorem 4.6 provides also a lower bound on $d_q(s)$. The fact that $d_q(s)$ grows at least linearly as a function of s for $s > q$ can be shown also by elementary means (see [45, Theorem 8.2.16]). It is an open problem whether we can ever have $t_q(s) < d_q(s)$. Extensive numerical data on lower and upper bounds for $t_q(s)$ can be found at the website `http://mint.sbg.ac.at`.

5 Duality Theory

The original approach to the construction of digital (t, m, s)-nets over $\mathbb{F}_q$ operates by means of their generating matrices (see Section 3), but there is an alternative approach which uses certain subspaces of the vector space $\mathbb{F}_q^{ms}$. The latter approach proceeds by the duality theory of Niederreiter and Pirsic [41] and establishes a close link between digital nets and linear error-correcting codes.

A crucial ingredient of this duality theory is the Niederreiter–Rosenbloom–Tsfasman weight, or NRT weight for short. It is named after the work of Niederreiter [25] and Rosenbloom and Tsfasman [51] and defined as follows.

Definition 5.1. Let $m \geq 1$ and $s \geq 1$ be integers. Put $v_m(\mathbf{a}) = 0$ if $\mathbf{a} = \mathbf{0} \in \mathbb{F}_q^m$, and for $\mathbf{a} = (a_1, \ldots, a_m) \in \mathbb{F}_q^m$ with $\mathbf{a} \neq \mathbf{0}$ let $v_m(\mathbf{a})$ be the largest value of j such that $a_j \neq 0$. Write a vector $\mathbf{A} \in \mathbb{F}_q^{ms}$ as the concatenation of s vectors of length m, i.e., $\mathbf{A} = (\mathbf{a}^{(1)}, \ldots, \mathbf{a}^{(s)}) \in \mathbb{F}_q^{ms}$ with $\mathbf{a}^{(i)} \in \mathbb{F}_q^m$ for $1 \leq i \leq s$. Then the *NRT weight* $V_m(\mathbf{A})$ of $\mathbf{A}$ is defined by

$$V_m(\mathbf{A}) = \sum_{i=1}^{s} v_m(\mathbf{a}^{(i)}) .$$

Remark 5.2. For $m = 1$ the NRT weight reduces to the Hamming weight on $\mathbb{F}_q^s$ which is basic in the theory of error-correcting codes.

Definition 5.3. The *minimum distance* $\delta_m(\mathcal{N})$ of a nonzero $\mathbb{F}_q$-linear subspace $\mathcal{N}$ of $\mathbb{F}_q^{ms}$ is given by

$$\delta_m(\mathcal{N}) = \min_{\mathbf{A} \in \mathcal{N} \setminus \{\mathbf{0}\}} V_m(\mathbf{A}) .$$

Now let P be a digital net with $m \times m$ generating matrices $C^{(1)}, \ldots, C^{(s)}$ over $\mathbb{F}_q$. We set up an $m \times ms$ matrix $M = M(C^{(1)}, \ldots, C^{(s)})$ over $\mathbb{F}_q$ as follows: for $1 \leq j \leq m$, the j-th row of M is obtained by concatenating the transposes of the j-th columns of $C^{(1)}, \ldots, C^{(s)}$. Let $\mathcal{M} \subseteq \mathbb{F}_q^{ms}$ be the row space of M and let $\mathcal{M}^{\perp}$ be its dual space, i.e.,

$$\mathcal{M}^{\perp} = \{\mathbf{A} \in \mathbb{F}_q^{ms} : \mathbf{A} \cdot \mathbf{M} = 0 \ \text{ for all } \ \mathbf{M} \in \mathcal{M}\} ,$$

where $\cdot$ denotes the standard inner product in $\mathbb{F}_q^{ms}$. Since the case of the dimension $s = 1$ is trivial for the construction of digital nets (it is obvious that the optimal quality parameter $t = 0$ can always be achieved in this case), we can assume that $s \geq 2$.

Theorem 5.4 (Niederreiter and Pirsic [41]). *Let $m \geq 1$ and $s \geq 2$ be integers. Then the point set P above is a digital (t, m, s)-net over $\mathbb{F}_q$ with quality parameter*

$$t = m + 1 - \delta_m(\mathcal{M}^{\perp}) .$$

Proof. In view of Theorem 3.4, it suffices to prove that

$$\varrho(C^{(1)}, \ldots, C^{(s)}) \geq \delta_m(\mathcal{M}^{\perp}) - 1 .$$

This inequality will follow if we can show that for any integers $0 \le d_1, \dots, d_s \le m$ with

$$\sum_{i=1}^{s} d_i = \delta_m(\mathcal{M}^\perp) - 1\,, \tag{5.1}$$

the row vectors $\mathbf{c}_j^{(i)}$, $1 \le j \le d_i$, $1 \le i \le s$, of the generating matrices are linearly independent over $\mathbb{F}_q$ (compare with Definition 3.3). Suppose, on the contrary, that these vectors were linearly dependent over $\mathbb{F}_q$, that is, that there exist coefficients $a_j^{(i)} \in \mathbb{F}_q$, not all 0, such that

$$\sum_{i=1}^{s} \sum_{j=1}^{d_i} a_j^{(i)} \mathbf{c}_j^{(i)} = \mathbf{0} \in \mathbb{F}_q^m\,.$$

Define $a_j^{(i)} = 0$ for $d_i < j \le m$, $1 \le i \le s$. Then

$$\sum_{i=1}^{s} \sum_{j=1}^{m} a_j^{(i)} \mathbf{c}_j^{(i)} = \mathbf{0} \in \mathbb{F}_q^m\,. \tag{5.2}$$

With $M = M(C^{(1)}, \dots, C^{(s)})$, the identity (5.2) can be written as

$$M\mathbf{A}^{\mathrm{T}} = \mathbf{0} \in \mathbb{F}_q^m\,, \tag{5.3}$$

where $\mathbf{A} = (\mathbf{a}^{(1)}, \dots, \mathbf{a}^{(s)}) \in \mathbb{F}_q^{ms}$ with

$$\mathbf{a}^{(i)} = (a_1^{(i)}, \dots, a_m^{(i)}) \in \mathbb{F}_q^m \quad \text{for } 1 \le i \le s\,.$$

The identity (5.3) shows that $\mathbf{A} \in \mathcal{M}^\perp$. Note that $\mathbf{A} \ne \mathbf{0} \in \mathbb{F}_q^{ms}$ and $v_m(\mathbf{a}^{(i)}) \le d_i$ for $1 \le i \le s$ by the definition of the $a_j^{(i)}$. Taking into account (5.1), we obtain

$$V_m(\mathbf{A}) = \sum_{i=1}^{s} v_m(\mathbf{a}^{(i)}) \le \sum_{i=1}^{s} d_i = \delta_m(\mathcal{M}^\perp) - 1\,,$$

which is a contradiction to the definition of $\delta_m(\mathcal{M}^\perp)$. □

We note that by construction $\dim(\mathcal{M}) \le m$, and so $\dim(\mathcal{M}^\perp) \ge ms - m$. This observation explains the dimensionality condition in the following straightforward consequence of Theorem 5.4.

Corollary 5.5 (Niederreiter and Pirsic [41]). *Let $m \ge 1$ and $s \ge 2$ be integers. Then from any $\mathbb{F}_q$-linear subspace $\mathcal{N}$ of $\mathbb{F}_q^{ms}$ with $\dim(\mathcal{N}) \ge ms - m$ we can construct a digital (t, m, s)-net over $\mathbb{F}_q$ with*

$$t = m + 1 - \delta_m(\mathcal{N})\,.$$

Corollary 5.5 serves as an important tool for the construction of good digital nets, as will be illustrated in Section 6.

An analog of the duality theory for digital nets was developed by Dick and Niederreiter [5] for the case of digital $(\mathbf{T}, s)$-sequences. Instead of a single subspace $\mathcal{N}$ of $\mathbb{F}_q^{ms}$ as in Corollary 5.5, we now need a sequence of subspaces satisfying a certain compatibility condition. The case $s = 1$ is again trivial (let the generating matrix $C^{(1)}$ be the $\infty \times \infty$ identity matrix over $\mathbb{F}_q$ to get the optimal quality parameter $t = 0$), and so we can assume that $s \geq 2$.

For each integer $m \geq 1$, let $\mathcal{N}_m$ be an $\mathbb{F}_q$-linear subspace of $\mathbb{F}_q^{ms}$ with $\dim(\mathcal{N}_m) \geq ms - m$. For $\mathbf{A} \in \mathbb{F}_q^{(m+1)s}$ we write $\mathbf{A} = (\mathbf{a}^{(1)}, \ldots, \mathbf{a}^{(s)})$ with $\mathbf{a}^{(i)} \in \mathbb{F}_q^{m+1}$ for $1 \leq i \leq s$. We put

$$\mathcal{A}_{m+1} = \{\mathbf{A} \in \mathcal{N}_{m+1} : v_{m+1}(\mathbf{a}^{(i)}) \leq m \text{ for } 1 \leq i \leq s\},$$

where v_{m+1} is given by Definition 5.1. Let $\mathcal{N}_{m+1,m} \subseteq \mathbb{F}_q^{ms}$ be the projection of $\mathcal{A}_{m+1}$ on the first m coordinates of each $\mathbf{a}^{(i)}$ for $1 \leq i \leq s$.

Definition 5.6. The sequence $(\mathcal{N}_m)_{m \geq 1}$ of $\mathbb{F}_q$-linear spaces is called a *dual space chain* over $\mathbb{F}_q$ if $\mathcal{N}_{m+1,m}$ is an $\mathbb{F}_q$-linear subspace of $\mathcal{N}_m$ with $\dim(\mathcal{N}_{m+1,m}) \geq \dim(\mathcal{N}_m) - 1$ for all $m \geq 1$.

Theorem 5.7 (Dick and Niederreiter [5]). *Let $s \geq 2$ be an integer. Then from any dual space chain $(\mathcal{N}_m)_{m \geq 1}$ over $\mathbb{F}_q$ we can construct a digital $(\mathbf{T}, s)$-sequence over $\mathbb{F}_q$ with*

$$\mathbf{T}(m) = m + 1 - \delta_m(\mathcal{N}_m) \quad \textit{for all } m \geq 1.$$

6 Special Constructions of Nets

The general principle enunciated in Lemma 4.4 reduces the problem of constructing $(t, m, s+1)$-nets to that of the construction of (t, s)-sequences. Since many concrete (t, s)-sequences are available (see Section 7), Lemma 4.4 yields a wealth of good nets. In the present section, we describe constructions of nets that are not derived from Lemma 4.4.

6.1 Polynomial Lattices

Let $\mathbb{F}_q((X^{-1}))$ be the field of formal Laurent series over $\mathbb{F}_q$ in the variable X^{-1}. Note that $\mathbb{F}_q((X^{-1}))$ contains the rational function field $\mathbb{F}_q(X)$ as a subfield. Let $m \geq 1$ and $s \geq 2$ be integers. Choose a polynomial $f \in \mathbb{F}_q[X]$ with $\deg(f) = m$ and an s-tuple $\mathbf{g} = (g_1, \ldots, g_s) \in \mathbb{F}_q[X]^s$ of polynomials with $\deg(g_i) < m$ for $1 \leq i \leq s$. Then expand each rational function $g_i(X)/f(X)$, $1 \leq i \leq s$, into a formal Laurent series

$$\frac{g_i(X)}{f(X)} = \sum_{k=1}^{\infty} u_k^{(i)} X^{-k} \in \mathbb{F}_q((X^{-1})) . \tag{6.1}$$

For $1 \le i \le s$, we set up the $m \times m$ matrix $C^{(i)} = (c_{jr}^{(i)})$ by putting

$$c_{jr}^{(i)} = u_{j+r}^{(i)} \in \mathbb{F}_q \quad \text{for } 1 \le j \le m\,, \;\; 0 \le r \le m-1\,.$$

A digital net over $\mathbb{F}_q$ with generating matrices $C^{(1)}, \dots, C^{(s)}$ is denoted by $P(\mathbf{g}, f)$ and called a *polynomial lattice*. In the case where f is irreducible over $\mathbb{F}_q$, this construction is equivalent to that in Niederreiter [25], and the general case was introduced in Niederreiter [30].

For the determination of the quality parameter of $P(\mathbf{g}, f)$, we proceed as in [30] and use the following definition which is instrumental in Theorem 6.2 below.

Definition 6.1. Let $s \ge 2$ and let $f \in \mathbb{F}_q[X]$ and $\mathbf{g} \in \mathbb{F}_q[X]^s$ be as above. Then the *figure of merit* $\varrho(\mathbf{g}, f)$ is defined by

$$\varrho(\mathbf{g}, f) = s - 1 + \min_{\mathbf{h}} \sum_{i=1}^{s} \deg(h_i)\,,$$

where the minimum is extended over all nonzero s-tuples $\mathbf{h} = (h_1, \dots, h_s) \in \mathbb{F}_q[X]^s$ with $\deg(h_i) < m$ for $1 \le i \le s$ and f dividing $\sum_{i=1}^{s} h_i g_i$ in $\mathbb{F}_q[X]$. Here we use the convention $\deg(0) = -1$.

Theorem 6.2 (Niederreiter [30]). *For $s \ge 2$, any polynomial lattice $P(\mathbf{g}, f)$ is a digital (t, m, s)-net over $\mathbb{F}_q$ with*

$$t = m - \varrho(\mathbf{g}, f)\,.$$

It is clear from Theorem 6.2 that in order to obtain a good (t, m, s)-net by this construction, i.e., a net with a small value of t, we need to find $\mathbf{g}$ and f with a large figure of merit $\varrho(\mathbf{g}, f)$. Depending on the size of m, s, and q, this can be done by exhaustive search or by random search, as e.g. in [9] and [16].

For $s = 2$ there is an explicit formula for $\varrho(\mathbf{g}, f)$ in terms of continued fractions. Without a serious loss of generality, we can assume that $g_1 = 1$ and $\gcd(g_2, f) = 1$. Then let

$$\frac{g_2}{f} = 1/(A_1 + 1/(A_2 + \cdots + 1/A_K))$$

be the continued fraction expansion of the rational function g_2/f, with partial quotients $A_k \in \mathbb{F}_q[X]$ satisfying $\deg(A_k) \ge 1$ for $1 \le k \le K$. Then for $\mathbf{g} = (g_1, g_2) = (1, g_2)$ we have

$$\varrho(\mathbf{g}, f) = m + 1 - \max_{1 \le k \le K} \deg(A_k)\,.$$

In particular, if g_2 and f are selected in such a way that $\deg(A_k) = 1$ for $1 \le k \le K$, then Theorem 6.2 implies that $P(\mathbf{g}, f)$ is a $(0, m, 2)$-net in base q. A convenient choice for implementations is a monomial $f(X) = X^m$ since then the formal Laurent series expansions in (6.1) are obtained trivially. For $f(X) = X^m \in \mathbb{F}_q[X]$, it was shown in [26] that there is always a choice of $g_2 \in \mathbb{F}_q[X]$ with $\gcd(g_2, f) = 1$ such that $\deg(A_k) = 1$ for $1 \le k \le K$ in the continued fraction expansion of $g_2(X)/X^m$.

If $s \ge 2$ and $f \in \mathbb{F}_q[X]$ with $\deg(f) = m \ge 1$ are fixed, then one may ask for choices of $\mathbf{g}$ such that the discrepancy of a resulting polynomial lattice $P(\mathbf{g}, f)$ becomes small. In the case where q is prime and $\mathbf{g}$ runs through all s-tuples $(g_1, \dots, g_s) \in \mathbb{F}_q[X]^s$ with $\gcd(g_i, f) = 1$ and $\deg(g_i) < m$ for $1 \le i \le s$, it was proved in [30] that on the average we get the discrepancy bound $D_N = O(N^{-1}(\log N)^s)$. In the special case $f(X) = X^m$, Larcher [15] showed that there exists a choice of $\mathbf{g}$ for which the discrepancy of $P(\mathbf{g}, f)$ satisfies $D_N = O\left(N^{-1}(\log N)^{s-1} \log\log(N+1)\right)$. An existence theorem with the same discrepancy bound was recently shown by Kritzer and Pillichshammer [14] for the case where q is a prime and where $f \in \mathbb{F}_q[X]$ has a positive degree and satisfies $f(0) \neq 0$.

A systematic method for the explicit construction of good polynomial lattices is the *component-by-component algorithm*, so named because the components of $\mathbf{g} = (g_1, \dots, g_s)$ are computed one by one in a recursive manner. Let $s \ge 2$ and $f \in \mathbb{F}_q[X]$ with $\deg(f) \ge 1$ be fixed. Start by choosing $g_1 = 1$. For $d \ge 2$, assume that $g_1, \dots, g_{d-1}$ have already been constructed. Then determine g_d which optimizes an *a priori* chosen quality measure for polynomial lattices $P((g_1, \dots, g_{d-1}, g_d), f)$. The algorithm ends when the s-tuple $(g_1, \dots, g_s)$ has been constructed. It was shown in [3] that, with q prime and a suitable choice of the quality measure, the component-by-component algorithm yields an s-dimensional polynomial lattice $P(\mathbf{g}, f)$ with discrepancy $D_N = O(N^{-1}(\log N)^s)$. Details on the implementation of this algorithm and further results on polynomial lattices can be found in [6, Chapter 10]. For a recent generalization of the component-by-component algorithm to so-called higher-order polynomial lattices, we refer to [1].

6.2 Hyperplane Nets

We now describe a general construction method based on the duality theory for digital nets, in particular Corollary 5.5. This method was introduced by Pirsic, Dick, and Pillichshammer [50]. Let $m \ge 1$ and $s \ge 2$ be integers and let $\mathbb{F}_{q^m}$ be the extension field of $\mathbb{F}_q$ of degree m. We can consider $\mathcal{Q} = \mathbb{F}_{q^m}^s$ as an ms-dimensional vector space over $\mathbb{F}_q$. Fix $\boldsymbol{\alpha} \in \mathcal{Q}$ with $\boldsymbol{\alpha} \neq \mathbf{0}$ and put

$$\mathcal{Q}_{\boldsymbol{\alpha}} = \{\boldsymbol{\beta} \in \mathcal{Q} : \boldsymbol{\alpha} \cdot \boldsymbol{\beta} = 0\},$$

where $\cdot$ is the standard inner product in $\mathcal{Q}$. Then $\mathcal{Q}_{\boldsymbol{\alpha}}$ is an $\mathbb{F}_q$-linear subspace of $\mathcal{Q}$ of dimension $ms - m$. Let $\sigma : \mathcal{Q} \to \mathbb{F}_q^{ms}$ be an isomorphism between vector spaces over $\mathbb{F}_q$ and let $\mathcal{N} = \sigma(\mathcal{Q}_{\boldsymbol{\alpha}}) \subseteq \mathbb{F}_q^{ms}$ be the image of $\mathcal{Q}_{\boldsymbol{\alpha}}$ under σ. Then $\dim(\mathcal{N}) =$

$ms - m$ as a vector space over $\mathbb{F}_q$. Hence, Corollary 5.5 yields a digital (t, m, s)-net over $\mathbb{F}_q$ which is called a *hyperplane net* over $\mathbb{F}_q$.

It was proved by Pirsic [49] that hyperplane nets include polynomial lattices as special cases. In analogy with Theorem 6.2, the quality parameter t of a hyperplane (t, m, s)-net over $\mathbb{F}_q$ is given by $t = m - \varrho(\boldsymbol{\alpha})$, where $\varrho(\boldsymbol{\alpha})$ is obtained by a suitable generalization of Definition 6.1 (see [47] and [6, Section 11.2]). Another result in [47] shows that for sufficiently large m and any $s \geq 2$ and q, there always exists a vector $\boldsymbol{\alpha} \in \mathcal{Q} = \mathbb{F}_{q^m}^s$ with

$$\varrho(\boldsymbol{\alpha}) \geq \lfloor m - (s-1)(\log_q m - 1) + \log_q (s-1)! \rfloor .$$

The component-by-component algorithm for polynomial lattices (see Section 6.1) was generalized to hyperplane nets by Pillichshammer and Pirsic [48], and this yields a hyperplane (t, m, s)-net over $\mathbb{F}_q$ with discrepancy $D_N = O(N^{-1}(\log N)^s)$, where $m \geq 1$, $s \geq 2$, and the prime power q are arbitrary.

An interesting special case of hyperplane nets is obtained by considering vectors $\boldsymbol{\alpha} \in \mathcal{Q} = \mathbb{F}_{q^m}^s$ of the form $\boldsymbol{\alpha} = (1, \alpha, \alpha^2, \ldots, \alpha^{s-1})$ with $\alpha \in \mathbb{F}_{q^m}$. Such a special hyperplane net is called a *cyclic digital net* over $\mathbb{F}_q$. The original definition of cyclic digital nets was given in an equivalent form by Niederreiter [35] and operates as follows. For integers $m \geq 1$ and $s \geq 2$, consider

$$\mathcal{P} = \{f \in \mathbb{F}_{q^m}[X] : \deg(f) < s\}$$

as a vector space over $\mathbb{F}_q$. Fix $\alpha \in \mathbb{F}_{q^m}$ and define the $\mathbb{F}_q$-linear subspace $\mathcal{P}_\alpha = \{f \in \mathcal{P} : f(\alpha) = 0\}$ of $\mathcal{P}$. Set up an $\mathbb{F}_q$-linear transformation $\tau : \mathcal{P} \to \mathbb{F}_q^{ms}$ in the following way. Write $f \in \mathcal{P}$ explicitly as

$$f(X) = \sum_{i=1}^{s} y_i X^{i-1} \quad \text{with } y_i \in \mathbb{F}_{q^m} \text{ for } 1 \leq i \leq s . \tag{6.2}$$

For each $i = 1, \ldots, s$, choose an ordered basis B_i of $\mathbb{F}_{q^m}$ over $\mathbb{F}_q$ and let $\mathbf{c}_i(f)$ be the coordinate vector of y_i in (6.2) with respect to B_i. Then define

$$\tau(f) = (\mathbf{c}_1(f), \ldots, \mathbf{c}_s(f)) \in \mathbb{F}_q^{ms} \quad \text{for all } f \in \mathcal{P} .$$

Let $\mathcal{N} = \tau(\mathcal{P}_\alpha) \subseteq \mathbb{F}_q^{ms}$ be the image of $\mathcal{P}_\alpha$ under τ. Then $\dim(\mathcal{N}) = ms - m$ as a vector space over $\mathbb{F}_q$, and so Corollary 5.5 yields a digital (t, m, s)-net over $\mathbb{F}_q$ which is in fact a cyclic digital net. For results concerning the existence of cyclic digital (t, m, s)-nets with a small quality parameter t or small discrepancy, we refer to [47] and [6, Chapter 11].

Remark 6.3. The construction of hyperplane nets suggests the following alternative approach to the general construction of digital nets over $\mathbb{F}_q$. Let $m \geq 1$ and $s \geq 2$ be integers. Choose an ms-tuple $B = (\beta_j^{(i)})_{1 \leq i \leq s,\, 1 \leq j \leq m}$ of elements of $\mathbb{F}_{q^m}$ which may be arranged also as an $s \times m$ matrix over $\mathbb{F}_{q^m}$. Consider the $\mathbb{F}_q$-linear transformation $L_B : \mathbb{F}_q^{ms} \to \mathbb{F}_{q^m}$ defined by

$$L_B(a_1^{(1)}, \dots, a_m^{(1)}, \dots, a_1^{(s)}, \dots, a_m^{(s)}) = \sum_{i=1}^{s} \sum_{j=1}^{m} a_j^{(i)} \beta_j^{(i)}$$

for all $a_j^{(i)} \in \mathbb{F}_q$, $1 \le i \le s$, $1 \le j \le m$. Let $\mathcal{N}_B \subseteq \mathbb{F}_q^{ms}$ be the kernel of L_B. Then $\mathcal{N}_B$ is an $\mathbb{F}_q$-linear subspace of $\mathbb{F}_q^{ms}$ with

$$\dim(\mathcal{N}_B) = ms - \operatorname{rank}(L_B) \ge ms - m .$$

Thus, by Corollary 5.5 we get a digital (t, m, s)-net over $\mathbb{F}_q$ with $t = m + 1 - \delta_m(\mathcal{N}_B)$. This construction is equivalent to the general construction of digital nets using generating matrices (see Section 3).

Remark 6.4. Let $\omega_1, \dots, \omega_m$ be an ordered basis of $\mathbb{F}_{q^m}$ over $\mathbb{F}_q$. Choose $\alpha_1, \dots, \alpha_s \in \mathbb{F}_{q^m}$ not all 0 and put $\beta_j^{(i)} = \alpha_i \omega_j$ for $1 \le i \le s$, $1 \le j \le m$. Then it is easily seen that in this special case the construction in Remark 6.3 is equivalent to the construction of a hyperplane net with $\boldsymbol{\alpha} = (\alpha_1, \dots, \alpha_s)$.

6.3 Nets Obtained from Global Function Fields

Powerful constructions of (t, m, s)-nets and $(\mathbf{T}, s)$-sequences are based on global function fields. Applications of global function fields to the construction of $(\mathbf{T}, s)$-sequences will be discussed in Section 7.2. Here we present a construction of digital (t, m, s)-nets using global function fields which is due to Niederreiter and Özbudak [39].

A *global function field* F over $\mathbb{F}_q$ is an algebraic function field of one variable with constant field $\mathbb{F}_q$, i.e., F is a finite extension of the rational function field over $\mathbb{F}_q$. We assume without loss of generality that $\mathbb{F}_q$ is the full constant field of F, which means that $\mathbb{F}_q$ is algebraically closed in F. Detailed information on global function fields can be found in the book of Stichtenoth [54].

Let F be a global function field with a full constant field $\mathbb{F}_q$. For a given dimension $s \ge 2$, choose s distinct places $P_1, \dots, P_s$ of F and put $e_i = \deg(P_i)$ for $1 \le i \le s$. For each $i = 1, \dots, s$, let ν_{P_i} be the normalized discrete valuation of F corresponding to P_i, let z_i be a local parameter at P_i, let F_{P_i} be the residue class field of P_i, and let $\psi_i : F_{P_i} \to \mathbb{F}_q^{e_i}$ be an $\mathbb{F}_q$-linear vector space isomorphism. Choose an arbitrary divisor G of F and put

$$a_i = \nu_{P_i}(G) \quad \text{for } 1 \le i \le s .$$

Write $\operatorname{div}(f)$ for the principal divisor of any $f \in F^*$ and let

$$\mathcal{L}(G) = \{f \in F^* : \operatorname{div}(f) + G \ge 0\} \cup \{0\}$$

be the Riemann–Roch space corresponding to G. Note that $\mathcal{L}(G)$ is a finite-dimensional vector space over $\mathbb{F}_q$.

For the moment, we fix i with $1 \le i \le s$. For $f \in \mathcal{L}(G)$ we have $\nu_{P_i}(f) \ge -a_i$, and so the local expansion of f at P_i has the form

$$f = \sum_{j=-a_i}^{\infty} f^{(j)}(P_i) z_i^j ,$$

where all $f^{(j)}(P_i) \in F_{P_i}$. Next we choose an integer $m \ge 1$. Let $m_i \ge 0$ and $0 \le r_i < e_i$ be the unique integers satisfying $m = m_i e_i + r_i$. For $f \in \mathcal{L}(G)$ we then define

$$\mathbf{c}_f^{(i)} = (\underbrace{0, \dots, 0}_{r_i}, \psi_i(f^{(-a_i+m_i-1)}(P_i)), \dots, \psi_i(f^{(-a_i)}(P_i))) \in \mathbb{F}_q^m .$$

Now we introduce the $\mathbb{F}_q$-linear transformation $\theta : \mathcal{L}(G) \to \mathbb{F}_q^{ms}$ by

$$\theta(f) = (\mathbf{c}_f^{(1)}, \dots, \mathbf{c}_f^{(s)}) \in \mathbb{F}_q^{ms} \quad \text{for } f \in \mathcal{L}(G) .$$

The image of θ is denoted by $\mathcal{N}_m(P_1, \dots, P_s; G)$. Note that, in general, the $\mathbb{F}_q$-linear subspace $\mathcal{N}_m(P_1, \dots, P_s; G)$ of $\mathbb{F}_q^{ms}$ depends also on the choice of the local parameters $z_1, \dots, z_s$ and on the choice of the $\mathbb{F}_q$-linear isomorphisms $\psi_1, \dots, \psi_s$, but we suppress this dependence in the notation for the sake of simplicity.

With m and $e_1, \dots, e_s$ as above and an integer $r \ge 0$, we put

$$\delta_m^*(e_1, \dots, e_s; r) = \min_{\mathbf{l}} \sum_{i=1}^{s} \max(0, m - (l_i + 1)e_i + 1) ,$$

where the minimum is extended over all $\mathbf{l} = (l_1, \dots, l_s) \in \mathbb{Z}^s$ with $\sum_{i=1}^{s} l_i e_i \le r$ and $0 \le l_i \le m_i$ for $1 \le i \le s$. According to [39], we have the following information on the vector space $\mathcal{N}_m(P_1, \dots, P_s; G)$.

Lemma 6.5. *Let G be a divisor of the global function field F with full constant field $\mathbb{F}_q$ and assume that* $\dim(\mathcal{L}(G)) \ge 1$ *and* $\deg(G) < ms - \sum_{i=1}^{s} r_i$. *Then the $\mathbb{F}_q$-linear subspace $\mathcal{N} = \mathcal{N}_m(P_1, \dots, P_s; G)$ of $\mathbb{F}_q^{ms}$ has minimum distance*

$$\delta_m(\mathcal{N}) \ge \delta_m^*(e_1, \dots, e_s; \deg(G))$$

and dimension

$$\dim(\mathcal{N}) = \dim(\mathcal{L}(G)) \ge \deg(G) + 1 - g ,$$

where g is the genus of F.

By choosing a divisor G of F with $\deg(G) = ms - m + g - 1$, imposing a suitable condition on m, and applying Corollary 5.5, the following result was derived in [39].

Theorem 6.6. *Let F be a global function field with full constant field $\mathbb{F}_q$ and of genus g. Let $P_1, \dots, P_s$ be $s \geq 2$ distinct places of F with degrees $e_1, \dots, e_s$, respectively. Let $m \geq 1$ be an integer and for $i = 1, \dots, s$ let r_i be the least residue of m modulo e_i. Assume that*

$$m \geq g + \sum_{i=1}^{s} r_i .$$

Then we can obtain a digital (t, m, s)-net over $\mathbb{F}_q$ with

$$t = m + 1 - \delta_m^*(e_1, \dots, e_s; ms - m + g - 1) .$$

Corollary 6.7. *Under the conditions of Theorem 6.6, we can obtain a digital (t, m, s)-net over $\mathbb{F}_q$ with*

$$t = g + \sum_{i=1}^{s} (e_i - 1) .$$

It was also shown in [39] that in some cases, and with a suitable choice of the $\mathbb{F}_q$-linear isomorphisms $\psi_i : F_{P_i} \to \mathbb{F}_q^{e_i}$, one can improve on Theorem 6.6 by getting a t-value that is smaller by 1. Another refinement is obtained in some cases by a good choice of the divisor G, which leads to the following result (see [39] and [46, Theorem 5.7.15]).

Theorem 6.8. *Let $s \geq 2$ be an integer. Let F be a global function field with full constant field $\mathbb{F}_q$ and of genus $g \geq 1$ such that F has at least s places of degree 1. If k and m are integers with $0 \leq k \leq g - 1$ and $m \geq \max(1, g - k - 1)$, then there exists a digital $(g - k - 1, m, s)$-net over $\mathbb{F}_q$ provided that*

$$\binom{s + m + k - g}{s - 1} A_k(F) < h(F) ,$$

where $A_k(F)$ is the number of positive divisors of F of degree k and $h(F)$ is the divisor class number of F.

Example 6.9. Let $q = 9$ and let F be the Hermitian function field over $\mathbb{F}_9$, that is, F is the finite extension of the rational function field $\mathbb{F}_9(X)$ defined by $F = \mathbb{F}_9(X, Y)$ with $Y^3 + Y = X^4$. Then $g = 3$, $h(F) = 4096$, and F has 28 places of degree 1. We apply Theorem 6.8 with $s = 28$, $k = 0$, $m = 5$, and we obtain a digital $(2, 5, 28)$-net over $\mathbb{F}_9$. The value $t = 2$ is the currently best value of the quality parameter for a $(t, 5, 28)$-net in base 9, according to the website `http://mint.sbg.ac.at` which contains an extensive database for parameters of (t, m, s)-nets.

7 Special Constructions of $(\mathbf{T}, s)$-Sequences

Finite fields are the most important tool for the construction of (t, s)-sequences and more generally of $(\mathbf{T}, s)$-sequences. All known constructions of good $(\mathbf{T}, s)$-sequences are based on the digital method described in Section 3.

7.1 Faure Sequences and Niederreiter Sequences

Chronologically the first constructions of digital (t, s)-sequences over $\mathbb{F}_q$ were those of Sobol' [53] (for $q = 2$), Faure [7] (for primes $q \geq s$), Niederreiter [27] (for prime powers $q \geq s$), and Niederreiter [28] (for arbitrary q and s). The first three constructions are special cases of the fourth construction, and so we describe only the latter construction.

For a given dimension $s \geq 1$, choose pairwise coprime polynomials $p_1, \ldots, p_s \in \mathbb{F}_q[X]$. Let $e_i = \deg(p_i) \geq 1$ for $1 \leq i \leq s$. Furthermore, choose polynomials $g_{iu} \in \mathbb{F}_q[X]$ with $\gcd(g_{iu}, p_i) = 1$ for $1 \leq i \leq s$ and $u \geq 1$. For integers $1 \leq i \leq s$, $u \geq 1$, and $0 \leq k < e_i$, consider the formal Laurent series expansion

$$\frac{X^k g_{iu}(X)}{p_i(X)^u} = \sum_{r=w}^{\infty} a^{(i)}(u, k, r) X^{-r-1} \tag{7.1}$$

with all $a^{(i)}(u, k, r) \in \mathbb{F}_q$ and an integer $w \leq 0$ which may depend on i, u, and k. Then define

$$c_{jr}^{(i)} = a^{(i)}(Q + 1, k, r) \quad \text{for } 1 \leq i \leq s, \ j \geq 1, \ r \geq 0,$$

where $j - 1 = Qe_i + k$ with integers $Q = Q(i, j)$ and $k = k(i, j)$ satisfying $0 \leq k < e_i$. For $1 \leq i \leq s$, set up the $\infty \times \infty$ matrix $C^{(i)} = (c_{jr}^{(i)})_{j \geq 1, r \geq 0}$. A digital sequence over $\mathbb{F}_q$ with generating matrices $C^{(1)}, \ldots, C^{(s)}$ is called a *Niederreiter sequence*.

Theorem 7.1 (Niederreiter [28]). *Any Niederreiter sequence based on the pairwise coprime nonconstant polynomials $p_1, \ldots, p_s \in \mathbb{F}_q[X]$ is a digital (t, s)-sequence over $\mathbb{F}_q$ with*

$$t = \sum_{i=1}^{s} (\deg(p_i) - 1) . \tag{7.2}$$

It was shown by Dick and Niederreiter [4] that the value of t in (7.2) is the least possible value of the quality parameter for a Niederreiter sequence.

In the special case where q is a prime, $1 \leq s \leq q$, and $p_i(X) = X - i + 1 \in \mathbb{F}_q[X]$ for $1 \leq i \leq s$, we obtain a digital $(0, s)$-sequence over $\mathbb{F}_q$ called a *Faure sequence*. The original construction by Faure [7] was given in terms of Pascal matrices, and it was pointed out in [32, Remark 4.52] that Faure sequences are special cases of Niederreiter sequences. An analogous construction of digital $(0, s)$-sequences over $\mathbb{F}_q$ using Pascal matrices was introduced for arbitrary prime powers q and dimensions $1 \leq s \leq q$ in Niederreiter [27]. This construction corresponds again to the choice of distinct monic linear polynomials $p_1, \ldots, p_s$ over $\mathbb{F}_q$ in Theorem 7.1. In view of Corollary 4.5, the condition $s \leq q$ is the best possible if one wants to obtain a $(0, s)$-sequence in base q.

A somewhat more general form of the rational functions in (7.1) was considered by Tezuka [55], [56, Section 6.1.2]. The denominators $p_i(X)^u$ in (7.1) are the same, whereas the numerators are more general but have to satisfy a certain linear independence condition. The resulting sequences are still digital (t, s)-sequences over $\mathbb{F}_q$ with t as in (7.2). A sequence of this type is called a *generalized Niederreiter sequence*. The special case of a generalized Niederreiter sequence where $q = 2$, $s \geq 1$ is an arbitrary dimension, $p_1(X) = X \in \mathbb{F}_2[X]$, and $p_2, \dots, p_s$ are distinct primitive polynomials over $\mathbb{F}_2$ is called a *Sobol' sequence* (see [53]).

If a prime power q and a dimension $s \geq 1$ are given, then the value of the quality parameter t in (7.2) is minimized by letting $p_1, \dots, p_s$ be s distinct monic irreducible polynomials over $\mathbb{F}_q$ of least degree. If with this choice, we put

$$T_q(s) = \sum_{i=1}^{s} (\deg(p_i) - 1) , \tag{7.3}$$

then this quantity depends only on q and s and is independent of the specific choice of $p_1, \dots, p_s$. It is easily seen that for fixed q the quantity $T_q(s)$ is of the order of magnitude $s \log s$ as $s \to \infty$ (compare with [28]). Let $U(s)$ denote the least value of t that is known to be achievable by Sobol' sequences for given s (recall that Sobol' sequences were constructed only for $q = 2$). Then $T_2(s) = U(s)$ for $1 \leq s \leq 7$ and $T_2(s) < U(s)$ for all $s \geq 8$.

7.2 Sequences Obtained from Global Function Fields

We have seen in Section 6.3 that global function fields are instrumental in the construction of (t, m, s)-nets. Now we discuss constructions of (t, s)-sequences and $(\mathbf{T}, s)$-sequences based on global function fields. In fact, these sequences yield substantial improvements on the Niederreiter sequences in Section 7.1. The idea of using global function fields for the construction of (t, s)-sequences goes back to Niederreiter [33] and the details were worked out in an improved form in several papers of Niederreiter and Xing. Therefore one speaks of the family of *Niederreiter–Xing sequences*. Systematic accounts of all these constructions were presented in [43] and [44]. A detailed exposition of two of these constructions is also available in Chapter 8 of the book [45].

We focus on the construction in [57] since it is the most flexible one. We use the same terminology and notation for global function fields as in Section 6.3. Let F be a global function field with full constant field $\mathbb{F}_q$ and genus g. Assume that F contains at least one place P_∞ of degree 1. Let D be a divisor of F with $\deg(D) = 2g$ and $\nu_{P_\infty}(D) = 0$. Furthermore, choose s distinct places $P_1, \dots, P_s$ of F with $P_i \neq P_\infty$ for $1 \leq i \leq s$. Consider the chain

$$\{0\} = \mathcal{L}(D - (2g+1)P_\infty) \subseteq \mathcal{L}(D - 2gP_\infty) \subseteq \cdots \subseteq \mathcal{L}(D - P_\infty) \subseteq \mathcal{L}(D)$$

of Riemann–Roch spaces. Note that $\dim(\mathcal{L}(D - P_\infty)) = g$ and $\dim(\mathcal{L}(D)) = g + 1$ by the Riemann–Roch theorem. At each step in the chain above, the two successive vector spaces over $\mathbb{F}_q$ are either identical or differ by 1 in dimension. Thus, there exist integers $0 = n_0 < n_1 < \cdots < n_g \leq 2g$ such that $\mathcal{L}(D - (n_u + 1)P_\infty)$ is a proper subspace of $\mathcal{L}(D - n_u P_\infty)$. Hence, we can choose

$$w_u \in \mathcal{L}(D - n_u P_\infty) \setminus \mathcal{L}(D - (n_u + 1)P_\infty) \quad \text{for } 0 \leq u \leq g\,.$$

It is easily seen that $\{w_0, w_1, \ldots, w_g\}$ is a basis of $\mathcal{L}(D)$.

Next, for each $i = 1, \ldots, s$, we consider the infinite chain

$$\mathcal{L}(D) \subset \mathcal{L}(D + P_i) \subset \mathcal{L}(D + 2P_i) \subset \cdots$$

of Riemann–Roch spaces. By starting from the basis $\{w_0, w_1, \ldots, w_g\}$ of $\mathcal{L}(D)$ and successively adding basis vectors at each step of the chain, we obtain for each integer $n \geq 1$ a basis

$$\{w_0, w_1, \ldots, w_g, h_1^{(i)}, h_2^{(i)}, \ldots, h_{n\,\deg(P_i)}^{(i)}\}$$

of $\mathcal{L}(D + nP_i)$. Now let z be a local parameter at P_∞. For $r = 0, 1, \ldots$, we put $z_r = z^r$ if $r \notin \{n_0, n_1, \ldots, n_g\}$ and $z_r = w_u$ if $r = n_u$ for some $u \in \{0, 1, \ldots, g\}$. Each $h_j^{(i)}$ with $1 \leq i \leq s$ and $j \geq 1$ has then a local expansion at P_∞ of the form

$$h_j^{(i)} = \sum_{r=0}^{\infty} a_{jr}^{(i)} z_r \quad \text{with all } a_{jr}^{(i)} \in \mathbb{F}_q\,.$$

Let $\mathbf{c}_j^{(i)}$ be the sequence obtained from the sequence $(a_{jr}^{(i)})_{r=0}^{\infty}$ by deleting the terms with $r = n_u$ for some $u \in \{0, 1, \ldots, g\}$. For $i = 1, \ldots, s$, the $\infty \times \infty$ generating matrix $C^{(i)}$ of a Niederreiter–Xing sequence is now the matrix whose j-th row is $\mathbf{c}_j^{(i)}$ for $j \geq 1$. The following theorem from [57] yields the quality parameter of this type of sequence.

Theorem 7.2. *Let F be a global function field with full constant field $\mathbb{F}_q$ and genus g which contains at least one place P_∞ of degree 1. Let D be a divisor of F with* $\deg(D) = 2g$ *and $\nu_{P_\infty}(D) = 0$ and let $P_1, \ldots, P_s$ be distinct places of F with $P_i \neq P_\infty$ for $1 \leq i \leq s$. Then any corresponding Niederreiter–Xing sequence is a digital (t, s)-sequence over $\mathbb{F}_q$ with*

$$t = g + \sum_{i=1}^{s} (\deg(P_i) - 1)\,.$$

In the special case where not only P_∞, but also $P_1, \ldots, P_s$ are places of F of degree 1, we obtain the following result.

Corollary 7.3. *For every prime power q and every dimension $s \geq 1$, there exists a digital $(V_q(s), s)$-sequence over $\mathbb{F}_q$, where*

$$V_q(s) = \min\{g \geq 0 : N_q(g) \geq s + 1\} \tag{7.4}$$

and $N_q(g)$ is the maximum number of places of degree 1 that a global function field with full constant field $\mathbb{F}_q$ and genus g can have.

Remark 7.4. The result in Corollary 7.3 can also be obtained by using the duality theory for digital sequences (see Section 5), as was shown in [5] (see also [6, Section 8.3]).

Since for any fixed q the integer $N_q(g)$ attains arbitrarily large values, the quantity $V_q(s)$ in (7.4) is well defined. Specific values of $V_q(s)$ can be derived from the data for $N_q(g)$ at the website `http://www.manypoints.org`. It was shown in [42], by using the class field theory of global function fields, that $V_q(s) = O(s)$ as $s \to \infty$ with an absolute implied constant. A detailed proof of this result can be found also in [45, Section 8.3].

We recall from Section 4 that $t_q(s)$ (or $d_q(s)$) is the least value of t for which there exists a (t, s)-sequence in base q (a digital (t, s)-sequence over $\mathbb{F}_q$). It follows from Corollary 7.3 that

$$t_q(s) \le d_q(s) \le V_q(s) .$$

Since $V_q(s) = O(s)$ as mentioned in the preceding paragraph, we conclude that $t_q(s) = O(s)$ and $d_q(s) = O(s)$ as $s \to \infty$ with an absolute implied constant. In view of Theorem 4.6, these asymptotic bounds are the best possible.

Note that the t-value $V_q(s)$ in (7.4) is obtained by just considering places of degree 1 in the construction of Niederreiter–Xing sequences. The full power of the construction of Niederreiter and Xing is realized by using places of arbitrary degrees as in Theorem 7.2. By optimizing the choices in Theorem 7.2, we are led to the quantity $X_q(s)$ defined as follows. For a global function field F containing at least one place of degree 1, we first define $L_s(F)$ by excluding one place of F of degree 1, letting $P_1, P_2, \ldots$ be a list of all other places of F arranged according to nondecreasing degrees, and putting

$$L_s(F) = \sum_{i=1}^{s} (\deg(P_i) - 1) .$$

Now define

$$X_q(s) = \min_F (g(F) + L_s(F)) , \tag{7.5}$$

where the minimum is extended over all global function fields F with a full constant field $\mathbb{F}_q$ and F containing at least one place of degree 1, and where $g(F)$ denotes the genus of F. It is trivial that $X_q(s) \le V_q(s)$. The following result is an immediate consequence of Theorem 7.2.

Corollary 7.5. *For every prime power q and every dimension $s \ge 1$, there exists a digital $(X_q(s), s)$-sequence over $\mathbb{F}_q$, where $X_q(s)$ is given by* (7.5).

In order to illustrate the dramatic improvement afforded by the construction of Niederreiter and Xing over previous constructions, we compare values of the quality parameter t obtained by three constructions in Table 7.1. The table refers to the most convenient base in practice, namely $q = 2$. The first row in Table 7.1 indicates the dimension s, the second row the t-value $U(s)$ for Sobol' sequences, the third row is the t-value $T_2(s)$ in (7.3) for binary Niederreiter sequences, and the fourth row is

Table 7.1: Table of t-values in base 2

s	1	2	3	4	5	6	7	8	9	10	11	12	13	14	15	20
$U(s)$	0	0	1	3	5	8	11	15	19	23	27	31	35	40	45	71
$T_2(s)$	0	0	1	3	5	8	11	14	18	22	26	30	34	38	43	68
$B_2(s)$	0	0	1	1	2	3	4	5	6	8	9	10	11	13	15	21

a known upper bound $B_2(s)$ on the t-value $X_2(s)$ in (7.5) for binary Niederreiter–Xing sequences. The linear increase of $X_2(s)$ as a function of s predicted by the theory can be seen very nicely in the last row of Table 7.1.

An approach to the construction of Niederreiter–Xing sequences using differentials of global function fields was described by Mayor and Niederreiter [21], but this approach yields the same values of the quality parameter as Theorem 7.2. The only improvements on Niederreiter–Xing sequences were obtained, in some special cases, in the work of Niederreiter and Özbudak [40]. For a given dimension $s \geq 2$, the construction in [40] uses a global function field F with full constant field $\mathbb{F}_q$ such that F has at least s distinct places of degree 1 and at least one place of degree 2. If g is the genus of F, then the construction in [40] yields a digital $(\mathbf{T}, s)$-sequence over $\mathbb{F}_q$ with $\mathbf{T}(m) = m$ for $1 \leq m \leq g$, $\mathbf{T}(m) = g$ for even $m \geq g+1$, and $\mathbf{T}(m) = g+1$ for odd $m \geq g+1$.

Example 7.6. Let q be an arbitrary prime power, let $s = q+1$, and let $F = \mathbb{F}_q(X)$ be the rational function field over $\mathbb{F}_q$. Then $g = 0$ and the assumptions in the above construction of Niederreiter and Özbudak are satisfied. Thus, we obtain a digital $(\mathbf{T}, q+1)$-sequence over $\mathbb{F}_q$ with $\mathbf{T}(m) = 0$ for even $m \geq 2$ and $\mathbf{T}(m) = 1$ for odd $m \geq 1$. On the other hand, Corollary 4.5 shows that the least possible t-value of a $(t, q+1)$-sequence in base q is ≥ 1, and the value $t = 1$ can be achieved by Theorem 7.1 with $s = q+1$. Since a $(1, q+1)$-sequence in base q is the same as a $(\mathbf{T}, q+1)$-sequence in base q with $\mathbf{T}(m) = 1$ for all $m \geq 1$, it is clear that the Niederreiter–Özbudak construction yields an improvement for the dimension $s = q+1$. Further examples of this type for other values of s can be found in [40, Section 5].

7.3 Sequences with Finite-Row Generating Matrices

The actual implementation of digital sequences involves, as the main step, the computation of the matrix-vector products $C^{(1)}\mathbf{n}, \dots, C^{(s)}\mathbf{n}$ in (3.5). The speed of this computation is related to the sparsity of the generating matrices $C^{(1)}, \dots, C^{(s)}$. The following case can be considered the most favorable one.

Definition 7.7. Let S be a digital sequence over $\mathbb{F}_q$ with generating matrices $C^{(1)}, \dots,$ $C^{(s)}$. Then S is a digital sequence with *finite-row generating matrices* if, for each $i = 1, \dots, s$, each row of $C^{(i)}$ contains only finitely many nonzero entries from $\mathbb{F}_q$.

In this context, it is convenient to say that a row of a finite-row generating matrix has *row length* k if the last nonzero entry of the row is the k-th entry. It was shown by Hofer and Larcher [11] that for digital $(0, s)$-sequences over $\mathbb{F}_q$, the best we can expect is that for every $i = 1, \dots, s$ and every $j \geq 1$, the j-th row of $C^{(i)}$ has row length $sj + 1 - \pi(i)$ for some permutation π of the set $\{1, \dots, s\}$. In this case, we say that the generating matrices have *shortest possible row lengths*.

It was proved in [11] that, for any prime q and any dimension $1 \leq s \leq q$, there exists a digital $(0, s)$-sequence over $\mathbb{F}_q$ with finite-row generating matrices having the shortest possible row lengths. The proof of this result is obtained by a suitable scrambling of the generating matrices of a given digital $(0, s)$-sequence over $\mathbb{F}_q$. If one starts from the Faure sequence in a prime base $q \geq s$ (see Section 7.1), then explicit formulas for the generating matrices of a digital $(0, s)$-sequence over $\mathbb{F}_q$ with finite-row generating matrices having the shortest possible row lengths can be given (see Hofer and Pirsic [12]).

The results above were generalized to arbitrary finite fields by Hofer [10]. This is achieved by an explicit construction that extends a method of Niederreiter [29] based on Hasse–Teichmüller derivatives (also called hyperderivatives). For an integer $k \geq 0$, the k-th *Hasse–Teichmüller derivative* is the $\mathbb{F}_q$-linear operator $H^{(k)}$ on $\mathbb{F}_q[X]$ which is determined by

$$H^{(k)}(X^r) = \binom{r}{k} X^{r-k} \quad \text{for all } r \in \mathbb{N}_0 .$$

Here we use a standard convention for binomial coefficients, namely $\binom{r}{k} = 0$ whenever $r < k$.

Let q be an arbitrary prime power and let s be an integer with $1 \leq s \leq q$. Choose s distinct elements $b_1, \dots, b_s$ of $\mathbb{F}_q$. For any integer $r \geq 0$, define the polynomial

$$f_r(X) = \prod_{i=1}^{s} (X - b_i)^{\lfloor (r+s-i)/s \rfloor} \in \mathbb{F}_q[X] .$$

Now put

$$c_{jr}^{(i)} = [H^{(j-1)}(f_r)](b_i) \in \mathbb{F}_q \quad \text{for } 1 \leq i \leq s, \; j \geq 1, \; r \geq 0 .$$

For $1 \leq i \leq s$, set up the $\infty \times \infty$ matrix $C^{(i)} = (c_{jr}^{(i)})_{j \geq 1, r \geq 0}$.

Theorem 7.8 (Hofer [10]). *Let q be a prime power and let s be an integer with $1 \leq s \leq q$. Let S be a digital sequence with generating matrices $C^{(1)}, \dots, C^{(s)}$ above. Then S is a digital $(0, s)$-sequence over $\mathbb{F}_q$ with finite-row generating matrices having the shortest possible row lengths.*

References

[1] J. Baldeaux, J. Dick, G. Leobacher, D. Nuyens, and F. Pillichshammer, Efficient calculation of the worst-case error and (fast) component-by-component construction of higher order polynomial lattice rules, *Numer. Algor.* **59** (2012), 403–431.

[2] A. T. Clayman, K. M. Lawrence, G. L. Mullen, H. Niederreiter, and N. J. A. Sloane, Updated tables of parameters of (t, m, s)-nets, *J. Combinatorial Designs* **7** (1999), 381–393.

[3] J. Dick, P. Kritzer, G. Leobacher, and F. Pillichshammer, Constructions of general polynomial lattice rules based on the weighted star discrepancy, *Finite Fields Appl.* **13** (2007), 1045–1070.

[4] J. Dick and H. Niederreiter, On the exact t-value of Niederreiter and Sobol' sequences, *J. Complexity* **24** (2008), 572–581.

[5] J. Dick and H. Niederreiter, Duality for digital sequences, *J. Complexity* **25** (2009), 406–414.

[6] J. Dick and F. Pillichshammer, *Digital Nets and Sequences: Discrepancy Theory and Quasi-Monte Carlo Integration*, Cambridge University Press, Cambridge, 2010.

[7] H. Faure, Discrépance de suites associées à un système de numération (en dimension s), *Acta Arith.* **41** (1982), 337–351.

[8] H. Faure and C. Lemieux, Improvements on the star discrepancy of (t, s)-sequences, *Acta Arith.* **154** (2012), 61–78.

[9] T. Hansen, G. L. Mullen, and H. Niederreiter, Good parameters for a class of node sets in quasi-Monte Carlo integration, *Math. Comp.* **61** (1993), 225–234.

[10] R. Hofer, A construction of digital $(0, s)$-sequences involving finite-row generator matrices, *Finite Fields Appl.* **18** (2012), 587–596.

[11] R. Hofer and G. Larcher, On existence and discrepancy of certain digital Niederreiter–Halton sequences, *Acta Arith.* **141** (2010), 369–394.

[12] R. Hofer and G. Pirsic, An explicit construction of finite-row digital $(0, s)$-sequences, *Uniform Distribution Theory* **6(2)** (2011), 13–30.

[13] P. Kritzer, Improved upper bounds on the star discrepancy of (t, m, s)-nets and (t, s)-sequences, *J. Complexity* **22** (2006), 336–347.

[14] P. Kritzer and F. Pillichshammer, Low discrepancy polynomial lattice point sets, preprint, 2012.

[15] G. Larcher, Nets obtained from rational functions over finite fields, *Acta Arith.* **63** (1993), 1–13.

[16] G. Larcher, A. Lauß, H. Niederreiter, and W.Ch. Schmid, Optimal polynomials for (t, m, s)-nets and numerical integration of multivariate Walsh series, *SIAM J. Numer. Analysis* **33** (1996), 2239–2253.

[17] G. Larcher and H. Niederreiter, Generalized (t, s)-sequences, Kronecker-type sequences and diophantine approximations of formal Laurent series, *Trans. Amer. Math. Soc.* **347** (1995), 2051–2073.

[18] K. M. Lawrence, A combinatorial characterization of (t, m, s)-nets in base b, *J. Combinatorial Designs* **4** (1996), 275–293.

[19] C. Lemieux, *Monte Carlo and Quasi-Monte Carlo Sampling*, Springer, New York, 2009.

[20] R. Lidl and H. Niederreiter, *Finite Fields*, Encyclopedia of Mathematics and Its Applications, Vol. 20, Cambridge University Press, Cambridge, 1997.

[21] D. J.S. Mayor and H. Niederreiter, A new construction of (t, s)-sequences and some improved bounds on their quality parameter, *Acta Arith.* **128** (2007), 177–191.

[22] G. L. Mullen, A. Mahalanabis, and H. Niederreiter, Tables of (t, m, s)-net and (t, s)-sequence parameters, *Monte Carlo and Quasi-Monte Carlo Methods in Scientific Computing* (H. Niederreiter and P. J.-S. Shiue, eds.), Lecture Notes in Statistics, Vol. 106, pp. 58–86, Springer, New York, 1995.

[23] G. L. Mullen and W.Ch. Schmid, An equivalence between (t, m, s)-nets and strongly orthogonal hypercubes, *J. Combinatorial Theory Ser. A* **76** (1996), 164–174.

[24] H. Niederreiter, Quasi-Monte Carlo methods and pseudo-random numbers, *Bull. Amer. Math. Soc.* **84** (1978), 957–1041.
[25] H. Niederreiter, Low-discrepancy point sets, *Monatsh. Math.* **102** (1986), 155–167.
[26] H. Niederreiter, Rational functions with partial quotients of small degree in their continued fraction expansion, *Monatsh. Math.* **103** (1987), 269–288.
[27] H. Niederreiter, Point sets and sequences with small discrepancy, *Monatsh. Math.* **104** (1987), 273–337.
[28] H. Niederreiter, Low-discrepancy and low-dispersion sequences, *J. Number Theory* **30** (1988), 51–70.
[29] H. Niederreiter, Quasi-Monte Carlo methods for multidimensional numerical integration, *Numerical Integration III* (H. Braß and G. Hämmerlin, eds.), International Series of Numerical Mathematics, Vol. 85, pp. 157–171, Birkhäuser, Basel, 1988.
[30] H. Niederreiter, Low-discrepancy point sets obtained by digital constructions over finite fields, *Czechoslovak Math. J.* **42** (1992), 143–166.
[31] H. Niederreiter, Orthogonal arrays and other combinatorial aspects in the theory of uniform point distributions in unit cubes, *Discrete Math.* **106/107** (1992), 361–367.
[32] H. Niederreiter, *Random Number Generation and Quasi-Monte Carlo Methods*, CBMS-NSF Regional Conference Series in Applied Mathematics, Vol. 63, SIAM, Philadelphia, 1992.
[33] H. Niederreiter, Factorization of polynomials and some linear-algebra problems over finite fields, *Linear Algebra Appl.* **192** (1993), 301–328.
[34] H. Niederreiter, Constructions of (t, m, s)-nets, *Monte Carlo and Quasi-Monte Carlo Methods 1998* (H. Niederreiter and J. Spanier, eds.), pp. 70–85, Springer, Berlin, 2000.
[35] H. Niederreiter, Digital nets and coding theory, *Coding, Cryptography and Combinatorics* (K. Q. Feng, H. Niederreiter, and C. P. Xing, eds.), Progress in Computer Science and Applied Logic, Vol. 23, pp. 247–257, Birkhäuser, Basel, 2004.
[36] H. Niederreiter, Constructions of (t, m, s)-nets and (t, s)-sequences, *Finite Fields Appl.* **11** (2005), 578–600.
[37] H. Niederreiter, Nets, (t, s)-sequences and codes, *Monte Carlo and Quasi-Monte Carlo Methods 2006* (A. Keller, S. Heinrich, and H. Niederreiter, eds.), pp. 83–100, Springer, Berlin, 2008.
[38] H. Niederreiter, Quasi-Monte Carlo methods, *Encyclopedia of Quantitative Finance* (R. Cont, ed.), pp. 1460–1472, John Wiley and Sons, Chichester, 2010.
[39] H. Niederreiter and F. Özbudak, Constructions of digital nets using global function fields, *Acta Arith.* **105** (2002), 279–302.
[40] H. Niederreiter and F. Özbudak, Low-discrepancy sequences using duality and global function fields, *Acta Arith.* **130** (2007), 79–97.
[41] H. Niederreiter and G. Pirsic, Duality for digital nets and its applications, *Acta Arith.* **97** (2001), 173–182.
[42] H. Niederreiter and C. P. Xing, Low-discrepancy sequences and global function fields with many rational places, *Finite Fields Appl.* **2** (1996), 241–273.
[43] H. Niederreiter and C. P. Xing, Quasirandom points and global function fields, *Finite Fields and Applications* (S. Cohen and H. Niederreiter, eds.), London Math. Society Lecture Note Series, Vol. 233, pp. 269–296, Cambridge University Press, Cambridge, 1996.
[44] H. Niederreiter and C. P. Xing, Nets, (t, s)-sequences and algebraic geometry, *Random and Quasi-Random Point Sets* (P. Hellekalek and G. Larcher, eds.), Lecture Notes in Statistics, Vol. 138, pp. 267–302, Springer, New York, 1998.
[45] H. Niederreiter and C. P. Xing, *Rational Points on Curves over Finite Fields: Theory and Applications*, London Math. Society Lecture Note Series, Vol. 285, Cambridge University Press, Cambridge, 2001.

[46] H. Niederreiter and C. P. Xing, *Algebraic Geometry in Coding Theory and Cryptography*, Princeton University Press, Princeton, 2009.

[47] F. Pillichshammer and G. Pirsic, The quality parameter of cyclic nets and hyperplane nets, *Uniform Distribution Theory* **4(1)** (2009), 69–79.

[48] F. Pillichshammer and G. Pirsic, Discrepancy of hyperplane nets and cyclic nets, *Monte Carlo and Quasi-Monte Carlo Methods 2008* (P. L'Ecuyer and A. B. Owen, eds.), pp. 573–587, Springer, Berlin, 2009.

[49] G. Pirsic, A small taxonomy of integration node sets, *Sitzungsber. Österr. Akad. Wiss. Math.-Naturw. Kl. II* **214** (2005), 133–140.

[50] G. Pirsic, J. Dick and F. Pillichshammer, Cyclic digital nets, hyperplane nets and multivariate integration in Sobolev spaces, *SIAM J. Numer. Analysis* **44** (2006), 385–411.

[51] M.Yu. Rosenbloom and M. A. Tsfasman, Codes for the m-metric, *Problems Inform. Transmission* **33** (1997), 45–52.

[52] R. Schürer, A new lower bound on the t-parameter of (t, s)-sequences, *Monte Carlo and Quasi-Monte Carlo Methods 2006* (A. Keller, S. Heinrich, and H. Niederreiter, eds.), pp. 623–632, Springer, Berlin, 2008.

[53] I. M. Sobol', Distribution of points in a cube and approximate evaluation of integrals (Russian), *Ž. Vyčisl. Mat. i Mat. Fiz.* **7** (1967), 784–802.

[54] H. Stichtenoth, *Algebraic Function Fields and Codes*, 2nd ed., Graduate Texts in Mathematics, Vol. 254, Springer, Berlin, 2009.

[55] S. Tezuka, Polynomial arithmetic analogue of Halton sequences, *ACM Trans. Modeling and Computer Simulation* **3** (1993), 99–107.

[56] S. Tezuka, *Uniform Random Numbers: Theory and Practice*, Kluwer Academic Publishers, Boston, 1995.

[57] C. P. Xing and H. Niederreiter, A construction of low-discrepancy sequences using global function fields, *Acta Arith.* **73** (1995), 87–102.

Alina Ostafe

Iterations of Rational Functions: Some Algebraic and Arithmetic Aspects

Abstract: In this survey we discuss several arithmetic and algebraic aspects of dynamical systems generated by rational functions, mostly over finite fields. We will mention some applications to justify the great importance of studying such dynamical systems and discuss mostly open questions which arise from these applications or are of purely theoretical interest.

Keywords: Iterations of Rational Functions, Degree Growth, Representation of Iterates, Exponential Sums, Pseudorandom Numbers, Periodic Structure, Intersection of Orbits, Diameter of Orbits, Stable Polynomials

2010 Mathematics Subject Classifications: 11K45,11L07, 11T06, 11T23, 11T55, 37P05, 37P25

Alina Ostafe: Department of Computing, Macquarie University, Sydney, Australia,
e-mail: alina.ostafe@mq.edu.au

1 Introduction

1.1 Background

Algebraic Dynamical Systems (ADS), that is, dynamical systems generated by iterations of polynomials and rational functions, is a classical area of mathematics with a rich history and a variety of results [5, 108, 113]. The ADS are exciting and challenging mathematical objects with intricate algebraic and number theoretic properties and very complex behavior. Their study requires very deep and diverse mathematical and computational methods. They are also valuable building blocks for various applications including Monte Carlo methods and cryptography [87, 111, 120]. Recently, surprising links with other natural sciences such as biology [63, 72, 118] and physics [6, 13, 54, 64, 106, 107, 123] have emerged.

During the preparation of this paper, A. O. was supported by the Swiss National Science Foundation Grants PBZHP2–133399 and PA00P2–139679. The author would like to thank Igor Shparlinski for proposing several problems and for important comments on the original draft, to Arne Winterhof for a careful read and valuable comments and to Andrew Hone, James Propp, Dylan Thurston for bringing to attention other examples of rational function systems with slow degree growth. The author is grateful to the referees for important comments.

The goal of this paper is to outline some algebraic and number theoretic properties of ADS over finite fields and show that they are very important for a variety of problems and applications. The great variety of results and research directions make it impossible to include all of them in one survey. Thus, here we mainly concentrate on several open questions, new and old, and hope this paper will attract more attention to them and lead to further progress.

1.2 Notation

We fix first some notations. From now on p always represents a prime, q a power of p, $\mathbb{F}_p$ the prime finite field with p elements, which we identify with $\{0, 1, \dots, p-1\}$, and $\mathbb{F}_q$ the finite field with q elements. Also, $\mathbb{F}$ always represents an arbitrary field unless otherwise specified, and we denote by $\overline{\mathbb{F}}$ its algebraic closure.

Let $F_1, \dots, F_m \in \mathbb{F}(X_1, \dots, X_m)$ be m rational functions in m variables over $\mathbb{F}$. For each $i = 1, \dots, m$ we define the k-th iteration of the rational function F_i by the recurrence relation

$$F_i^{(0)} = X_i, \quad F_i^{(k)} = F_i(F_1^{(k-1)}, \dots, F_m^{(k-1)}), \quad k = 1, 2, \dots. \tag{1.1}$$

In this paper we discuss several algebraic properties of dynamical systems generated by rational functions, we refer to [5, 87, 108, 111, 113] for a background on ADS. A very good reference to this subject is also the recent survey [112] which covers also topics included in the present survey and more motivation for studying ADS.

1.3 Iterations

We define the vectors $\vec{u}_n = (u_{n,1}, \dots, u_{n,m}) \in \mathbb{F}^m$ by the recurrence relation

$$u_{n+1,i} = F_i(u_{n,1}, \dots, u_{n,m}), \quad n = 0, 1, \dots, \quad i = 1, \dots, m, \tag{1.2}$$

with some *initial vector* $\vec{u}_0 = (u_{0,1}, \dots, u_{0,m}) \in \mathbb{F}^m$.

Using the following vector notation

$$\vec{F} = (F_1(X_1, \dots, X_m), \dots, F_m(X_1, \dots, X_m)),$$

we have the recurrence relation

$$\vec{u}_{n+1} = \vec{F}(\vec{u}_n), \quad n = 0, 1, \dots. \tag{1.3}$$

In particular, for any $n \geq 0$ and $i = 1, \dots, m$ we have

$$u_{n,i} = F_i^{(n)}(\vec{u}_0) = F_i^{(n)}(u_{0,1}, \dots, u_{0,m})$$

or

$$\vec{u}_n = \vec{F}^{(n)}(\vec{u}_0)\,,$$

provided that $\vec{u}_n$ has been generated by (1.3).

Clearly, if we work over a finite field of q elements, the above sequence (1.3) of vectors $\{\vec{u}_n\}$ is eventually periodic with some period $\tau \le q^m$, that is, for some integer $s \ge 0$,

$$\vec{u}_{n+\tau} = \vec{u}_n\,, \quad n \ge s\,. \tag{1.4}$$

We always assume that s and τ are chosen to minimize the sum

$$T = s + \tau \le p^m\,.$$

Thus, in particular, T is the *trajectory length* of the iterations of the initial vector $\vec{u}_0$ and, hence, the vectors $\vec{u}_0, \dots, \vec{u}_{T-1}$ are pairwise distinct.

In the next sections we exemplify a few applications in pseudorandom number generation and discuss other theoretical problems in the study of rational function dynamical systems.

2 Distribution of Elements, Degree Growth and Representation

2.1 Exponential Sums and Linear Combinations of Iterates

It is very well known that most of the pseudorandom number generators (PRNGs) used in Monte Carlo methods and cryptography are based on the iteration of rational functions, see [39, 50, 81, 82, 85, 86, 111, 112, 120] and references therein. However, a "randomly" chosen system of such functions is expected to yield a rather poor generator, with a short cycle length. Here we discuss the properties of such systems that lead to better generators. Surprisingly, these constructions bring together several notions of intrinsic interest to the theory of polynomial rings over finite fields, such as algebraic entropy and automorphisms. We outline some of the constructions considered and present several open questions.

In this section we restrict ourselves to prime fields $\mathbb{F}_p$, and we also assume that the sequence generated by (1.2) is purely periodic, that is $s = 0$ in (1.4), however, similar results hold over arbitrary finite fields and for arbitrary s.

Also, when we work with rational functions, one should take care of possible zeros occurring in the denominator, that is, one has to decide how to compute $u_{n+1,i}$ if $\vec{u}_n$ is a pole of $F_i \in \mathbb{F}_p(X_1, \dots, X_m)$. One can "define" (see [34, 85, 86, 112]) these functions separately on the set of their poles, for example $0^{-1} = 0$.

Amongst many randomness measures [81, 82], a good pseudorandom sequence should have good distribution properties, that is, the elements in the sequence

$$\left\{\left(\frac{u_{n,1}}{p},\ldots,\frac{u_{n,m}}{p}\right)\right\}, \quad n \geq 0, \tag{2.1}$$

in the unit interval $[0,1)^m$ derived from the sequence $\{\vec{u}_n\}$, $n \geq 0$, generated by (1.2) in $\mathbb{F}_p$ should be *uniformly distributed*. Typically, the distribution properties of a sequence are derived from bounds of exponential sums with elements of this sequence. The relation is made explicit in the celebrated *Erdős–Turan–Koksma inequality*, see [29, Theorem 1.21], and, thus, reduces the problem of studying the distribution of the vectors (2.1) to estimating the following exponential sum

$$S_{\vec{a}}(N) = \sum_{n=0}^{N-1} \mathbf{e}_p\Big(\sum_{i=1}^{m} a_i u_{n,i}\Big), \tag{2.2}$$

where

$$\mathbf{e}_p(z) = \exp(2\pi i z/p), \quad \vec{a} = (a_1,\ldots,a_m) \in \mathbb{Z}^m.$$

The technique suggested in [84, 85], which was later slightly improved in [88], leads to estimating exponential sums with

$$L_{\vec{a},k,l}(X_1,\ldots,X_m) = \sum_{i=1}^{m} a_i\big(F_i^{(k)}(X_1,\ldots,X_m) - F_i^{(l)}(X_1,\ldots,X_m)\big). \tag{2.3}$$

So, we reduce the problem to estimating exponential sums with some polynomial or rational function argument for which the classical Weil bound (see [74, Chapter 5]) immediately implies:

Lemma 2.1. *For any nonconstant rational function $F \in \mathbb{F}_p(X_1,\ldots,X_m)$ of total degree D we have the bound*

$$\sum_{x_1,\ldots,x_m=1}^{p}{}^{*} \mathbf{e}_p(F(x_1,\ldots,x_m))) = O(Dp^{m-1/2}),$$

where $\sum^*$ means that the poles of F are excluded from the range of summation.

Obtaining nontrivial estimates of the sums (2.2) (our goal is to obtain a significant saving over the trivial bound, that is a power of p), reduces to studying several algebraic properties of the iterates $F_1^{(k)},\ldots,F_m^{(k)}$, $k \geq 1$, see also [87, 112]:

(i) *Linear independence* of the iterates $F_1^{(k)},\ldots,F_m^{(k)}$, $k \geq 1$, with 1, which ensures that $L_{\vec{a},k,l}(X_1,\ldots,X_m)$ is nonconstant, as otherwise we have the trivial bound.

(ii) *Slow degree growth* of the iterates $F_1^{(k)},\ldots,F_m^{(k)}$, $k \geq 1$, such that the rational function $L_{\vec{a},k,l}(X_1,\ldots,X_m)$ is of small degree (so that Lemma 2.1 is nontrivial for this function).

(iii) *Long trajectory length*: the estimates of $S_{\vec{a}}(N)$ are in general nontrivial if the period of the sequence we generate is large enough, close to the *maximal period* p^m.
(iv) The highest form of $L_{\vec{a},k,l}(X_1,\dots,X_m)$ should have a *low dimensional locus of singularity* (to apply the Deligne bound or other similar bounds [28, 69], which give much stronger estimates than Lemma 2.1).

An alternative approach is to construct $F_1,\dots,F_m$ of a special form thus to *replace using the Weil bound* altogether with other techniques to obtain even better bounds avoiding the need of controlling the degree growth.

In the *univariate* ($m = 1$) case, many constructions of polynomial or rational function generators were considered: linear polynomials [81, 82], inversive generator [85, 86], power generator [19, 37–39, 73, 121], Dickson polynomials [50] and Rédei functions [58], see also the surveys [120, 125].

However, for a nonlinear polynomial $f \in \mathbb{F}_p[X]$ of degree $d \geq 2$, it has been shown [84] that if $1 \leq N \leq \tau$, where $\tau \leq p$ is the period of the sequence generated by f and (1.2), we have the following estimate

$$S_a(N) = O(N^{-1/2}p^{1/2}(\log p)^{-1/2}),$$

which was slightly improved in [88].

We note that the estimate is nontrivial only in a very small range $p \geq N \geq p/\log p$, and, thus, τ should be close to the maximal period p, which raises also the question how to construct sequences generated by nonlinear polynomials which achieve maximal period. As explained in [87], the reason behind this is that the degree of iterated nonlinear univariate polynomials always grows exponentially, and thus, the saving over the trivial bound is only logarithmic. This motivates us to try to explore the multivariate case and see if this brings in new effects on the algebraic properties of polynomial iterates.

However, studying the linear independence of polynomial iterates in the *multivariate* setting is in general a hard problem as the degree growth and the representation under iterations of generic polynomials cannot be easily controlled. For the univariate case, the linear independence of the iterates is clear as we always have a monotonic degree growth, which is exponential in the number of iterations. In the multivariate case, efforts have been made to construct systems for which this algebraic property is controlled.

2.2 Generic Multivariate Polynomials

In the papers [55, 56] the authors considered polynomial systems $\mathcal{F} = \{F_1,\dots,F_m\}$, $F_1,\dots,F_m \in \mathbb{F}[X_1,\dots,X_m]$, over an arbitrary field $\mathbb{F}$ such that

$$\deg F_1 \geq 2, \quad \text{and} \quad F_i(X_1,\dots,X_m) = X_{i-1}, \quad i = 2,\dots,m.$$

For these classes of polynomials, for some special choices of the polynomial F_1, the degrees of the iterations of the polynomial F_1 grow strictly monotonically, which is essential for the approach of [55, 56]:

(i) F_1 has a *dominating term*, that is,

$$F_1 = a_{d_1\dots d_m} X_1^{d_1} \dots X_m^{d_m} + G\,,$$

where $a_{d_1\dots d_m} \in \mathbb{F}^*$, $G \in \mathbb{F}[X_1,\dots,X_m]$ and $\deg_{X_i} G < d_i$, $i = 1,\dots,m$;

(ii) F_1 is *non-quasi-linear* in X_m, that is, $\deg_{X_m} F_1 > 0$ and F_1 is not of the form $aX_m + G$, where $a \in \mathbb{F}^*$ and $G \in \mathbb{F}[X_1,\dots,X_{m-1}]$;

(iii) F_1 has a dominating variable X_1, that is, $\deg F_1 = d$,

$$F_1 = a_d X_1^d + a_{d-1} X_1^{d-1} + \dots + a_0\,,$$

where $a_i \in \mathbb{F}[X_2,\dots,X_m]$, $i = 0,\dots,d-1$, $a_d \in \mathbb{F}^*$.

For these classes of polynomials, the authors of [55, 56] proved that the iterations $F_1^{(k)},\dots,F_m^{(k)}$ are linearly independent for any k, and thus the polynomial $L_{\vec{a},k,l}$ defined by (2.3) is nonconstant.

For $\mathbb{F} = \mathbb{F}_p$, the distribution of the elements in the sequence generated by the above systems is given [55, 56].

In [94] the authors suggest a new approach, based on some combinatorial arguments, which avoids the need to verify the property of monotonic degree growth as in all previous approaches. It applies to arbitrary polynomial systems, such that their iterations on $\mathbb{F}_p^m$-vectors generate sufficiently long trajectories. We remark that this condition is anyway needed for applications to pseudorandom number generation and cryptography. The following result is proved in [94, Lemma 1]:

Lemma 2.2. *Let $\mathcal{F} = \{F_1,\dots,F_m\} \subset \mathbb{F}_p[X_1,\dots,X_m]$ be a system of m polynomials over $\mathbb{F}_p$ of degree at most D. Assume that for some initial vector $\vec{u}_0 \in \mathbb{F}_p^m$ the sequence of vectors $\{\vec{u}_n\}$ given by* (1.2) *has the trajectory of length T. Then, for any nonnegative integers $k < \ell \le \lceil T/p^{m-1}\rceil - 1$ and any nonzero $\vec{a} = (a_1,\dots,a_m) \in \mathbb{F}_p^m$, $L_{\vec{a},k,l}$ defined by* (2.3) *is a nonconstant polynomial of degree*

$$\deg L_{\vec{a},k,l} = O(D^\ell)\,.$$

However, for the polynomial systems $F_1,\dots,F_m \in \mathbb{F}_p[X_1,\dots,X_m]$, $m > 1$, described in this section, the same exponential degree growth is expected, which gives again very weak estimates as in the univariate case: for $1 \le N \le \tau$, where $\tau \le p^m$ is the period of the sequence generated by $F_1,\dots,F_m$ and (1.2), we have the following estimate

$$S_{\vec{a}}(N) = O(N^{-1/2} p^{m/2} (\log p)^{-1/2})\,.$$

2.3 Systems with Slow Degree Growth

One important characteristics of the dynamical system generated by $F_1, \dots, F_m \in \mathbb{F}(X_1, \dots, X_m)$ is the degree growth of the functions (1.1). It is of great interest for the theory of dynamical systems and has been studied in a number of works, see, for example, [13, 59, 123] and references therein. It is also important for applications to PRNGs [55, 56, 84, 95, 120] as remarked above and also for cryptography to get sequences of provable high linear complexity (see [100, 120, 125]).

To give a flavor of the results that can be obtained, if one succeeds to construct systems with a slow degree growth, that is polynomial in the number of iterations, we have the following estimate for the exponential sum (2.2), see [52, 95, 96]:

Lemma 2.3. *Assume that for some real $\gamma \geq 0$ and integers $m \geq s \geq 1$, $\nu \geq 1$ and $k_0 \geq 0$, a system $\mathcal{F} = \{F_1, \dots, F_m\}$ of m rational functions in $\mathbb{F}_p(X_1, \dots, X_m)$ satisfies the following condition. For every integers $k_1, \ell_1, \dots, k_\nu, \ell_\nu \geq k_0$ such that the components of the vectors*

$$(k_1 \dots, k_\nu) \quad \text{and} \quad (\ell_1 \dots, \ell_\nu)$$

are not permutations of each other, and nonzero vector $\vec{a} = (a_1, \dots, a_s) \in \mathbb{F}_p^s$, the linear combination

$$L_{\vec{a},k_1,\ell_1,\dots,k_\nu,\ell_\nu} = \sum_{i=1}^{s} a_i \sum_{j=1}^{\nu} \left(F_i^{(k_j)} - F_i^{(\ell_j)}\right),$$

is a nonconstant rational function over $\mathbb{F}_p$ of degree

$$\deg L_{\vec{a},k_1,\ell_1,\dots,k_\nu,\ell_\nu} = O(k^\gamma).$$

Then, for any sequence $\{\vec{u}_n\}$ generated by (1.2) that is purely periodic with an arbitrary period τ and $N \leq \tau$, we have

$$\max_{\vec{a} \in \mathbb{F}_p^s \setminus \{\vec{0}\}} \left| \sum_{n=0}^{N-1} e_p\left(\sum_{i=1}^{s} a_i u_{n,i}\right) \right| |S_s(\vec{a}; N)| = O(p^\alpha N^{1-\beta}),$$

where

$$\alpha = \frac{2m(\gamma + \nu) - \nu}{4\nu(\gamma + \nu)} \quad \text{and} \quad \beta = \frac{1}{2\nu}$$

and the implied constant depends only on γ, m and ν.

We note that in this case the saving is a power of p and the result is nontrivial whenever $N \geq p^{\alpha/\beta}$.

In [52, 92, 95, 98] several types of multivariate rational function systems with slow degree growth over a finite field $\mathbb{F}_p$ have been constructed and studied. However, the degree growth for these systems behave the same over any field $\mathbb{F}$, as we present them below.

Let $\mathcal{A} = \{A_1, \ldots, A_m\}$ be an arbitrary polynomial automorphism in $\mathbb{F}[X_1, \ldots, X_m]$, that is, there exists a polynomial system $\mathcal{A}^{-1} = \{A_1^{-1}, \ldots, A_m^{-1}\}$ such that

$$\mathcal{A}^{-1} \circ \mathcal{A} = (X_1, \ldots, X_m) .$$

We consider systems of the form

$$\mathcal{R} = \{R_1, \ldots, R_m\} = \mathcal{A}^{-1} \circ \mathcal{F} \circ \mathcal{A} , \tag{2.4}$$

where $\mathcal{F}$ is defined having the "triangular" form

$$\begin{aligned} F_1(X_1, \ldots, X_m) &= X_1^{e_1} G_1(X_2, \ldots, X_m) + H_1(X_2, \ldots, X_m) , \\ &\vdots \\ F_{m-1}(X_1, \ldots, X_m) &= X_{m-1}^{e_{m-1}} G_{m-1}(X_m) + H_{m-1}(X_m) , \\ F_m(X_1, \ldots, X_m) &= g_m X_m^{e_m} + h_m, \end{aligned} \tag{2.5}$$

with $e_1, \ldots, e_m \in \{-1, 1\}$, $G_i, H_i \in \mathbb{F}[X_{i+1}, \ldots, X_m]$, $i = 1, \ldots, m-1$, and $g_m, h_m \in \mathbb{F}$, $g_m \neq 0$.

The systems (2.5) have been further investigated in [91, 92, 96, 100, 101].

We also define

$$\begin{aligned} G_i^{(\ell)}(X_{i+1}, \ldots, X_m) &= G_i(F_{i+1}^{(\ell-1)}, \ldots, F_m^{(\ell-1)}) , \\ H_i^{(\ell)}(X_{i+1}, \ldots, X_m) &= H_i(F_{i+1}^{(\ell-1)}, \ldots, F_m^{(\ell-1)}). \end{aligned} \tag{2.6}$$

For the classes of rational function systems (2.5), it has been shown in [95, 98] that the degrees of the iterations of F_i, $i = 1, \ldots, m$, grow significantly slower (we note that the proof for the degree growth in [95, 98] holds over any field $\mathbb{F}$) than the exponential growth expected for the iterations of a "generic" system of m rational functions in m variables. However, to control the degree growth, further conditions were needed. Let $F_1, \ldots, F_m$ be rational functions defined by (2.5). From now on we consider the system (2.5) satisfying the following conditions for F_i for any $i = 1, \ldots, m$:

(i) if $e_i = 1$, as in [95, 96], we assume that the polynomial G_i has a unique leading monomial $X_{i+1}^{s_{i,i+1}} \ldots X_m^{s_{i,m}}$, that is,

$$G_i = g_i X_{i+1}^{s_{i,i+1}} \ldots X_m^{s_{i,m}} + \widetilde{G}_i ,$$

where $g_i \in \mathbb{F}^*$ and $\widetilde{G}_i \in \mathbb{F}[X_{i+1}, \ldots, X_m]$ with

$$\deg_{X_j} \widetilde{G}_i < s_{i,j} , \quad \deg_{X_j} H_i \le s_{i,j} , \quad j = i+1, \ldots, m ; \tag{2.7}$$

(ii) if $e_i = -1$, we assume that the polynomial H_i has a unique leading monomial $X_{i+1}^{s_{i,i+1}} \ldots X_m^{s_{i,m}}$, that is,

$$H_i = h_i X_{i+1}^{s_{i,i+1}} \ldots X_m^{s_{i,m}} + \widetilde{H}_i ,$$

where $h_i \in \mathbb{F}^*$ and $\widetilde{H}_i \in \mathbb{F}[X_{i+1}, \ldots, X_m]$, and

$$\deg_{X_j} \widetilde{H}_i < s_{i,j} , \quad \deg_{X_j} G_i < 2 s_{i,j} , \quad j = i+1, \ldots, m . \tag{2.8}$$

In [52, 95] the follwing result was proved:

Lemma 2.4. *Let $\mathcal{R}$ and $\mathcal{F}$ be defined by* (2.4) *and* (2.5) *satisfying the conditions* (2.7) *and* (2.8) *and such that $s_{i,i+1} \neq 0$, $i = 1, \dots, m-1$. If $X_1^{e_{i,1}} \dots X_m^{e_{i,m}}$ is the lexicographically highest monomial of A_i^{-1}, then*

$$\deg A_i^{-1} \circ \mathcal{F}^{(k)} = \sum_{j=1}^{m} e_{i,j} \left(\frac{1}{(m-j)!} s_{j,j+1} \dots s_{m-1,m} k^{m-j} + O(k^{m-1-j}) \right),$$

for $i = 1, \dots, m$.

We note that for $\mathcal{A} = \{X_1, \dots, X_m\}$, we recover the formula obtained in [95, Lemma 1].

Several other examples of rational maps are known achieving a slow degree growth. We give several examples that were mentioned in [59]:

- It can be shown that the degree of the k-th iterate of the system

  $$F_1 = X_2, \quad F_2 = \frac{X_2^2 + 1}{X_1}$$

  is $2k$, see [59, 61] and references therein.
- The rational system

  $$F_1 = X_2, \quad F_2 = X_3, \quad F_3 = X_4, \quad F_4 = \frac{X_2 X_4 + X_3^2}{X_1},$$

  which induces the so-called *Somos-4 recurrence*, has the degree of the k-th iterate like k^2, see [41, 59].

  The previous two examples are said to be *Laurent polynomials*, that is, they are rational functions that can be written as a polynomial divided by a monomial. Below we give an example of a system that achieves a quadratic degree growth, but it's not a Laurent polynomial.
- The system

  $$F_1 = X_2, \quad F_2 = X_3, \quad F_3 = X_4, \quad F_4 = \frac{X_4(X_1 X_4 - X_2 X_3)}{X_1 X_3 - X_2^2}$$

  has the degree at the k-th iteration given by $\lfloor (2k^2 + 6k + 9)/5 \rfloor$, see [59].

  In particular, we can use Lemma 2.3 with $\gamma = 1$ for the first example and with $\gamma = 2$ for the second and third examples.

Problem 2.5. *There are many interesting examples with even slower degree growth than the one of the systems* (2.5), *but, unfortunately, appear only as isolated examples, and not as classes of polynomials. For cryptographic purposes it is important to have parametric families of such systems as in this case the parameters are usually assumed to be secret. Moreover, we would like to have such constructions in any dimension.*

So far we have encountered two types of growth of the degree of rational function systems: polynomial and exponential. It is certainly interesting to understand whether these are the only two possible types of growth. In this sense, Hasselblatt and Propp asked the following questions [59, Questions 9.5 and 9.6], see also [115]:

Problem 2.6. *Can the degree growth of a rational function be subexponential, but superpolynomial, in the number of iterates?*

Problem 2.7. *If the degree growth of a rational function is bounded by a polynomial in the number of iterates k, can it exhibit an intermediate behavior, such as $\sqrt{k}$.*

No such examples are known.

2.4 Exponential Degree Growth, but Sparse Representation

Having a slow degree growth for our systems is not the only property that will give good candidates for PRNGs. For example, one can consider rational function systems with sparse representation and still obtain good estimates for exponential sums, even without any information on the degree growth, which is the main drawback in all previous approaches. For applications to PRNGs, it is desirable to construct classes of polynomials such that under iterations the representation of the polynomial $L_{\vec{a},k,l}$ defined by (2.3) is "small". For this purpose, in [99] a multidimensional analogue of the classical power generator was introduced, that does not have any embedded homogeneous properties.

Let $F_1, \dots, F_m \in \mathbb{F}_p[X_1, \dots, X_m]$ be defined by

$$\begin{aligned} F_1 &= (X_1 - h_1)^{e_1} G_1 + h_1 , \\ &\vdots \\ F_{m-1} &= (X_{m-1} - h_{m-1})^{e_{m-1}} G_{m-1} + h_{m-1} , \\ F_m &= g_m (X_m - h_m)^{e_m} + h_m, \end{aligned} \tag{2.9}$$

where $G_i \in \mathbb{F}_p[X_{i+1}, \dots, X_m]$, $i = 1, \dots, m-1$, and $g_m, h_i \in \mathbb{F}_p$, $g_m \neq 0$, $e_i \in \mathbb{N}$, $i = 1, \dots, m$. For this new class of triangular polynomial systems, one has the following representation of iterates [99]:

Lemma 2.8. *Let $F_1, \dots, F_m \in \mathbb{F}_p[X_1, \dots, X_m]$ be defined by (2.9). Then,*

$$F_i^{(k)} = (X_i - h_i)^{e_i^k} G_{i,k} + h_i , \quad F_m^{(k)} = g_m^{1+e_m+\dots+e_m^{k-1}} (X_m - h_m)^{e_m^k} + h_m ,$$

where, for $i = 1, \dots, m-1$ and $k = 1, 2, \dots$, we define

$$G_{i,k} = G_i^{e_i^{k-1}} \left(G_i^{(2)}\right)^{e_i^{k-2}} \dots G_i^{(k)} ,$$

with $G_i^{(k)}$ defined by (2.6).

This class of systems extends the class of nonlinear PRNGs for which a power saving is possible in estimates on their measure of uniformity of distribution for sequences (1.2). The authors achieved this result by using the sparsity of the polynomial $L_{\vec{a},k,l}$ defned by (2.3) and reducing the exponential sum to estimating binomial sums and a recent result of Cochrane and Pinner [21], obtaining, for $N \le \tau$, where τ is the period of $\{\vec{u}_n\}$ generated by (1.2) and (2.9), the estimate

$$S_{\vec{a}}(N) = O\left(N^{1/2} p^{m/2-3/184}\right) .$$

This bound is achieved under the assumption that

$$\min_{i=1,\dots,m} t_{p-1}(e_i) \ge p^{3/46} ,$$

where e_i are the exponents appearing in the system (2.9) and $t_{p-1}(e_i)$ is the multiplicative order of e_i modulo $p-1$.

2.5 Representation of Iterates

Another natural problem is to say something about the representation of iterations of rational functions over finite fields, and in particular, for applications, it is important to control the "size" of the representation. Questions of this type are very hard and only a few results are known.

In the univariate case, over a field of characteristic zero, it is proved in [40] (see also [128, 129]) that the minimum number of terms necessary to express an iterate $f^{(n)}$ of a rational function f tends to infinity with n, provided f is not of an explicitly described special shape.

We denote by T_d the Chebyshev polynomial of degree d defined by

$$T_d(x + x^{-1}) = x^d + x^{-d} .$$

Theorem 2.9. *Let $\mathbb{F}$ be a field of characteristic 0 and $f \in \mathbb{F}(X)$ of degree $d \ge 3$. Suppose that f is not conjugate (with respect to the group action given by $PGL_2(\mathbb{F})$ on $\mathbb{F}(X)$) to $\pm X^d$ or to $\pm T_d(X)$. Then, for any integer $n \ge 3$, we cannot express $f^{(n)}$ as a ratio of two polynomials having altogether less than $((n-2)\log d - \log 2016)/\log 5$ terms.*

For univariate polynomials over finite fields, no results of this type are known, but it is clear that at least for some special classes of polynomials (e.g., linearised polynomials) or rational functions, such a result does not hold.

Problem 2.10. *What types of growth for the number of terms exist for iterates of rational functions over finite fields? Classify these functions on their type of growth.*

Moreover, in the multivariate case one can also show that an analogue of the result [40] does not hold anymore, and this happens over any field, not necessarily over finite fields, as the next example shows.

Let $m = 3$ and

$$\begin{aligned} F_1 &= X_1 - 2X_2(X_1X_3 + X_2^2) - X_3(X_1X_3 + X_2^2)^2 \\ F_2 &= X_2 + X_3(X_1X_3 + X_2^2) \\ F_3 &= X_3. \end{aligned}$$

Then, for any $k \geq 1$,

$$\begin{aligned} F_1^{(k)} &= X_1 - 2kX_2(X_1X_3 + X_2^2) - k^2X_3(X_1X_3 + X_2^2)^2, \\ F_2^{(k)} &= X_2 + kX_3(X_1X_3 + X_2^2), \quad F_3^{(k)} = X_3. \end{aligned}$$

This example appears in the literature as the *Nagata automorphism* [80]. In this case, the degree growth is constant, as the number of distinct monomials at every iteration, over any field of characteristic 0. Over a field of characteristic p, the degree at the k-th iterate is 2 whenever $(k, p) = 1$ and 1 otherwise.

Another way of measuring the complexity of the iterations of a polynomial $F \in \mathbb{F}[X_1, \ldots, X_m]$ over a field $\mathbb{F}$ with the polynomials $F_1, \ldots, F_m \in \mathbb{F}[X_1, \ldots, X_m]$ is in terms of the so-called *additive complexity*, which is the smallest number of "+" signs in the formulas evaluating these polynomials. For example, the polynomial

$$F(X, Y) = (X^2 + 2Y)^{1000}(3X + Y^3)^{1000} + (X^{100} + Y^{200})^{10}$$

is of total degree 5000 and, thus, has a very long representation via the list of coefficients. However, it is of additive complexity 4 and, thus, has a very concise representation as in the above (which also makes its evaluation at any point very efficient).

For a rational function $F = F_1/F_2 \in \mathbb{F}(X_1, \ldots, X_m)$, $F_1, F_2 \in \mathbb{F}[X_1, \ldots, X_m]$, we define the additive complexity as the maximum of the additive complexities of F_1 and F_2, respectively.

Problem 2.11. *Find general classes of polynomials or rational functions such that under iterations the additive complexity grows polynomially in the number of iterations (linearly, if possible).*

2.6 Deligne and Dwork-Regular Polynomials

Much better estimates for the exponential sum (2.2) would be possible if one constructs polynomial systems in which certain polynomials are Deligne for which Deligne type bounds given by [28] can be applied.

Lemma 2.12. *Let $F \in \mathbb{F}_p[X_1,\dots,X_m]$ be a polynomial with m variables, degree $d \geq 1$, which can be written as,*

$$F = F_d + \cdots + F_0\,, \quad \text{with } F_i \text{ homogeneous of degree } i.$$

Suppose that the following two conditions hold:
(i) *The degree of the polynomial F is prime to p;*
(ii) *The locus $F_d = 0$ is a nonsingular hypersurface in the projective space of dimension $m-1$.*

In this case we call F to be a Deligne polynomial. Then we have the estimate,

$$\left| \sum_{x_1,\dots,x_m \in \mathbb{F}_p} \mathbf{e}_p(F(x_1,\dots,x_m)) \right| \leq (d-1)^m p^{m/2}\,.$$

However, we note that this bound also depends on the degree of the polynomial F, or in our case the degree of the polynomial $L_{\vec{a},k,l}$ defined by (2.3), and, thus, on the degree growth of the iterations $F_1^{(k)},\dots,F_m^{(k)}$, $k \geq 1$.

Problem 2.13. *Construct polynomials $F_1,\dots,F_m \in \mathbb{F}_p[X_1,\dots,X_m]$ such that the leading term of $L_{\vec{a},k,l}$ defined by* (2.3) *describes a nonsingular hypersurface in $\mathbb{P}^{m-1}$, and the degree of the iterates $F_1^{(k)},\dots,F_m^{(k)}$, $k \geq 1$, does not grow faster than a polynomial in k.*

A homogeneous polynomial $F \in \mathbb{F}_p[X_1,\dots,X_m]$, $m \geq 2$, is called *Dwork-regular* with respect to the coordinates $X_1,\dots,X_m$ if the variety defined by the vanishing of F and $X_i \partial F/\partial X_i$ is empty, see [31, 70].

A polynomial $F \in \mathbb{F}_p[X_1,\dots,X_m]$ of degree d is *affine Dwork-regular* if the homogenization of F, that is,

$$F(X_0,X_1,\dots,X_m) = X_0^d F(X_1/X_0,\dots,X_m/X_0)$$

is Dwork-regular with respect to $X_0,\dots,X_m$.

Problem 2.14. *Similarly, construct systems such that $L_{\vec{a},k,l}$ defined by* (2.3) *is affine Dwork-regular* [31] *for which there exist Deligne-type estimates* [70].

2.7 Distribution in Prime and Polynomial Times

So far we discussed only the distribution of consecutive elements in the sequences we generate. However, we can also consider the distribution of the sequence (1.2) at prime moments of time $n = \ell$.

Problem 2.15. *Study the distribution in prime times of the sequence generated by* (1.2), *that is $\{\vec{u}_\ell\}$, when $\ell = 2, 3, \dots$ runs through consecutive primes.*

In turn, this is equivalent (see [29]) to studying exponential sums

$$T_{\vec{a}}(N) = \sum_{\substack{\ell \le N \\ \ell \text{ prime}}} \mathbf{e}_p\Big(\sum_{i=1}^{m} a_i u_{\ell,i}\Big), \quad \vec{a} = (a_1, \dots, a_m) \in \mathbb{Z}^m . \tag{2.10}$$

Note that the results of [10, 11, 14, 44] have an interpretation as results on the behavior at prime moments of time of the dynamical systems generated by the linear transformation $x \mapsto gx$ on $\mathbb{F}_p$, that is, of the sequence hg^{ℓ}, where ℓ runs through the primes up to N.

The standard technique of estimating the sums (2.10), that is based on the Vaughan identity [122], (see also [27, Chapter 24]), reduces the problem of estimating the sums (2.10) to some single (such as (2.2) for $m = 1$) and bilinear sums and in particular leads to the following problem:

Problem 2.16. *Given positive integer numbers H, K and a sequence* (1.2), *estimate the sums*

$$\sum_{h=1}^{H} \sum_{k=1}^{K} \alpha_h \beta_k \mathbf{e}_p\Big(\sum_{i=1}^{m} a_i u_{hk,i}\Big), \quad \vec{a} = (a_1, \dots, a_m) \in \mathbb{Z}^m ,$$

for arbitrary sequences $(\alpha_h)_{h=1}^{H}$ and $(\beta_k)_{k=1}^{K}$ of complex numbers.

Another interesting problem is to study the distribution at polynomial times:

Problem 2.17. *Study the distribution in polynomial times of the sequence generated by* (1.2), *that is, $\{\vec{u}_{f(n)}\}$, $n = 1, 2, \dots$, for some $f \in \mathbb{Z}[X]$.*

3 Structure of Rational Function Maps

3.1 Trajectory Length and Periodic Structure

Motivated by applications to pseudorandom number generation where having a large trajectory is essential, in [98] the authors considered the question about the length of trajectories generated by iterations (1.3) over a finite field $\mathbb{F}_q$. We remark that in this case a trajectory falls into a cycle if $\vec{u}_t = \vec{u}_s$ for some integers $t > s \ge 0$.

We note that Silverman [114] has considered a question about periods of general polynomial systems but in a somewhat dual situation when the initial value is fixed and the iterations are considered over a family of finite fields. The results of [114], which are further generalized in [2], apply to very general systems, however the estimates are only logarithmic rather than a power of the field size.

The following result was proved in [98, Theorem 4]:

Theorem 3.1. *Let $F_1, \dots, F_m \in \mathbb{F}_q(X_1, \dots, X_m)$ be rational functions defined by* (2.5) *satisfying the conditions* (2.7) *and* (2.8) *and such that $s_{i,i+1} \ne 0$, $i = 1, \dots, m-1$.*

Then, for any $T \geq 1$ for all but $O(T^3 q^{m-1})$ initial vectors $\vec{u}_0 \in \mathbb{F}_q^m$, the trajectory length of the iterations (1.3) *exceeds* T.

In [93, 98], the authors consider constructions from rational function systems (2.5) which generate sequences of the largest possible period p^m. In particular, they give an exact description, which leads to an immediate construction of such systems. Although this condition is not needed for having good PRNGs, it is a very important question for the theory of dynamical systems.

Theorem 3.2. *Let $\mathcal{F} = \{F_1, \ldots, F_m\}$ be a system of polynomials over $\mathbb{F}_p$ defined by* (2.5)*. Then the sequence $\{\vec{u}_n\}$ generated by* (1.3) *is purely periodic with period $\tau = p^m$ if and only if the following conditions are satisfied*

(i) *for every $i < m$ with $e_i = 1$, we have*

$$\prod_{\vec{v} \in \mathbb{F}_p^{m-i}} G_i(\vec{v}) = 1 \quad \text{and} \quad \sum_{\vec{v} \in \mathbb{F}_p^{m-i}} R_i(\vec{v}) \neq 0\,;$$

(ii) *for every $i < m$ with $e_i = -1$, we have:*

(a) *if $R_{i,p^{m-i}-1}(\vec{u}_0) = 0$, then*

$$R_{i,p^{m-i}}(\vec{u}_0) = S_{i,p^{m-i}-1}(\vec{u}_0) \quad \text{and} \quad S_{i,p^{m-i}}(\vec{u}_0) S_{i,p^{m-i}-1}(\vec{u}_0) \neq 0\,;$$

(b) *if $R_{i,p^{m-i}-1}(\vec{u}_0) \neq 0$, then*

$$X^2 - \frac{R_{i,p^{m-i}}(\vec{u}_0)}{R_{i,p^{m-i}-1}(\vec{u}_0)} X - \frac{\prod_{\vec{v} \in \mathbb{F}_p^{m-i}} G_i(\vec{v})}{R_{i,p^{m-i}-1}(\vec{u}_0)}$$

is a primitive polynomial over $\mathbb{F}_p$, where

$$F_i^{(k)} = \frac{X_i R_{i,k} + S_{i,k}}{X_i R_{i,k-1} + S_{i,k-1}}\,,$$

where $R_{i,k}, S_{i,k}$ are defined by the recurrence relations

$$R_{i,k} = G_i^{(k)} R_{i,k-2} + H_i^{(k)} R_{i,k-1}\,,$$
$$S_{i,k} = G_i^{(k)} S_{i,k-2} + H_i^{(k)} S_{i,k-1},$$

for $k \geq 1$, with the initial rational functions

$$R_{i,0} = 1, \quad S_{i,0} = 0\,, \quad R_{i,1} = H_i\,, \quad S_{i,1} = G_i\,;$$

(iii) *if $e_m = 1$, then $g_m = 1$;*

(iv) *if $e_m = -1$, then $X^2 - h_m X - g_m$ is a primitive polynomial over $\mathbb{F}_p$.*

Theorem 3.2 is interesting because it shows that studying the maximal period for the highly nonlinear systems (2.5) reduces to studying the maximal period achieved by some certain univariate linear or inversive congruential generator. However, the period length of the linear [71, 81] and inversive [18, 34] generators is very well understood.

Also, in [93], in the case $e_i = 1$ for all $i = 1, \ldots, m$, under certain conditions, the period of sequences generated by the iterations of the system (2.5) is given in terms of the order of the values of the polynomials G_i in the initial vector.

However, no such results are known yet for the systems (2.9). We believe that, as for the systems (2.5), one can reduce the problem to studying the maximal period of the classical univariate power generator [38] using a combination of the ideas of [38] and [110].

Problem 3.3. *Give necessary and sufficient conditions for other classes of multivariate rational function dynamical systems (in particular* (2.9)*) to achieve a maximal period.*

Also, as mentioned above, in [2, 114] nontrivial, but rather weak, lower bounds on the period of a general ADS modulo almost all primes p are given. By the *Birthday Paradox* one expects that the orbit length is of order $q^{m/2}$ (for a sufficiently large q). Indeed, for a rational function system $\mathcal{F}$, it is natural to expect that the map induced by $\vec{x} \mapsto \mathcal{F}(\vec{x})$ behaves like a random map on $\mathbb{F}_q^m$, for which the trajectory length is of this order, see [35] for a detailed treatment of cycle structure of random maps on finite sets. For example, the Pollard integer factorization algorithm (where a quadratic polynomial $f(X) = X^2 + c$ is iterated in a residue ring, see [25, Section 5.2.1]) is based on this assumption.

Problem 3.4. *Find classes of rational function dynamical systems such that they behave like random maps modulo infinitely many primes.*

Problem 3.5. *Improve the result of* [114] *(with a logarithmic lower bound on the "typical" period length) for special types of (multivariate) rational function dynamical systems, for example, for the systems with the polynomial degree growth.*

Another interesting parameter to investigate is the period t_f of a polynomial $f \in \mathbb{F}_q[X]$ as a map from $\mathbb{F}_q$ to $\mathbb{F}_q$. That is, t_f is the smallest t such that for any periodic point $x \in \mathbb{F}_q$ we have $f^{(t)}(x) = x$. Thus, t_f is the least common multiple of all cycle lengths of f. It is interesting to compare this parameter with the result of [109] for random maps.

3.2 Graph of Rational Function Maps

Let $\mathcal{F} = \{F_1, \ldots, F_m\}$ be a rational function system that induces a rational mapping from $\mathbb{F}_q^m$ to $\mathbb{F}_q^m$. The *functional graph* of $\mathcal{F}$ is the directed graph $G(\mathcal{F}) = (V, E)$, where $V = \mathbb{F}_q$ is the set of vertices and E is the set of edges $(\vec{v}_1, \vec{v}_2) \in \mathbb{F}_q^{2m}$ with $\vec{v}_2 = \mathcal{F}(\vec{v}_1)$. Each connected component of $G(\mathcal{F})$ consists of a directed cycle, sometimes called *limit cycle*, and to each vertex in the cycle is attached a *tree*. The number of edges leaving every vertex is called the *out-degree* and is always one, and the number of edges entering a vertex is called the *in-degree*. In the univariate case the in-degree of every vertex in the graph defined by a polynomial f is always between zero and the degree of f.

Understanding the structure of the graph of a rational map is a long standing open question and only a few results are known in this direction. When $\mathcal{F}$ is a linear system, then the number of limit cycles and their length, as well as the structure of the trees, is determined by the factorization of the characteristic polynomial of the matrix representation of the system, see [32, 60, 78].

Problem 3.6. *What is the distribution of the in-degrees in a "typical" functional graph?*

Problem 3.7. *What is the expected number of trees, attached to cycles, in a "typical" functional graph?*

Problem 3.8. *What is the expected number of connected components in the graph of a "typical" rational map?*

We note that for $m = 1$, there are some recent results [36] in which some (rather weak) lower and upper bounds for the average number of connected components of graphs associated to polynomials or rational functions of a given degree are given. Moreover, the authors also find lower bounds on the average number of periodic points of such graphs.

In [22, 24] a special class of monomial systems over a finite field $\mathbb{F}_q$ which have only fixed points as limit cycles is described, in which case the system is called a *fixed point system*. For $q = 2$, the case of so-called *Boolean* monomial systems, a sufficient condition for such systems to have only fixed points as limit cycles is given in [23]. In [24], the authors associate with each monomial system $\mathcal{F}$ a Boolean system $T(\mathcal{F})$ and a linear system $L(\mathcal{F})$ over $\mathbb{Z}_{q-1}$. The main result of [24] is that $\mathcal{F}$ is a fixed point system if and only if $T(\mathcal{F})$ and $L(\mathcal{F})$ are fixed point systems.

However, the author is not aware of similar results for more general polynomial systems or for rational function systems.

3.3 Common Composites and Intersection of Orbits

Let $f, g \in \mathbb{F}[X]$ be nonconstant polynomials over a field $\mathbb{F}$. We say that f_1 and f_2 have a *common composite* if there exist $u, v \in \mathbb{F}[X]$ such that

$$u(f_1(X)) = v(f_2(X)) .$$

There are necessary and sufficient conditions for two polynomials to have a common composite as long as the degree of the common composite is not divisible by the characteristic p of $\mathbb{F}$, see [12, 15] and references therein. The following result is proved in [12, Theorem 1.1]:

Theorem 3.9. *Two polynomials $f_1, f_2 \in \mathbb{F}[X]$ have a common composite if and only if there is a nonempty finite subset A of $\overline{\mathbb{F}}$ which admits a function $\ell : A \to \mathbb{Z}$ such that*
- *for $a \in A$ and $i \in \{1,2\}$, $\ell(a)/m_i(a)$ is a positive integer, where $m_i(a)$ denotes the multiplicity of $X = a$ as a root of $f_i(X) - f_i(a)$;*
- *for $i \in \{1,2\}$, $a \in A$ and $b \in \overline{\mathbb{F}}$, if $f_i(a) = f_i(b)$, then $b \in A$ and $\ell(a)/m_i(a) = \ell(b)/m_i(b)$.*

Moreover, in the same paper, the authors prove that, if $f_1, f_2 \in \mathbb{F}[X]$ have a common composite, then they have a common composite of degree $\mathrm{lcm}(\deg(f_1), \deg(f_2))p^s$ for some $s \geq 0$ (if the characteristic $p = 0$, then we use the convention $0^0 = 1$).

The authors of [12] also give several other results, examples and counterexamples, as well as algorithms to decide if two polynomials have a common composite of degree less than any fixed bound. However, no results are known for the multivariate case.

Problem 3.10. *Let $F_1, F_2 \in \mathbb{F}[X_1, \ldots, X_m]$. Give necessary and sufficient conditions for F_1, F_2 to have a common composite in $\mathbb{F}[X_1, \ldots, X_m]$. There are several ways to interpret this. One way is to find polynomials $u, v \in \mathbb{F}[X]$ such that*

$$u(F_1) = v(F_2)\,.$$

Of course, one can also ask when two polynomial systems $\mathcal{F}_1$ and $\mathcal{F}_2$ in m variables have a common composite, that is, there exist $U, V \in \mathbb{F}[X_1, \ldots, X_m]$ such that

$$U(\mathcal{F}_1) = V(\mathcal{F}_2)\,.$$

For a polynomial $f \in \mathbb{F}[X]$ and $x \in \mathbb{F}$, we define the orbit set of f in x as

$$\mathrm{Orb}_x(f) = \{f^{(n)}(x) \colon n = 1, 2, \ldots\}\,. \tag{3.1}$$

Little is known about the relation between orbits of different polynomials and the only few results known are only for polynomials over $\mathbb{C}$, see [47, 48]. For example, in [48, Theorem 1.1] the authors proved the following result:

Theorem 3.11. *Let $f_1, f_2 \in \mathbb{C}[X]$ be nonlinear polynomials and $x, y \in \mathbb{C}$. Let $\mathrm{Orb}_x(f_1)$ and $\mathrm{Orb}_y(f_2)$ be the orbit sets defined by (3.1) of f_1 and f_2 in the points x and y, respectively. If $\mathrm{Orb}_x(f_1) \cap \mathrm{Orb}_y(f_2)$ is infinite, then f_1 and f_2 have a common iterate, that is, there exist $n, m \geq 1$ with $f_1^{(n)} = f_2^{(m)}$.*

Studying problems of this type involve deep number theoretic tools such as counting integral points on curves, classifying diophantine equations with infinite many S-integral solutions, where S is a set of primes, and several results on decomposition of polynomials, see [47, 48].

We also mention that in [9] the authors combine complex-analytic and arithmetic tools to show that for any fixed $a, b \in \mathbb{C}$ and any integer $d \geq 2$, the set of $c \in \mathbb{C}$ for which both a and b are preperiodic for $X^d + c$ is infinite if and only if $a^d = b^d$.

However, there are no known analogues in the literature of Theorem 3.11 over finite fields.

Problem 3.12. *Is it true that, if for two nonlinear polynomials $f_1, f_2 \in \mathbb{F}_q[X]$ of fixed degrees and some initial values $(x, y) \in \mathbb{F}_q^2$ the intersection of the orbits $\mathrm{Orb}_x(f_1)$ and $\mathrm{Orb}_y(f_2)$ is much larger than the "expected" value (which is large enough too), say*

$$\#\mathrm{Orb}_x(f_1) \cap \mathrm{Orb}_y(f_2) \geq q^{\delta} \frac{\#\mathrm{Orb}_x(f_1)\#\mathrm{Orb}_y(f_2)}{q} \geq q^{\eta}$$

for some fixed $\eta > \delta > 0$ and sufficiently large q, then there exist integers $n, m \geq 1$ such that $f_1^{(n)} \equiv f_2^{(m)} \mod (X^q - X)$?

It is possible that in Problem 3.12 one may need further conditions.

Problem 3.13. *Let $f_1, f_2 \in \mathbb{Z}[X]$ and $x, y \in \mathbb{Z}$. Is it true that, if the intersection of the orbit sets $\mathrm{Orb}_x(f_1)$ and $\mathrm{Orb}_y(f_2)$ is "big" enough modulo almost all primes, then there exist $n, m \geq 1$ such that $f_1^{(n)} = f_2^{(m)}$ over $\mathbb{Z}$?*

One can also generalise the previous two questions to the multivariate case.

Problem 3.14. *Is it true that, if for two nonlinear polynomial systems $\mathcal{F}_1$ and $\mathcal{F}_2$ defined by m polynomials in m variables over $\mathbb{F}_q$ of fixed degrees and some initial values $\vec{x}, \vec{y} \in \mathbb{F}_q^m$ the intersection of the orbit sets $\mathrm{Orb}_{\vec{x}}(\mathcal{F}_1)$ and $\mathrm{Orb}_{\vec{y}}(\mathcal{F}_2)$ is "big" enough, then there exist $n, m \geq 1$ such that $\mathcal{F}_1^{(n)} = \mathcal{F}_2^{(m)} \mod (X_1^q - X_1, \dots, X_m^q - X_m)$? As mentioned above, one might need some further conditions on the polynomial systems. Similarly, consider Problem 3.13 for the multivariate case.*

4 Geometric Properties of Orbits

4.1 Diameter of Orbits

In [20, 57] the authors considered orbits of the dynamical system generated by $f \in \mathbb{F}_p(X)$, that is, sequences

$$u_n = f(u_{n-1}), \quad n = 1, 2, \dots, \tag{4.1}$$

with some initial value $u_0 = u \in \mathbb{F}_p$ and orbit size $\tau_u \leq p$.

Given an initial value $u \in \mathbb{F}_p$, we consider how far the sequence (4.1) propagates in N steps, that is, we study the *diameter*

$$L_u(N) = \max_{0 \leq n \leq N} |u_n - u| .$$

Trivially, for any $u \in \mathbb{F}_p$ and $N < \tau_u$, we have $L_u(N) \geq N$. Using certain estimates of exponential sums, it has been shown in [57], that, when $f \in \mathbb{F}_p(X)$ is an arbitrary rational function,

$$L_u(N) = p^{1+o(1)},$$

provided that $N \geq p^{1/2+\varepsilon}$ for some fixed $\varepsilon > 0$. However, it is well known that the "birthday paradox" usually leads to orbits of length of order $p^{1/2}$, thus, obtaining nontrivial estimates for such short orbits of this length is an important open question.

For linear fractional transformations $(aX + b)/(cX + d) \in \mathbb{F}_p(X)$, the authors of [57] obtained a bound which is nontrivial for essentially any N, using a new technique based on results of additive combinatorics due to Bourgain [14].

Theorem 4.1. *For any $\varepsilon > 0$ there exists an absolute constant $\delta > 0$ such that, for every linear fractional function $(aX + b)/(cX + d) \in \mathbb{F}_p(X)$ with $ad \neq bc$, $c \neq 0$, and initial value $u \in \mathbb{F}_p$,*

$$L_u(N) \gg N^{1+\delta}$$

provided that $N \leq \min(\tau_u, p^{1-\varepsilon})$.

Moreover, in [20], the authors combined standard exponential sum techniques with results on arithmetic combinatorics to obtain results on the expansion of orbits generated by univariate nonlinear polynomials $f \in \mathbb{F}_p[X]$.

Theorem 4.2. *Let $f \in \mathbb{F}_p[X]$ be a polynomial of degree $d \geq 2$ and let $\{u_n\}$ be the sequence generated by (4.1) with the initial value $u \in \mathbb{F}_p$. Then, for $\tau_u \geq N \geq 1$ we have*

$$L_u(N) \gg \min\left\{N^{2\kappa(d)/(1+2\kappa(d))+o(1)} p^{1/(1+2\kappa(d))}, N^{1+(d-1)/(2\kappa(d)-d+1)+o(1)}\right\},$$

where $\kappa(d)$ is the smallest integer κ such that for $k \geq \kappa$ there exists a constant $C(k, d)$ depending only on k and d such that the number of solutions to the system of equations

$$x_1^\nu + \cdots + x_k^\nu = x_{k+1}^\nu + \cdots + x_{2k}^\nu, \quad \nu = 1, \ldots, d,$$

in positive integers $x_1, \ldots, x_{2k} \leq H$, is $O(H^{2k-d(d+1)/2+o(1)})$ as $H \to \infty$.

The classical result of Hua [62, Theorem 15] on the Vinogradov mean value theorem implies that for $d \geq 11$,

$$\kappa(d) \leq \left\lfloor d^2(3 \log d + \log\log d + 4)\right\rfloor - 11,$$

see also [122, Theorem 7.4]. Furthermore, explicit numerical estimates on $\kappa(d)$ for $2 \leq d \leq 10$ can be found in [62, Chapter IV]. By a very recent striking result of Wooley [127] we have

$$\kappa(d) \leq d^2 - 1$$

for any $d \geq 3$.

Recently, M.-C. Chang [17] proved, using a different approach and techniques, a better lower bound for a quadratic polynomial $f = aX^2 + bX + c \in \mathbb{F}_p[X]$ with some initial value $u \in \mathbb{F}_p$ and $N \le \tau_u$,

$$L_u(N) \gg \min\left\{Np^c, \frac{1}{\log p}N^{4/5}p^{1/5}, N^{1/13 \log\log N}\right\},$$

for some constant $c > 0$.

Using an appropriate modification of previous techniques (distribution of elements in the sequence, exponential sums over finite fields, additive combinatorics [16]), it would be interesting to extend such results to the multivariate case. Let $F_1, \dots, F_m$ be a system of m-variate rational functions and $\vec{u}_0 \in \mathbb{F}_p^m$ an initial vector. We consider how far the sequence (1.2) propagates in N steps, that is, we study

$$L_{\vec{u}}(N) = \max_{0 \le n \le N} \|u_n - u\|,$$

where, for $(v_1, \dots, v_m) \in \mathbb{F}_p^m$, $\|(v_1, \dots, v_m)\| = \max_{i=1,\dots,m} v_i$.

Problem 4.3. *Give lower bounds for $L_{\vec{u}}(N)$ for different classes of multivariate rational function systems.*

4.2 Convex Hull of Trajectories

Let $\mathcal{F} = \{F_1, \dots, F_m\}$ be a system of m-variate rational functions over $\mathbb{F}_p$ and $\vec{u}_0 \in \mathbb{F}_p^m$ an initial vector. Let the sequence $\{\vec{u}_n\}$, $n \ge 0$, be defined by (1.2) and define $C_{\mathcal{F}}(N)$ to be the convex hull of the vectors $\vec{u}_1, \dots, \vec{u}_N$ in $[0, p-1]^m$.

Problem 4.4. *Give upper and lower bounds on the number of vertices of $C_{\mathcal{F}}(N)$.*

The following question is yet another generalization, together with Problem 4.3, of the problem of estimating the diameter of the trajectory of a univariate algebraic dynamical system.

Problem 4.5. *Give upper and lower bounds on the volume of $C_{\mathcal{F}}(N)$.*

As the first step, one can investigate these quantities numerically and compare them with the corresponding expectations for a set of uniformly distributed random points in a unit cube. This has been extensively studied in discrete geometry.

5 Stability, Absolute Irreducibility and Coprimality

5.1 Motivation

Here we describe some results and problems concerning the irreducibility of iterates. As it was shown in [91, 96], new stronger results about the distribution of the elements in the sequences we generate can be obtained if more information becomes available about the algebraic structure of linear combinations of $F_1^{(k)}, \dots, F_m^{(k)}$, in particular about the irreducibility of these combinations.

5.2 Stable Univariate Polynomials

Let $\mathbb{F}$ be a field. For a polynomial $f \in \mathbb{F}[X]$ we have the sequence of iterations defined by (1.1) for $m = 1$, that is

$$f^{(0)} = X, \quad f^{(n)} = f(f^{(n-1)}), \quad n \geq 1.$$

Following [4, 7, 66, 67], we say that $g \in \mathbb{F}[X]$ is *f-stable* if $g(f^{(n)})$ is irreducible for all $n \geq 0$. We say that f is *stable* if f is f-stable, that is, all iterates $f^{(n)}$ are irreducible over $\mathbb{F}$.

Studying the stability of a polynomial is an exciting problem which has attracted a lot of attention. However, only a few results are known and the problem is far from being well understood. At the heart of all known results lies the classical Capelli's Lemma, see [33].

Lemma 5.1. *Let $\mathbb{F}$ be a field, $f, g \in \mathbb{F}[X]$, and let $\beta \in \overline{\mathbb{F}}$ be any root of g. Then $g(f)$ is irreducible over $\mathbb{F}$ if and only if both g is irreducible over $\mathbb{F}$ and $f - \beta$ is irreducible over $\mathbb{F}(\beta)$.*

We mention first the quadratic polynomial case. Let $\mathbb{F}$ be a field of characteristic different from 2. As in [67], for a quadratic polynomial $f(X) = aX^2 + bX + c \in \mathbb{F}[X]$, $a \neq 0$, we define $\gamma = -b/2a$ as the unique critical point of f (that is, the zero of the derivative f') and consider the set defined by (3.1) from which we take out the element $f(\gamma)$. Using a slight abuse of notation, we still denote this set by $\mathrm{Orb}(f)$, that is

$$\mathrm{Orb}(f) = \{f^{(n)}(\gamma) \colon n = 2, 3, \dots\}. \tag{5.1}$$

It is shown in [65–67] that critical orbits play a very important role in the dynamics of polynomial iterations. The following result is well known [7, 8, 67] and follows directly from Lemma 5.1:

Theorem 5.2. *Let $\mathbb{F}$ be a field of characteristic different from 2, $f, g \in \mathbb{F}[X]$, $f = aX^2 + bX + c$ and $\gamma = -b/2a$ the unique critical point of f. Suppose that g is such that $g(f^{(n-1)})$ is of degree d and irreducible over $\mathbb{F}$ for some $n \geq 1$. Then $g(f^{(n)})$ is irreducible over $\mathbb{F}$ if $(-a)^d g(f^{(n)}(\gamma))$ is not a square in $\mathbb{F}$. If $\mathbb{F}$ is finite, then we have an "if and only if" statement.*

In particular, f is stable if the adjusted orbit

$$\overline{\mathrm{Orb}}(f) = \{-af(\gamma)\} \cup \mathrm{Orb}(f)$$

contains no squares. If $\mathbb{F}$ is finite, then we have an "if and only if" statement, that is, f is stable if and only if $\chi(f^{(n)}(\gamma)) = -1$, $n = 2, \dots, t_f$, where χ is the quadratic character of $\mathbb{F}$ and t_f is the smallest value of t such that $f^{(t)}(\gamma) = f^{(s)}(\gamma)$ for some positive integer $s < t$.

The first part of Theorem 5.2 can easily be extended to more general polynomials of even degree, see [1].

Theorem 5.2 tells us that the stability of quadratic polynomials over $\mathbb{F}_q$ (for an odd q, $q = p^s$, p prime) can be tested in at most q steps by simply examining $-af(\gamma)$ and each element of $\mathrm{Orb}(f)$. In [97], using Theorem 5.2 and methods from analytic number theory, we significantly reduced this bound down to $q^{3/4}$.

However, the case of an arbitrary polynomial $f \in \mathbb{F}_q[X]$ is not yet settled. The only known result in this case was proved in [51] and is based on Lemma 5.1, but also on new techniques based on resultants of polynomials together with the Stickelberger's theorem [117].

Theorem 5.3. *Let $q = p^s$, p be an odd prime, and $f \in \mathbb{F}_q[X]$ a stable polynomial of degree $d \geq 2$ with leading coefficient a_d, nonconstant derivative f', $\deg f' = k \leq d-1$ and a_{k+1} the coefficient of X^{k+1} in f. Let us suppose that γ_i, $i = 1, \dots, k$, are the roots of the derivative f'. Then*

(i) *if $d = \deg f$ is even,*

$$S_1 = \left\{ a_d^k \prod_{i=1}^{k} f^{(n)}(\gamma_i) \mid n > 1 \right\} \cup \left\{ (-1)^{\frac{d}{2}} a_d^k \prod_{i=1}^{k} f(\gamma_i) \right\}$$

contains only nonsquares in $\mathbb{F}_q$;

(ii) *if $d = \deg f$ is odd,*

$$S_2 = \left\{ (-1)^{\frac{(d-1)}{2}+k} (k+1) a_{k+1} a_d \prod_{i=1}^{k} f^{(n)}(\gamma_i) \mid n \geq 1 \right\}$$

contains only squares in $\mathbb{F}_q$.

Theorem 5.3 has immediately led to a nontrivial estimate [51] for the length of the sets S_1 and S_2 that are analogues of the orbit sets (5.1) of a stable polynomial with the additional property that its derivative is irreducible.

We should also remark that several other results regarding the stability of higher degree polynomials of the form $X^n - b \in \mathbb{F}[X]$, for some field $\mathbb{F}$, are given in [26]. In particular, in [26] the authors prove that $X^n - b \in \mathbb{F}[X]$ is stable over $\mathbb{F}$ in the following cases: (i) $\mathbb{F} = \mathbb{Q}$ and $b \in \mathbb{Z}$; (ii) $\mathbb{F} = \mathbb{Q}(t)$ and $b \in \mathbb{Z}[t]$; (iii) $\mathbb{F} = \mathbb{K}(t)$ and $b \in \mathbb{K}[t]$ for $\mathbb{K}$ an arbitrary closed field; (iv) $\mathbb{F} = \mathbb{K}(t)$, $b \in \mathbb{K}(t)$, $b \notin \mathbb{K}$, $n \geq 3$ and $\mathbb{K}$ an arbitrary field of characteristic 0. Moreover, in [26] the authors prove that, if there exist $n \geq 3$ and $b \in \mathbb{Q}$ such that $X^n - b$ is not stable over $\mathbb{Q}$, then there exists a triple of integers (x, y, z) such that $xyz \neq 0$, $\gcd(x, y, z) = 1$, satisfying $x^p + y^p = z^r$, where p is a prime dividing n and $r \geq 3$ is even.

Standard heuristics, based on the density of irreducible polynomials suggest that one should expect that there are very few stable nonlinear polynomials over finite fields while almost all polynomials over $\mathbb{Z}$ should be stable. In [49, 51] estimates of the number of stable polynomials over the finite field $\mathbb{F}_q$ of odd characteristic are given. More precisely, Gomez and Nicolás [49] have proved that there are $O(q^{5/2}(\log q)^{1/2})$ stable quadratic polynomials over $\mathbb{F}_q$ for an odd prime power q. For arbitrary degree polynomials, the following estimate was proved in [51].

Theorem 5.4. *The number of stable polynomials $f \in \mathbb{F}_q[X]$ of degree d is*

$$O\left(q^{d+1-1/\log(2d^2)}\right).$$

Over $\mathbb{Q}$, it is proved in [1] that almost all monic quadratic polynomials $f \in \mathbb{Z}[X]$ are stable.

We note that in Theorem 5.3 only a necessary condition for the stability of a polynomial f over $\mathbb{F}_q$, $q = p^s$, p prime, was given. However, no necessary and sufficient condition is known for the stability of arbitrary polynomials over a finite field. In fact, over $\mathbb{Q}$ even the quadratic case is not fully settled.

Problem 5.5. *Can the stability of $f \in \mathbb{F}[X]$, where $\mathbb{F} = \mathbb{F}_q$ or $\mathbb{F} = \mathbb{Q}$, be tested in finitely many steps?*

Moreover, we note that the results of [51] hold only over a field of odd characteristic.

Problem 5.6. *Study the stability of $f \in \mathbb{F}_{2^s}[X]$, $s \geq 1$, of degree $d \geq 3$.*

Another interesting question is to look only at the special class of polynomials of degree a multiple of the characteristic of the field. We note that in [1] it is shown that there are no stable quadratic polynomials over finite fields of characteristic two. One might expect that this is the case over any field of characteristic two, which is not true as it is also shown in [1] where an example of a stable quadratic polynomial over a function field of characteristic two is given.

Problem 5.7. *Do there exist stable polynomials of degree a multiple of the characteristic p of $\mathbb{F}_q$?*

5.3 On the Growth of the Number of Irreducible Factors

In [66, Conjecture 1] was conjectured the following:

Problem 5.8. *Let $f \in \mathbb{Z}[X]$ be monic and quadratic and suppose that 0 is not periodic. Show that f is eventually stable, that is, some iterate of f is a product of f-stable polynomials.*

As noted in [66], if 0 is a periodic point, then the number of irreducible factors goes to infinity with the number of iterations (at least as a linear function in the number of iterations) and the reason behind this is that there exists $n \geq 1$ such that X is a factor of $f^{(n)}$. The simplest example is to consider $f(X) = Xg(X) \in \mathbb{Z}[X]$ or the family $f(X) = X^2 + kX - (k+1)$, $k \in \mathbb{Z}$, given in [66].

In [53] we prove that [66, Conjecture 1] does not hold over finite fields. Indeed, we combine the method of Gomez and Nicolás [49] with some new ideas to show that for almost all polynomials $f \in \mathbb{F}_q[X]$ the number $r_n(f)$ of irreducible divisors of the n-th iterate $f^{(n)}$ grows at least linearly with n if n is of order up to $\log q$.

Theorem 5.9. *If q is odd, then for any fixed $\varepsilon > 0$ for all but $o(q^{d+1})$ polynomials $f \in \mathbb{F}_q[X]$ of degree d, we have*

$$r_n(f) \geq (0.5 + o(1))\, n\,,$$

when $n \to \infty$ and $L \geq n$, where

$$L = \left\lceil \left(\frac{1}{2\log d} - \varepsilon \right) \log q \right\rceil .$$

Our tools to prove this are resultants of iterated polynomials, the Stickelberger's Theorem [117] and estimates of certain character sums. In particular, we use the following formula for the discriminant of polynomial iterates [53, Lemma 3]:

Lemma 5.10. *Let $f \in \mathbb{F}_q[X]$ be a polynomial of degree $d \geq 2$ with leading coefficient f_d and nonconstant derivative f' of degree $k \leq d-1$. Suppose that γ_i, $i = 1, \ldots, k$, are the roots of the derivative f'. Then, for $n \geq 1$, we have*

$$\operatorname{Disc} f^{(n)} = (-1)^{d\left(\frac{d(d-1)}{2}+k\right)} f_d^{\frac{d^n-1}{d-1}\left((k-1)d^n + k\frac{d^n-d}{d-1} + 2d\right)} \left((k+1)f_{k+1}\right)^{d^n}$$
$$\operatorname{Disc} f^{(n-1)d} \prod_{i=1}^{k} f^{(n)}(\gamma_i).$$

We also note that a similar computation was obtained in [68, Lemma 3.1 and Theorem 3.2] for iterated rational functions.

Recall that, as $D \to \infty$, the expected number of irreducible divisors of a random polynomial over $\mathbb{F}_q$ of degree D is asymptotic to $\log D$ (uniformly in q), see [105, Section 2.3] for relevant references and more precise results. So, since $\deg f^{(n)} = d^n$ where $d = \deg f$, the linear in n lower bound in Theorem 5.9 appears to be right by the order of magnitude.

Problem 5.11. *Extend the bound of Theorem 5.9 to any n (beyond the current threshold $n = O(\log q)$).*

Although we do not know how to obtain such a result, we can construct some examples of polynomials for which r_n grows linearly (which, as we have mentioned, appears to the expected rate of growth). Indeed, take any quadratic polynomial $f(X) =$

$X^2 + 2aX + a^2 - a \in \mathbb{F}_q[X]$ with $a \in \mathbb{F}_q$ and set $\gamma = -a$. Clearly $f(\gamma) = \gamma$, thus $f^{(n)}(\gamma) = \gamma$ for any $n = 1, 2, \ldots$. We now get from Lemma 5.10 that

$$\mathrm{Disc} f^{(n)} = (-1)^{n-1}\gamma .$$

So, if -1 is a nonsquare in $\mathbb{F}_q$ (for example, for a prime $q = p \equiv 3 \pmod 4$), then $\mathrm{Disc} f^{(n)}$ is a square or a nonsquare depending only on the parity of n. Therefore, for this polynomial we have $r_n(f) \geq n$ for any $n \geq 1$. A concrete example is given by $f(X) = X^2 + X + 2 \in \mathbb{F}_3[X]$ (we take $a = 2$ in the above construction).

For n of order larger than $\log q$, we use a different technique related to Mason's proof of the ABC-conjecture in its polynomial version, see [76, 116], to prove a lower bound on the largest degree $D_n(f)$ of the irreducible divisors of $f^{(n)}$, see [53, Theorem 11].

The approaches and some results used to derive lower bounds on $r_n(f)$ and $D_n(f)$ are readily combined in [53, Theorem 12] to obtain the lower bound $n^{1+o(1)}$ as $n \to \infty$ (uniformly over q) on the largest degree of square-free divisors of $f^{(n)}$.

5.4 Stable Multivariate Polynomials

Studying multivariate generalizations of this problem is also an interesting problem. We can generalize the notion of stability from the previous section to multivariate polynomials in different ways, for example, as the absolute irreducibility of the iterations of multivariate polynomials.

Let $F \in \mathbb{F}[X_1, \ldots, X_m]$ and $\mathcal{F} = \{F_1, \ldots, F_m\}$ be a polynomial system in $\mathbb{F}[X_1, \ldots, X_m]$. We say that F is *$\mathcal{F}$-stable* if

$$F^{(n)} = F(F_1^{(n-1)}, \ldots, F_m^{(n-1)}) \tag{5.2}$$

is absolutely irreducible for all $n \geq 1$.

Problem 5.12. *Let $F \in \mathbb{F}[X_1, \ldots, X_m]$ and $\mathcal{F} = \{F_1, \ldots, F_m\}$ be a polynomial system in $\mathbb{F}[X_1, \ldots, X_m]$. Give necessary and sufficient conditions for which F is $\mathcal{F}$-stable?*

One might start by looking first at the generalization of the Eisenstein criterion to the multivariate case, which is due to Williams [124]:

Theorem 5.13. *Let $F \in \mathbb{F}_q[X_1, \ldots, X_m]$ be such that, if F is regarded as a polynomial in some indeterminate X_i for some $1 \leq i \leq m$ of degree $d < q$, there exists an absolutely irreducible polynomial $P \in \mathbb{F}_q[X_1, \ldots, X_{i-1}, X_{i+1}, \ldots, X_m]$ with the properties*

$$P \nmid F_d\,, \quad P \mid F_r\,, \ r = 0, \ldots, d-1\,, \quad P^2 \nmid F_0\,,$$

where F_r, $r = 0, \ldots, d$, denotes the coefficient of X_i^r. Then F is absolutely irreducible in $\mathbb{F}_q[X_1, \ldots, X_m]$.

We note that for $m = 1$, if $f \in \mathbb{Q}[X]$ is an Eisenstein polynomial, then so is every iterate $f^{(n)}$, $n \geq 2$. However, this is not true anymore for the multivariate case.

Advances in studying this type of problem have recently been made due to results connecting geometric and algebraic properties of polynomials. In particular, the Eisenstein irreducibility criterion for univariate and multivariate polynomials has been studied via *Newton polytopes*, see [30, 43, 89, 90] and references therein. This gives a geometric criterion to construct infinite classes of absolutely irreducible multivariate polynomials over an arbitrary field, see [43]. The notion of Newton polytopes associated with polynomials was introduced by Ostrowski [102], who proves that the factorization of polynomials implies the decomposition of polytopes, see [103, 104].

Let $F \in \mathbb{F}[X_1, \ldots, X_m]$ be a nonconstant polynomial over an arbitrary field $\mathbb{F}$. The *Newton polytope*, denoted by P_F, associated to the polynomial

$$F = \sum a_{i_1 \ldots i_m} X_1^{i_1} \ldots X_m^{i_m}$$

is the convex hull in $\mathbb{R}^m$ of the set of support vectors of F, that is, of

$$\mathrm{Supp}(F) = \{(i_1 \ldots i_m) \in \mathbb{N}^m : a_{i_1 \ldots i_m} \neq 0\} .$$

A polytope in $\mathbb{R}^m$ is called *integral* if all of its vertices are points with integer coordinates. An integral polytope C is called *integrally decomposable* if there exist integral polytopes A and B with at least two points each such that $C = A + B = \{a + b : a \in A, b \in B\}$. Otherwise, C is called *integrally indecomposable.*

In [43] the relation between decomposability of polytopes and factorization of polynomials is established:

Theorem 5.14 (Irreducibility Criterion). *Let $F \in \mathbb{F}[X_1, \ldots, X_m]$ be a nonzero polynomial over a field $\mathbb{F}$, not divisible by any X_i, $i = 1, \ldots, m$. If the Newton polytope P_F of F is integrally indecomposable, then F is absolutely irreducible over $\mathbb{F}$.*

Gao also gave general constructions of indecomposable polytopes, and, thus, of classes of absolutely irreducible polynomials, see [43, Theorems 4.2 and 4.11], from which we mention just one.

Theorem 5.15. *Let Q be any integral polytope in $\mathbb{R}^n$ contained in a hyperplane H and $\vec{v}$ is an integral point lying outside H. Suppose that $\vec{v}_1, \ldots, \vec{v}_k$ are all the vertices of Q. Then the convex hull* $\mathrm{conv}(\vec{v}, Q)$ *is integrally indecomposable if and only if*

$$\gcd(\vec{v} - \vec{v}_1, \ldots, \vec{v} - \vec{v}_k) = 1 .$$

Using this result one can give infinitely many examples of absolutely irreducible polynomials, see [43]. For example, if

$$F = aX^n + bY^r + cX^uY^v + \sum c_{i,j}X^iY^j \in \mathbb{F}_q[X, Y], \quad a, b, c \in \mathbb{F}_q^*,$$

is such that P_F is the triangle with vertices $(n, 0), (0, r), (u, v)$, then F is absolutely irreducible if and only if $\gcd(r, n, u, v) = 1$.

Problem 5.16. *Let $\mathbb{F}$ be an arbitrary field, $F \in \mathbb{F}[X_1, \dots, X_m]$ and $\mathcal{F} = \{F_1, \dots, F_m\}$ a polynomial system in $\mathbb{F}[X_1, \dots, X_m]$. Study the $\mathcal{F}$-stability of F via Newton polytopes.*

5.5 Coprimality of Iterates

It is well known [45, 71] that given two random multivariate polynomials, they are almost always relatively prime. Thus one might expect the following algebraic behavior of ADS whose study is also very interesting:

Problem 5.17. *Let $F, G \in \mathbb{F}[X_1, \dots, X_m]$ be such that $\gcd(F, G) = 1$. Let $F_1, \dots, F_m \in \mathbb{F}[X_1, \dots, X_m]$. Describe the conditions for which $\gcd(F^{(k)}, G^{(k)}) = 1$ for all $k \geq 2$, where $F^{(k)}$ and $G^{(k)}$ are defined by* (5.2).

We remark that Problem 5.17 was initially studied in [46] where it is proved that $\gcd(F^{(k)}, G^{(k)}) = 1$ for all $k \geq 2$ whenever $F_i \in \mathbb{F}_q[X_i]$, $i = 1, \dots, m$, but their approach does not apply to more general polynomials.

Moreover, Allem and Trevisan [3] also used geometric properties of Newton polytopes to obtain a criterion for the coprimality of two multivariate polynomials. They proved in [3, Theorem 2] that, if two polytopes have no common parallel edges, then they have no common integral summands, which gives a coprimality criterion for multivariate polynomials:

Theorem 5.18. *Let $F, G \in \mathbb{F}[X_1, \dots, X_m]$ over a field $\mathbb{F}$. If the Newton polytopes P_F and P_G have no parallel edges, then $\gcd(F, G) = 1$.*

This motivates us to further study properties of Newton polytopes of iterations of multivariate polynomials and solve the problem:

Problem 5.19. *Let $F, G \in \mathbb{F}[X_1, \dots, X_m]$. If P_F and P_G have no parallel edges, describe the conditions for which $P_{F^{(k)}}$ and $P_{G^{(k)}}$ have no parallel edges for any $k \geq 2$.*

6 More Problems

6.1 Multiplicative Independence

The study of polynomial dynamical systems plays a very important role in solving different problems over finite fields. One such example is finding elements of high order in $\mathbb{F}_q$, $q = p^s$. Giving an efficient algorithm to find or test primitive elements in finite fields is a well known open problem. The only algorithms for testing or constructing high order elements require factoring $q - 1$, which is known to be not solvable in polynomial time. Thus, in [42] the author asks the question if it is possible to find elements in $\mathbb{F}_q$ that can be proved being primitive or having high orders without the knowledge of how $q - 1$ factors. He succeeds in giving a new method to solve this problem and, thus, obtains the following result:

Theorem 6.1. *Let $n > 1$ and $m = q^{\lfloor \log_q n \rfloor}$. Let $g \in \mathbb{F}_q[X]$ with $\deg g \le 2\log_q n$ and g not of the form $aX^k, aX^{p^l} + b$ with $a, b \in \mathbb{F}_q$, $k, l \ge 0$. Let $\alpha \in \mathbb{F}_{q^n}$ have degree n and be a root of $X^m - g$. Then α has order at least*

$$n^{\frac{\log_q n}{4\log_q(2\log_q n)} - \frac{1}{2}}.$$

The surprising fact behind this result is that the proof relies in proving some properties of the dynamical system defined by a certain polynomial $f \in \mathbb{F}_q[X]$. More precisely, it relies on the *multiplicative independence* of iterates of $f \in \mathbb{F}_q[X]$, see [42]:

Theorem 6.2. *Let $f \in \mathbb{F}_q[X]$ be not a monomial or a binomial of the form $aX^{p^l} + b$. Then the polynomials $f, f^{(2)}, \dots, f^{(n)}, \dots$ are multiplicatively independent, that is, if*

$$f^{k_1}(f^{(2)})^{k_2} \dots (f^{(n)})^{k_n} = 1$$

for any integers $n \ge 1, k_1, k_2, \dots, k_n$, then $k_1 = \dots = k_n = 0$.

As noted in [112], there are several possible interpretations of what a multivariate analogue of Theorem 6.2 may look like, but this direction has not been studied so far. One of them is:

Problem 6.3. *Let $F \in \mathbb{F}_q[X_1, \dots, X_m]$. Study the multiplicative independence of the iterations $F^{(k)}$ of F with $F_1, \dots, F_m \in \mathbb{F}[X_1, \dots, X_m]$ as defined by* (5.2).

6.2 Complete Polynomials

Let $f \in \mathbb{F}_q[X]$ be a univariate polynomial over $\mathbb{F}_q$. In [83] the authors study classes of univariate polynomials over finite fields which are *complete*, that is, permutation polynomials $f \in \mathbb{F}_q[X]$ with the additional property that $X + f(X)$ is also a permutation of $\mathbb{F}_q$. In [126] the author considers dynamical systems generated by univariate polynomials and generalises the property of being complete to studying *k-complete* mapping polynomials f, such that both f and

$$f_k = X + f + \dots + f^{(k-1)}$$

are permutation polynomials for some $k \ge 2$. For $k = 2$ we recover the definition of f being a complete mapping considered in [79, 83]. We note that complete maps were initially introduced in [75] where they were used in the construction of Latin squares.

Let K be a positive integer. We call a permutation polynomial $f \in \mathbb{F}_q$ to be *K-strong complete* if f is k-complete for any $2 \le k \le K$.

Problem 6.4. *Characterise the classes of polynomials which are k-complete or K-strong complete, for some $k, K > 2$.*

References

[1] O. Ahmadi, F. Luca, A. Ostafe, and I. E. Shparlinski, On stable quadratic polynomials, *Glasgow Math. J.* **54(2)** (2012), 359–369.

[2] A. Akbary and D. Ghioca, Periods of orbits modulo primes, *J. Number Theory* **129(11)** (2009), 2831–2842.

[3] L. E. Allem and V. Trevisan, GCD of multivariate polynomials via Newton polytopes, *Applied Math. and Comp.* **217** (2011), 8377–8386.

[4] N. Ali, Stabilité des polynômes, *Acta Arith.* **119** (2005), 53–63.

[5] V. Anashin and A. Khrennikov, *Applied algebraic dynamics*, Walter de Gruyter, Berlin, 2009.

[6] A. d'Auriac, J.-Ch. Maillard, and C.-M. Viallet, On the complexity of some birational transformations, *J. Phys. A* **39** (2006), 3641–3654.

[7] M. Ayad and D. L. McQuillan, Irreducibility of the iterates of a quadratic polynomial over a field, *Acta Arith.* **93** (2000), 87–97.

[8] M. Ayad and D. L. McQuillan, Corrections to: Irreducibility of the iterates of a quadratic polynomial over a field, *Acta Arith.* **93** (2000), 87–97; *Acta Arith.* **99** (2001), 97.

[9] M. Baker and L. Demarco, Preperiodic points and unlikely intersections, *Duke Math. J.* **159** (2011), 1–29.

[10] W. Banks, A. Conflitti, J. B. Friedlander, and I. E. Shparlinski, Exponential sums over Mersenne numbers, *Compos. Math.* **140** (2004), 15–30.

[11] W. Banks, J. B. Friedlander, M. Z. Garaev, and I. E. Shparlinski, Exponential and character sums with Mersenne numbers, *J. Aust. Math. Soc.* **92(1)** (2012), 1–13.

[12] R. Beals, J. Wetherell, and M. Zieve, Polynomials with a common composite, *Israel J. Math.* **174** (2009), 93–117.

[13] E. Bedford and T. T. Truong, Degree complexity of birational maps related to matrix inversion, *Comm. Math. Phys.* **298** (2010), 357–368.

[14] J. Bourgain, Mordell's exponential sum estimate revisited, *J. Amer. Math. Soc.* **18** (2005), 477–499.

[15] A. Bremner and P. Morton, Polynomial relations in characteristic p, *Quart. J. Math. Oxford*, **29(2)** (1978), 335–347.

[16] B. Bukh and J. Tsimerman, Sum-product estimates for rational functions, *Proc. Lond. Math. Soc.* **104** (2012), 1–26.

[17] M.-C. Chang, Expansions of quadratic maps in prime fields, *Proc. Amer. Math. Soc.*, to appear.

[18] W. S. Chou, The period lengths of inversive pseudorandom vector generations, *Finite Fields Appl.* **1** (1995), 126–132.

[19] W.-S. Chou and I. E. Shparlinski, On the cycle structure of repeated exponentiation modulo a prime, *J. Number Theory* **107** (2004), 345–356.

[20] J. Cilleruelo, M. Z. Garaev, A. Ostafe, and I. E. Shparlinski, On the concentration of points of polynomial maps and applications, *Math. Zeit.* **272(3–4)** (2012), 825–837.

[21] T. Cochrane and C. Pinner, Explicit bounds on monomial and binomial exponential sums, *Quart. J. Math.* **62** (2011), 323–349.

[22] O. Colón-Reyes, *Monomial dynamical systems over finite fields*, Ph.D. thesis, Virginia Tech, 2005.

[23] O. Colón-Reyes, R. Laubenbacher, and B. Pareigis, Boolean monomial dynamical systems, *Annals of Combinatorics* **8** (2004), 425–439.

[24] O. Colón-Reyes, A. S. Jarrah, R. Laubenbacher, and B. Sturmfels, Monomial dynamical systems over finite fields, *Complex Systems* **16(4)** (2006), 333–342.

[25] R. Crandall and C. Pomerance, *Prime numbers: A computational perspective*, 2nd ed., Springer-Verlag, New York, 2005.

[26] Lynda Danielson and Burton Fein, On the irreducibility of the iterates of $x^n - b$, *Proc. Amer. Math. Soc.* **130(6)** (2002), 1589–1596.

[27] H. Davenport, *Multiplicative number theory*, Springer-Verlag, New York, 2000.

[28] P. Deligne, La conjecture de Weil. I, *Inst. Hautes Études Sci. Publ. Math.* **43** (1974), 273–307.

[29] M. Drmota and R. Tichy, *Sequences, discrepancies and applications*, Springer-Verlag, Berlin, 1997.

[30] G. Dumas, Sur queques cas d'irreductibilite des polynomes à coefficients rationnnels, *J. Math. Pures Appl.* **2** (1906), 191–258.

[31] B. Dwork, On the Zeta function of a hypersurface, *Publ. Math. IHES* **12** (1962), 5–68.

[32] B. Elspas, *The theory of autonomous linear sequential networks*, pp. 45–60, IRE Transaction on Circuit Theory, 1959.

[33] B. Fein and M. Schacher. Properties of iterates and composites of polynomials, *J. London Math. Soc.* **54(3)** (1996), 489–497.

[34] M. Flahive and H. Niederreiter, *On inversive congruential generators for pseudorandom numbers*, Finite Fields, Coding Theory, and Advances in Commun. and Comp., pp. 75–80, Marcel Dekker, New York, 1992.

[35] P. Flajolet and A. M. Odlyzko, Random mapping statistics, *Lecture Notes in Comput. Sci.* **434** (1990), 329–354.

[36] R. Flynn and D. Garton, *Graph components and dynamics over finite fields*, http://arxiv.org/abs/1108.4132v1, preprint, 2011.

[37] J. B. Friedlander, J. Hansen and I. E. Shparlinski, On character sums with exponential functions, *Mathematika* **47** (2000), 75–85.

[38] J. B. Friedlander, C. Pomerance and I. E. Shparlinski, Period of the power generator and small values of Carmichael's function, *Math. Comp.* **70** (2001), 1591–1605; see also **71** (2002), 1803–1806.

[39] J. Friedlander and I. Shparlinski, On the distribution of the power pgenerator, *Math. Comp.* **70** (2001), 1575–1589.

[40] C. Fuchs and U. Zannier, Composite rational functions expressible with few terms, *J. Europ. Math. Soc.* **14** (2012), 175–208.

[41] D. Gale, The strange and surprising saga of the Somos sequences, *Math. Intell.* **13** (1991), 49–50.

[42] S. Gao, Elements of provable high orders in finite fields, *Proc. Amer. Math. Soc.* **127** (1999), 1615–1623.

[43] S. Gao, Absolute irreducibility of polynomials via Newton polytopes, *J. of Algebra* **237(2)** (2001), 501–520.

[44] M. Z. Garaev and I. E. Shparlinski, The large sieve inequality with exponential functions and the distribution of Mersenne numbers modulo primes, *Intern. Math. Research Notices* **39** (2005), 2391–2408.

[45] J. von zur Gathen and J. Gerhard. *Modern computer algebra*, Cambridge University Press, 1999.

[46] J. von zur Gathen, J. Gutierrez, and R. Rubio, Multivariate polynomial decomposition, *Appl. Algebra Engrg. Comm. Comput.* **14(1)** (2003), 11–31.

[47] D. Ghioca, T. Tucker, and M. Zieve, Intersections of polynomial orbits, and a dynamical Mordell–Lang conjecture, *Invent. Math.* **171** (2008), 463–483.

[48] D. Ghioca, T. Tucker, and M. Zieve, Linear relations between polynomial orbits, *Duke Math. J.* **161(7)** (2012), 1379–1410.

[49] D. Gomez and A. P. Nicolás, An estimate on the number of stable quadratic polynomials, *Finite Fields and Their Appl.* **16** (2010), 401–405.

[50] D. Gomez-Perez, J. Gutierrez, and I. E, Shparlinski, Exponential sums with Dickson polynomials, *Finite Fields Appl.* **12** (2006), 16–25.
[51] D. Gomez-Perez, A. P. Nicolás, A. Ostafe, and D. Sadornil, Stable polynomials over finite fields, *Revista Matemática Iberoamericana*, to appear.
[52] D. Gomez-Perez, A. Ostafe, and I. Shparlinski, Algebraic entropy, automorphisms and sparsity of algebraic dynamical systems and pseudorandom number generators, *Math. Comp.*, to appear.
[53] D. Gomez-Perez, A. Ostafe, and I. E. Shparlinski, On irreducible divisors of iterated polynomials, *Revista Matemática Iberoamericana*, to appear.
[54] B. Grammaticos, R. G. Halburd, A. Ramani, and C.-M. Viallet, How to detect the integrability of discrete systems, *J. Phys. A* **42** (2009), 454002, p. 30.
[55] F. Griffin, H. Niederreiter, and I. E. Shparlinski, *On the distribution of nonlinear recursive congruential pseudorandom numbers of higher orders*, Lect. Notes in Comp. Sci. 1719, pp. 87–93, Springer-Verlag, Berlin, 1999.
[56] J. Gutierrez and D. Gomez-Perez, *Iterations of multivariate polynomials and discrepancy of pseudorandom numbers*, Lect. Notes in Comp. Sci. 2227, pp. 192–199, Springer-Verlag, Berlin, 2001.
[57] J. Gutierrez and I. E. Shparlinski, Expansion of orbits of some dynamical systems over finite fields, *Bull. Aust. Math. Soc.* **82** (2010), 232–239.
[58] J. Gutierrez and A. Winterhof, Exponential sums of nonlinear congruential pseudorandom number generators with Rédei functions, *Finite Fields Appl.* **14** (2008), 410–416.
[59] B. Hasselblatt and J. Propp, Degree growth of monomial maps, *Ergodic Theory and Dynamical Systems* **27** (2007), 1375–1397.
[60] A. Hernández-Toledo, Linear finite dynamical systems, *Communications in Algebra* **33** (2005), 2977–2989.
[61] A. Hone, Singularity confinement for maps with the Laurent property, *Phys. Lett. A.* **361** (2007), 341–345.
[62] L. K. Hua, *Additive theory of prime numbers*, Amer. Math. Soc., Providence, RI, 1965.
[63] A. S. Jarrah and R. Laubenbacher, *On the algebraic geometry of polynomial dynamical systems*, Emerging applications of algebraic geometry, IMA Vol. Math. Appl. 149, pp. 109–123, Springer, New York, 2009.
[64] D. Jogia, J. A. G. Roberts, and F. Vivaldi, An algebraic geometric approach to integrable maps of the plane, *J. Phys. A: Math. Gen.* **39** (2006), 1133–1149.
[65] R. Jones, Iterated Galois towers, associated martingales, and the p-adic Mandelbrot set, *Compositio Math.* **43** (2007), 1108–1126.
[66] R. Jones, The density of prime divisors in the arithmetic dynamics of quadratic polynomials, *J. Lond. Math. Soc.* **78** (2008), 523–544.
[67] R. Jones and N. Boston, Settled polynomials over finite fields, *Proc. Amer. Math. Soc.* **140(4)** (2012), 1849–11863.
[68] R. Jones and M. Manes, Galois theory of quadratic rational functions, *Comment. Math. Helv.*, to appear.
[69] N. Katz, Estimates for "singular" exponential sums, *International Mathematics Research Notices* **16** (1999), 875–899.
[70] N. Katz, On a question of Browning and Heath-Brown, *Analytic number theory*, Cambridge Univ. Press, Cambridge (2009), 267–288.
[71] D. E. Knuth, *The art of computer programming, Seminumerical algorithms*, vol. 2, Addison-Wesley, 1973.
[72] R. Laubenbacher and B. Stigler, A computational algebra approach to the reverse engineering of gene regulatory networks, *J. Theoret. Biol.* **229** (2004), 523–537.

[73] G. Martin and C. Pomerance, The iterated Carmichael λ-function and the number of cycles of the power generator, *Acta Arith.* **118** (2005), 305–335.

[74] R. Lidl and H. Niederreiter, *Finite fields*, Cambridge Univ. Press, Cambridge, 1997.

[75] H. B. Mann, The construction of orthogonal latin squares, *Ann. Math. Statist.* **13** (1942), 418–423.

[76] R. C. Mason, *Diophantine Equations over Functions Fields*, Cambridge, Cambridge Univ. Press, 1984.

[77] J. H. McKay and S. S.-S. Wang, A chain rule for the resultant of two polynomials, *Archiv der Mathematik* **53** (1989), 347–351.

[78] D. Milligan and M. Wilson, The behavior of affine boolean sequential networks, *Connection Science* **5** (1993), 153–167.

[79] G. L. Mullen and H. Niederreiter, Dickson polynomials over finite fields and complete mappings, *Canad. Math. Bull.* **30** (1987), 19–27.

[80] M. Nagata, *On automorphism group of* $k[x, y]$, Department of Mathematics, Kyoto University, Lectures in Mathematics, No. 5. Kinokuniya Book-Store Co., Ltd., Tokyo, 1972.

[81] H. Niederreiter, Quasi-Monte Carlo methods and pseudo-random numbers, *Bull. Amer. Math. Soc.* **84** (1978), 957–1041.

[82] H. Niederreiter, *Random number generation and Quasi-Monte Carlo methods*, SIAM Press, 1992.

[83] H. Niederreiter and K. H. Robinson, Complete mappings of finite fields, *J. Austral. Math. Soc. Ser. A* **33(2)** (1982), 197–212.

[84] H. Niederreiter and I. E. Shparlinski, On the distribution and lattice structure of nonlinear congruential pseudorandom numbers, *Finite Fields and Their Appl.* **5** (1999), 246–253.

[85] H. Niederreiter and I. E. Shparlinski, On the distribution of inversive congruential pseudorandom numbers in parts of the period, *Math. Comp.* **70** (2001), 1569–1574.

[86] H. Niederreiter and I. E. Shparlinski, On the average distribution of inversive pseudorandom numbers, *Finite Fields and Their Appl.* **8** (2002), 491–503.

[87] H. Niederreiter and I. E. Shparlinski, *Dynamical systems generated by rational functions*, Lect. Notes in Comp. Sci. 2643, pp. 6–17, Springer-Verlag, Berlin, 2003.

[88] H. Niederreiter and A. Winterhof, Exponential sums for nonlinear recurring sequences, *Finite Fields Appl.* **14** (2008), 59–64.

[89] O. Ore, Zur Theorie der Irreduzibilitätskriterien, *Math. Zeit.* **18** (1923), 278–288.

[90] O. Ore, Zur Theorie der Eisensteinschen Gleichungen, *Math. Zeit.* **20** (1924), 267–279.

[91] A. Ostafe, *Multivariate permutation polynomial systems and pseudorandom number generators*, Finite Fields and Their Appl., pp. 144–154, 2010.

[92] A. Ostafe, *Pseudorandom vector sequences derived from triangular polynomial systems with constant multipliers*, Lect. Notes in Comp. Sci., WAIFI 2010, pp. 62–72, Springer-Verlag, Berlin, 2010.

[93] A. Ostafe, Pseudorandom vector sequences of maximal period generated by triangular polynomial dynamical systems, *Designs, Codes and Cryptography* **63(1)** (2012), 59–72.

[94] A. Ostafe, E. Pelican, and I. E. Shparlinski, On pseudorandom numbers from multivariate polynomial systems, *Finite Fields and Their Appl.* **16** (2010), 320–328

[95] A. Ostafe and I. E. Shparlinski, On the degree growth in some polynomial dynamical systems and nonlinear pseudorandom number generators, *Math. Comp.* **79** (2010), 501–511.

[96] A. Ostafe and I. E. Shparlinski, Pseudorandom numbers and hash functions from iterations of multivariate polynomials, *Cryptography and Communications* **2** (2010), 49–67.

[97] A. Ostafe and I. E. Shparlinski, On the length of critical orbits of stable quadratic polynomials, *Proc. Amer. Math. Soc.* **138** (2010), 2653–2656.

[98] A. Ostafe and I. E. Shparlinski, *Degree growth, linear independence and periods of a class of rational dynamical systems*, Arithmetic, Geometry, Cryptography and Coding Theory 2010, Contemp. Math., to appear.

[99] A. Ostafe and I. E. Shparlinski, On the power generator of pseudorandom numbers and its multivariate analogue, *J. Complexity* **28(2)** (2012), 238–249.

[100] A. Ostafe, I. E. Shparlinski, and A. Winterhof, On the generalized joint linear complexity profile of a class of nonlinear pseudorandom multisequences, *Adv. Math. Comm.* **4** (2010), 369–379.

[101] A. Ostafe, I. E. Shparlinski, and A. Winterhof, Multiplicative character sums of a class of nonlinear recurrence vector sequences, *Intern. J. Number Theory* **7(6)** (2011), 1557–1571.

[102] A. M. Ostrowski, Über die Bedeutung der Theorie der konvexen Polyeder für die formale Algebra, *Jahresberichte Deutsche Math.* **30** (1921), 98–99.

[103] A. M. Ostrowski, On multiplication and factorization of polynomials, I. Lexicographic orderings and extreme aggregates of terms, *Aequationes Math.* **13** (1975), 201–228.

[104] A. M. Ostrowski, On multiplication and factorization of polynomials, II. Irreducibility discussion, *Aequationes Math.* **14** (1976), 1–32.

[105] D. Panario, *What do random polynomials over finite fields look like?*, Lect. Notes in Comp. Sci. 2948, pp. 89–108, Springer-Verlag, Berlin, 2003.

[106] J. A. G. Roberts and F. Vivaldi, Arithmetical method to detect integrability in maps, *Phys. Rev. Lett.* **90** (2003), 034102.

[107] J. A. G. Roberts and F. Vivaldi, A combinatorial model for reversible rational maps over finite fields, *Nonlinearity* **22** (2009), 1965–1982.

[108] K. Schmidt, *Dynamical systems of algebraic origin*, Progress in Math., vol. 128. Birkhäuser Verlag, Basel, 1995.

[109] E. Schmutz, Period lengths for iterated functions, *Combinatorics, Probability and Computing* **20** (2011), 289–298.

[110] M. Sha, On the cycle structure of repeated exponentiation modulo a prime power, *Fibonacci Quart.* **49(4)** (2011), 340–347.

[111] I. E. Shparlinski, On some dynamical systems in finite fields and residue rings, *Discr. and Cont. Dynam. Syst., Ser. A* **17** (2007), 901–917.

[112] I. E. Shparlinski, *Algebraic dynamical systems over finite fields*, Handbook of Finite Fields (G. Mullen and D. Panario, eds.), CRC Press, in press.

[113] J. H. Silverman, *The arithmetic of dynamical systems*, Springer, New York, 2007.

[114] J. H. Silverman, Variation of periods modulo p in arithmetic dynamics, *New York J. Math.* **14** (2008), 601–616.

[115] J. H. Silverman, Dynamical degrees, arithmetic degrees, and canonical heights for dominant rational self-maps of projective space, *Ergodic Theory and Dynamical Systems* (2012), 1–32, http://arxiv.org/abs/1111.5664v1.

[116] N. Snyder, An alternate proof of Mason's theorem, *Elemente Math.* **55** (2000), 93–94.

[117] L. Stickelberger, *Über eine neue Eigenschaft der Diskriminanten algebraischer Zahlkörper*, Verh. 1 Internat. Math. Kongresses, 1897, pp. 182–193, Leipzig, 1898.

[118] B. Stigler, *Polynomial dynamical systems in systems biology*, Modeling and simulation of biological networks, Proc. Sympos. Appl. Math., vol. 64, pp. 53–84, Amer. Math. Soc., Providence, RI, 2007.

[119] R. G. Swan, Factorization of polynomials over finite fields, *Pacific Journal of Mathematics* **12** (1962), 1099–1106.

[120] A. Topuzoğlu and A. Winterhof, *Pseudorandom sequences*, Topics in Geometry, Coding Theory and Cryptography, pp. 135–166, Springer-Verlag, Berlin, 2006.

[121] T. Vasiga and J. O. Shallit, On the iteration of certain quadratic maps over $GF(p)$, *Discr. Math.* **277** (2004), 219–240.

[122] R. C. Vaughan, An elementary method in prime number theory, *Acta Arith.* **37** (1980), 111–115.

[123] C.-M. Viallet, Algebraic dynamics and algebraic entropy, *Int. J. Geom. Methods Mod. Phys.* **5** (2008), 1373–1391.

[124] K. S. Williams, Eisenstein's criteria for absolute irreducibility over a finite field, *Canad. Math. Bull.* **9** (1966), 575–580.

[125] A. Winterhof, *Recent results on recursive nonlinear pseudorandom number generators*, Lect. Notes in Comp. Sci. 6338, Sequences and Their Applications Đ SETA 2010, pp. 113–124, Springer-Verlag, Berlin, 2010.

[126] A. Winterhof, *Generalizations of complete mappings of finite fields and some applications* (submitted), 2012.

[127] T. D. Wooley, *Vinogradov's mean value theorem via efficient congruencing, II*, http://arxiv.org/abs/1112.0358, preprint, 2011.

[128] U. Zannier, On the number of terms of a composite polynomial, *Acta Arith.* **127** (2007), 157–167.

[129] U. Zannier, On composite lacunary polynomials and the proof of a conjecture of Schinzel, *Invent. Math.* **174** (2008), 127–138.

Igor E. Shparlinski

Additive Combinatorics over Finite Fields: New Results and Applications

Abstract: We give a survey of recently emerged directions in additive combinatorics over finite fields. We describe a variety of concrete results and outline their applications to a broad spectrum of other problems.

Keywords: Additive Combinatorics, Finite Fields, Character Sums, Polynomials

2010 Mathematics Subject Classifications: 11B30, 11C08, 11C20, 11T23, 11T30

Igor E. Shparlinski: Department of Computing, Macquarie University, Sydney, Australia, e-mail: igor.shparlinski@mq.edu.au

1 Introduction

Additive combinatorics over finite fields is a very old area, with such celebrated results as the *Cauchy–Davenport theorem*, see [193, Theorem 5.4]. The *Erdős-Heilbronn conjecture* (see [89]), is another keystone in the area. This conjecture, established by Dias da Silva & Hamidoune [82], has eventually led to a very general statement by Alon [5] which is now known as *Combinatorial Nullstellensatz*. So the area has never been dormant.

Yet, the new era in additive combinatorics over finite fields starts with the pioneering work of Bourgain, Katz & Tao [36], that essentially shows that a set of elements in a prime finite field cannot behave simultaneously like an arithmetic progression (that is, not to grow under pairwise element addition) and a geometric progression (that is, not to grow under pairwise element multiplication). Since that time, this area has enjoyed a lot of attention, has been developed in various directions and has had a wide range of important applications, see [15, 29, 30, 35, 96, 98, 119, 155, 165, 169, 172] and references therein. Here we describe a series of recently emerged directions and present some concrete results. However, the main emphasis is made on new applications to a scope of seemingly unrelated problems from number theory and the theory of finite fields, including applications to such classical areas as *character sums*

The author is very grateful to Khodakhast Bibak, Javier Cilleruelo, Moubariz Garaev, Alex Iosevich, Misha Rudnev and Arne Winterhof for the careful reading of the manuscript and many helpful comments. During the preparation of this paper, the author was supported in part by the Australian Research Council Grant DP1092835.

and *polynomials*. We also indicate some possible directions for future research. Note that applications to *theoretical computer science* are as exciting and important, but we refer to [15] for an exhaustive survey of such applications. We also note the very recent surveys of Bourgain [23], Garaev [98], Green [112] and Shkredov [172] that also explain the underlying ideas and various applications.

The most interesting and well-studied case is the classical *sum-product problem*. For sets of real numbers this question has been introduced and studied by Erdős & Szemerédi [90]. The goal is to show that for an arbitrary finite set $\mathcal{A}$ (in a certain ring) at least one of the sets

$$\{a_1 + a_2 \colon a_1, a_2 \in \mathcal{A}\} \quad \text{and} \quad \{a_1 a_2 \colon a_1, a_2 \in \mathcal{A}\} \tag{1.1}$$

is of size substantially larger than the size of $\mathcal{A}$.

In the case of a set in a finite field, this is roughly the result of Bourgain, Katz & Tao [36]. The initial result has been improved and generalized in numerous directions. Here we outline some of them. We also give a short survey of recently emerged applications.

Unfortunately, due to the space constraints, we are not able to describe the methods behind the results (although in some cases we mention some basic underlying ideas). We only mention that one of the basic notions that appears in this area is the notion of *multiplicative energy* $E_\times(\mathcal{A})$ of a set $\mathcal{A}$:

$$E_\times(\mathcal{A}) = \#\{a_1 a_2 = a_3 a_4 \colon a_1, a_2, a_3, a_4 \in \mathcal{A}\}\,. \tag{1.2}$$

One can also define the *additive energy* $E_+(\mathcal{A})$ in a similar way. Analogues of these notions have also been defined and studied for two distinct sets, see [22, 34, 108]. Furthermore, Schoen & Shkredov [167] have successfully used a "cubic" generalization of the energy.

We also have had to leave out such exciting areas of additive combinatorics in finite fields as

- the *Erdős distance problem* [83, 94, 117, 130, 132, 144, 145] as well as its modification in some other settings (distinct volumes, configurations, and so on defined by arbitrary sets in $\mathbb{F}_q^n$) and metrics [14, 64, 142, 195, 200, 202, 203, 205];
- the *Kakeya problem* and other related problems about the directions defined by arbitrary sets in vector spaces over a finite field, see [84–86, 88, 128, 131, 151];
- estimating the size of the sets in a finite field that avoid arithmetic or geometric progressions, sum sets and similar linear and non-linear relations; in particular these results include finite field analogues of the Roth and Szemerédi theorems, see [1, 6, 8, 12, 77, 81, 112, 113, 153, 154, 158, 181];
- estimating the size of the sets in vector spaces over a finite field that define only some restrictive geometric configurations such as integral distances, acute angles, and pairwise orthogonal systems, see [75, 133, 134, 183, 194, 204];
- distribution of the values of determinants and permanents of matrices with entries from general sets, see [74, 196, 197];

and several others.

Additive combinatorics in $\mathbb{Z}$, $\mathbb{Q}$, $\mathbb{R}$, $\mathbb{C}$, polynomials and matrix rings, function fields and other infinite algebraic domains is a beautiful and very active area as well, see [17, 51–53, 79, 80, 86, 172, 190, 191, 193] and references therein. However, here we limit the scope of this survey to the case of sets in finite fields. In fact, most of the time we only consider the case of prime fields. Indeed, since a subfield of a composite field is closed under both addition and multiplication, it is clear that the sets of finite fields containing a massive subfield behaves similarly to both arithmetic and geometric progressions under addition and multiplication, respectively (see, however [107, 136, 140, 157, 171] for some results for arbitrary finite fields).

We conclude with a general comment that additive combinatorics presents a rare (maybe unique) example of an area where finite field results are actually trailing behind their prototypes that are already known over the integers. In fact, quite unusually one expects stronger results in the case of real numbers. For example, the famous conjecture of Erdős & Szemerédi [90] asserts that in the case of $\mathcal{A} \subseteq \mathbb{R}$, one of the sets (1.1) is of size $(\#\mathcal{A})^{2+o(1)}$, with the currently strongest result towards this conjecture being what is due to Solymosi [189] that gives the exponent $4/3$ instead of 2. On the other hand, for sets in finite fields, the best possible result and the currently known estimates are much more modest, see Section 3.1 for details.

2 Notation

For a prime power q, we denote by $\mathbb{F}_q$ the finite field of q elements.

Let $\mathbb{Z}_T$ denote the residue ring modulo T and let $\mathbb{Z}_T^*$ be its unit group.

Given m sets $\mathcal{A}_1, \dots, \mathcal{A}_m \subseteq \mathbb{F}_q$ and a rational function

$$F(X_1, \dots, X_m) \in \mathbb{F}_q(X_1, \dots, X_m) ,$$

we define the set

$$F(\mathcal{A}_1, \dots, \mathcal{A}_m) = \{F(a_1, \dots, a_m) \colon (a_1, \dots, a_m) \in (\mathcal{A}_1 \times \dots \times \mathcal{A}_m) \setminus \mathcal{P}_F\} ,$$

where $\mathcal{P}_F$ is the set of poles of F.

In particular, for an integer k, we use $k\mathcal{A}$ and $\mathcal{A}^k$ to denote k-fold sum and product sets, respectively.

If $\mathcal{A}$ is a finite set, $\#\mathcal{A}$ represents the number of elements of $\mathcal{A}$.

Throughout the paper, the implied constants in the symbols "O", "$\ll$" and "$\gg$" may depend on the real positive parameter ε. Recall that the notations $U \ll V$ and $V \gg U$ are equivalent to $U = O(V)$.

Furthermore, the letter p always denotes a prime; q always denotes a prime power; k, m and n (as well as K, M and N) always denote positive integers.

By $\mathbf{e}_p(u)$ we mean as usual $\exp(2\pi i u/p)$.

3 Estimates from Arithmetic Combinatorics

3.1 Classical Sum-Product Problem

The pioneering result of Bourgain, Katz & Tao [36] asserts that for any fixed $\varepsilon > 0$ there exists some $\delta > 0$ such that for an arbitrary set $\mathcal{A} \subseteq \mathbb{F}_p$ of cardinality $p^{\varepsilon} \le \#\mathcal{A} \le p^{1-\varepsilon}$ we have

$$\max\{\#(\mathcal{A}^2), \#(2\mathcal{A})\} \gg (\#\mathcal{A})^{1+\delta}.$$

It is obvious that the upper bound $\#\mathcal{A} \le p^{1-\varepsilon}$ is necessary for the sets $\mathcal{A}^2$, $2\mathcal{A}$ to have a room to grow. However, there is no obvious reason for the set $\mathcal{A}$ not to be too small. In fact, it has been shown by Bourgain, Glibichuk & Konyagin [35] that the lower bound $\#\mathcal{A} \ge p^{\varepsilon}$ can be simply dropped. The results of [35, 36] have triggered a quest for getting the best possible explicit dependence of δ on ε. The work of Garaev [96] has essentially closed this question for large sets. We present the [96, Theorem 1] result in a more general form (which has also been briefly mentioned in [96], since it may be more useful for applications (see Theorem 3.4 below).

Theorem 3.1. *For arbitrary sets $\mathcal{A}, \mathcal{B}, C \subseteq \mathbb{F}_p$, with $0 \notin \mathcal{B}$, we have*

$$\#(\mathcal{A} \cdot \mathcal{B}) \cdot \#(\mathcal{A} + C) \ge \tfrac{3}{8} \min\left\{p\#\mathcal{A}, \frac{(\#\mathcal{A})^2\#\mathcal{B}\#C}{p}\right\}.$$

The method of Garaev [96], which has its roots in the work of Elekes [87] on the sum-product problem over the reals, is based on bounds of certain multiple exponential sums. Cilleruelo [66] has shown that in many cases there is an elementary combinatorial approach that leads to similar results. For example, Cilleruelo [66, Theorem 3.1] gives a purely combinatorial proof of Theorem 3.1, see also [66, Theorem 3.2 and 3.3] for combinatorial proofs of other results of this type from [100, 119].

Although typically we do not explain the underlying techniques, the method of Garaev [96] is so delightfully simple and robust (and so applicable in many other situations) that we now outline a proof of Theorem 3.1 as it is given in [11] (see also the proof of Theorem 3.28 that is based on the idea of [96] as well).

We consider the solutions of the equation

$$s \cdot \frac{1}{b} + c = t, \quad b \in \mathcal{B},\ c \in C,\ s \in \mathcal{S},\ t \in \mathcal{T}, \tag{3.1}$$

where

$$\mathcal{S} = \mathcal{A} \cdot \mathcal{B} \quad \text{and} \quad \mathcal{T} = \mathcal{A} + C.$$

As in [96] we note that for any triplet $(a, b, c) \in \mathcal{A} \times \mathcal{B} \times C$ there is a unique solution, namely $s = ab$, $t = a + c$. So for the number J of solutions to (3.1) we have $J \ge \#\mathcal{A}\#\mathcal{B}\#C$. On the other hand, using the orthogonality relation between exponential functions, we write

$$J = \frac{1}{p}\sum_{\lambda=0}^{p-1}\sum_{b\in\mathcal{B}}\sum_{c\in\mathcal{C}}\sum_{s\in\mathcal{S}}\sum_{t\in\mathcal{T}} \mathbf{e}_p(\lambda(sb^{-1}+c-t))$$

$$\le \frac{\#\mathcal{B}\#\mathcal{C}\#\mathcal{S}\#\mathcal{T}}{p} + \frac{1}{p}\sum_{\lambda=1}^{p-1}\left|\sum_{b\in\mathcal{B}}\sum_{s\in\mathcal{S}}\mathbf{e}_p(\lambda sb^{-1})\right|\left|\sum_{c\in\mathcal{C}}\mathbf{e}_p(\lambda c)\right|\left|\sum_{t\in\mathcal{T}}\mathbf{e}_p(\lambda t)\right|.$$

Applying the classical bound for bilinear exponential sums, see (4.1) below, and then the Cauchy inequality, we obtain

$$J \le \frac{\#\mathcal{B}\#\mathcal{C}\#\mathcal{S}\#\mathcal{T}}{p} + \frac{\sqrt{p\#\mathcal{B}\#\mathcal{S}}}{p}\left(\sum_{\lambda=1}^{p-1}\left|\sum_{c\in\mathcal{C}}\mathbf{e}_p(\lambda c)\right|^2\right)^{1/2}\left(\sum_{\lambda=0}^{p-1}\left|\sum_{t\in\mathcal{T}}\mathbf{e}_p(\lambda t)\right|^2\right)^{1/2}.$$

Extending the summation to $\lambda = 0$ and using the orthogonality relation between exponential functions again, we see that the remaining sums satisfy

$$\sum_{\lambda=0}^{p-1}\left|\sum_{c\in\mathcal{C}}\mathbf{e}_p(\lambda c)\right|^2 \le p\#\mathcal{C} \quad\text{and}\quad \sum_{\lambda=0}^{p-1}\left|\sum_{t\in\mathcal{T}}\mathbf{e}_p(\lambda t)\right|^2 \le p\#\mathcal{T}.$$

This yields the bound

$$\#\mathcal{A}\#\mathcal{B}\#\mathcal{C} \le \frac{\#\mathcal{B}\#\mathcal{C}\#\mathcal{S}\#\mathcal{T}}{p} + \sqrt{p\#\mathcal{B}\#\mathcal{C}\#\mathcal{S}\#\mathcal{T}}, \tag{3.2}$$

which implies Theorem 3.1 (in fact, with the constant $(3-\sqrt{5})/2 \ge 3/8$).

For example, if $\mathcal{A} = \mathcal{B} = \mathcal{C}$ and $\#\mathcal{A} \gg p^{2/3}$, then Theorem 3.1 gives the bound

$$\max\{\#(\mathcal{A}^2), \#(2\mathcal{A})\} \gg \sqrt{\#\mathcal{A}p},$$

which Garaev [96] has shown to be optimal.

For smaller sets the situation is less understood and is far from being settled. Various improvements of the initial result of [36] are due to Bourgain & Garaev [29], Garaev [95], Hart, Iosevich & Solymosi [118], Katz & Shen [141], Li [155] and Rudnev [165] and many others. The current state of affairs has been conveniently summarized by Bukh & Tsimerman [49] which we now reproduce here.

Theorem 3.2. *For an arbitrary set $\mathcal{A} \subseteq \mathbb{F}_p$, we have*

$$\max\{\#\left(\mathcal{A}^2\right), \#(2\mathcal{A})\} \gg \begin{cases} (\#\mathcal{A})^{12/11+o(1)}, & \text{if } \#\mathcal{A} \le p^{1/2}, \\ (\#\mathcal{A})^{7/6}p^{-1/24+o(1)}, & \text{if } p^{1/2} \le \#\mathcal{A} \le p^{35/68}, \\ (\#\mathcal{A})^{10/11}p^{1/11+o(1)}, & \text{if } p^{35/68} \le \#\mathcal{A} \le p^{13/24}, \\ (\#\mathcal{A})^2 p^{-1/2}, & \text{if } p^{13/24} \le \#\mathcal{A} \le p^{2/3}, \\ (\#\mathcal{A})^{1/2}p^{1/2}, & \text{if } \#\mathcal{A} \ge p^{2/3}. \end{cases}$$

The bounds of Theorem 3.2 are due to the work of Garaev [96] (the last two bounds), Li [155] (the second and third bounds) and Rudnev [165] (the first bound). Rudnev [165, Remark 2] also mentions the bound

$$\max\{\#(\mathcal{A}/\mathcal{A}), \#(\mathcal{A} \pm \mathcal{A})\} \gg (\#\mathcal{A})^{12/11}$$

for a set $\mathcal{A} \subseteq \mathbb{F}_p$ with $\#\mathcal{A} \le p^{1/2}$ (for any choice of the sign). Clearly for large sets $\mathcal{A} \subseteq \mathbb{F}_p$, one can get a nontrivial estimate from Theorem 3.1 taken with $\mathcal{B} = \mathcal{A}$ and $C = \mathcal{A}^{-1}$.

Explicit lower bounds on $\max\{\#\mathcal{A} + \mathcal{B}), \#\mathcal{A} \cdot \mathcal{B})\}$, for sets $\mathcal{A}, \mathcal{B} \subseteq \mathbb{F}_p$, are given by Garaev [97] and Shen [168], see also [98, Theorem 3.1].

We also present a beautiful result of Bourgain & Glibichuk [34, Lemma 5] that applies to arbitrary fields:

Theorem 3.3. *For arbitrary sets $\mathcal{A}, \mathcal{B} \subseteq \mathbb{F}_q$, such that $\#\mathcal{A} > 1$ and $\mathcal{B}$ is not contained in a proper subfield of $\mathbb{F}_q$,*

$$\max\{\#(\mathcal{A} + \mathcal{A} \cdot \mathcal{B}), \#(\mathcal{A} - \mathcal{A} \cdot \mathcal{B})\} \ge 2^{-1/4} \min\{\#\mathcal{A}(\#\mathcal{B})^{1/7}, (\#\mathcal{A})^{6/7} q^{1/7}\}.$$

In their seminal paper, Erdős & Szemerédi [90] have also introduced a sum-product problem associated with a graph. Namely, let $\mathcal{A} = \{a_1, \dots, a_N\}$ be an N-element set in an arbitrary ring. Consider a graph $\mathcal{G}$, with the vertex set $\mathcal{V} = \{1, \dots, N\}$ and the edge set $\mathcal{E}$. Then instead of (1.1) one can consider more general sets

$$\{a_u + a_v : [uv] \in \mathcal{E}\} \quad \text{and} \quad \{a_u a_v : [uv] \in \mathcal{E}\},$$

where for $u, v \in \mathcal{V}$ we use $[uv]$ to denote the edge from u to v.

Alon, Angel, Benjamini & Lubetzky [7] have recently achieved very interesting results on this problem; obtaining their finite field analogues is certainly an important and challenging problem.

3.2 Multifold Sum-Product Problem

A more general question about k-fold sum and product sets $\mathcal{A}^k$ and $k\mathcal{A}$ has also been studied (albeit not so actively as the classical case of $k = 2$), see [11, 55, 78, 107, 110, 118, 156] and references therein. Motivated by some new applications, the following estimate has been derived in [11] from Theorem 3.1 if one applies it inductively with $\mathcal{B} = \mathcal{A}^{k-1}$ and $C = (k-1)\mathcal{A}$:

Theorem 3.4. *For an arbitrary subset $\mathcal{A} \subseteq \mathbb{F}_p^*$ and integer $k \ge 1$, we have*

$$\#(\mathcal{A}^k) \cdot \#(k\mathcal{A}) \ge \min\left\{cp\#\mathcal{A}, \frac{c^{k-1}(\#\mathcal{A})^{2k}}{p^{k-1}}\right\},$$

where $c = 3/8$.

Furthermore, it is shown by Balog, Broughan & Shparlinski [11] the method of Garaev [96] and can be used for multifold sum-product problems in a more direct way, rather than inductively as in Theorem 3.4 as well. In particular, the following estimate is derived in [11] from a bound of exponential sums of Bourgain & Garaev [29, Theorem 1.2] (see also Theorem 4.4 below).

Theorem 3.5. *For arbitrary sets $\mathcal{A}, \mathcal{B}, \mathcal{C}, \mathcal{D} \subseteq \mathbb{F}_p^*$, we have*

$$\max\{\#(\mathcal{A}\cdot\mathcal{B}\cdot\mathcal{C}), \#(\mathcal{A}+\mathcal{D})\} \gg \min\{\sqrt{p\#\mathcal{A}}, (\#\mathcal{A})^{16/21}(\#\mathcal{B}\#\mathcal{C})^{1/7}(\#\mathcal{D})^{8/21}p^{-40/189+o(1)}\}.$$

The bound of Theorem 3.4 is nontrivial only if $\#\mathcal{A} \gg p^{1/2}$ (however, in this case it is more precise than Theorem 3.2).

Open Question 3.6. *Obtain a lower bound on $\max\{\#(\mathcal{A}^k), \#(k\mathcal{A})\}$ for $k \geq 3$ that is stronger than that of Theorem 3.2 for sets of any size.*

We note that naively one may suggest that for any set $\mathcal{A} \subseteq \mathbb{F}_p^*$ of cardinality $\#\mathcal{A} \geq p^{\varepsilon}$ for some ε and any $\delta > 0$, there is k, depending only on ε and $\delta > 0$ for which

$$\max\{\#(\mathcal{A}^k), \#(k\mathcal{A})\} \gg p^{1-\delta}.$$

This, however, is wrong as the following example shows, see [55]:

Let $\mathcal{H}$ be a multiplicative subgroup of $\mathbb{F}_p^*$ of order $\#\mathcal{H} \sim p^{3/4+o(1)}$ (there are infinitely many primes that have such a subgroup, see [92]).

By the pigeon-hole principle, there exists $s \in \mathbb{F}_p$ such that for the set

$$\mathcal{A} = \mathcal{H} \cap \{s+1, \ldots, s+\lceil p^{3/4} \rceil\}$$

we have

$$\#\mathcal{A} \geq \frac{\#\mathcal{H}p^{3/4}}{p} \sim p^{1/2+o(1)}.$$

However, that uniformly over all integers $k \geq 1$ we have

$$\max\{\#(\mathcal{A}^k), \#(k\mathcal{A})\} \leq kp^{3/4+o(1)}.$$

In particular, for a fixed k, the sizes of both sets are limited by $p^{3/4+o(1)}$.

3.3 Sum-Inversion Estimates

Bourgain [19, Theorem 4.1] has given a bound for the following *sum-inversion problem*:

Theorem 3.7. *For any $\varepsilon > 0$ there is some $\delta > 0$ so that for an arbitrary set $\mathcal{A} \subseteq \mathbb{F}_p^*$ with $\#\mathcal{A} < p^{1-\varepsilon}$ we have*

$$\max\{\#(2\mathcal{A}), \#(2\mathcal{A}^{-1})\} \gg (\#\mathcal{A})^{1+\delta}.$$

The proof of Theorem 3.7 in [19] is based on a finite field version of the Szemerédi–Trotter incidence theorem, see Section 3.5. Using a recent explicit estimate of Helfgott & Rudnev [127, Theorem 2] or its improvement due to Jones [137, 138], which we present as Theorem 3.15 below, in the argument of [19] one can easily derive a result with an explicit dependence of δ on ε in Theorem 3.7.

Using the method of Garaev [96], the following estimate, generalizing that of [50] is shown by Balog, Broughan & Shparlinski [11].

Theorem 3.8. *For arbitrary sets $\mathcal{A}, \mathcal{B}, C \subseteq \mathbb{F}_p^*$, we have*

$$\#(\mathcal{A} + \mathcal{B}) \cdot \#\left(\mathcal{A}^{-1} + C\right) \geq \tfrac{1}{6} \min\left\{p\#\mathcal{A}, \frac{(\#\mathcal{A})^2 \#\mathcal{B}\#C}{p}\right\}.$$

Taking $\mathcal{B} = \mathcal{A}^{-1}$ and $C = \mathcal{A}$ in Theorem 3.8, we derive:

Corollary 3.9. *For an arbitrary set $\mathcal{A} \subseteq \mathbb{F}_p^*$, we have*

$$\#\left(\mathcal{A} + \mathcal{A}^{-1}\right) \geq 6^{-1/2} \min\left\{\sqrt{p\#\mathcal{A}}, \frac{(\#\mathcal{A})^2}{\sqrt{p}}\right\}.$$

Furthermore, using an inductive argument, the following analogue of Theorem 3.4 is given in [11]:

Theorem 3.10. *For an arbitrary set $\mathcal{A} \subseteq \mathbb{F}_p^*$, we have*

$$\#(k\mathcal{A}) \cdot \#\left(k\mathcal{A}^{-1}\right) \geq \min\left\{cp\#\mathcal{A}, \frac{c^{k-1}(\#\mathcal{A})^{2k}}{p^{k-1}}\right\},$$

where $c = 1/6$.

As in Section 3.2 we note that there are sets $\mathcal{A} \subseteq \mathbb{F}_p$ of size $\#\mathcal{A} \geq p^{1/2}$ and such that uniformly over all integers $k \geq 1$ we have

$$\max\left\{\#(k\mathcal{A}), \#(k\mathcal{A}^{-1})\right\} \leq kp^{3/4+o(1)}. \tag{3.3}$$

Indeed, taking $\mathcal{H}$ to be the set of inverses modulo p of integers $h = 1, \ldots, \lceil p^{3/4} \rceil$ we see that by the pigeon-hole principle, there exists $s \in \mathbb{F}_p$ such that for the set

$$\mathcal{A} = \mathcal{H} \cap \left\{s + 1, \ldots, s + \left\lceil p^{3/4} \right\rceil\right\}$$

we have

$$\#\mathcal{A} \geq \frac{\#\mathcal{H} p^{3/4}}{p} \geq p^{1/2}.$$

which yields (3.3).

3.4 Equations over Finite Fields with Variables from Arbitrary Sets

It is certainly natural to ask about the extreme cases, when a certain function applied to several arbitrary sets (or to a Cartesian product of one set) gives the whole finite field. This can be re-cast as a question about solvability of equations with variables from arbitrary sets.

A series of very general results of this kind is due to Glibichuk [107] (see also [109]).

For example, as in [107], we say that a set $\mathcal{A} \subseteq \mathbb{F}_q$ is *special* if for some $\alpha \in \mathbb{F}_q^*$ and a proper subfield $\mathbb{F}_r \subseteq \mathbb{F}_q$, $\mathbb{F}_r \neq \mathbb{F}_q$, we have

$$\{\alpha a : a \in \mathcal{A}\} \subseteq \mathbb{F}_r .$$

Otherwise, we say that $\mathcal{A}$ is *nonspecial*. Then, by [107, Theorem 6] we have:

Theorem 3.11. *For any nonspecial set $\mathcal{A} \subseteq \mathbb{F}_q$ with $\#A > q^{1/(n-\delta)}$, arbitrary integer $n \geq 2$ and positive $\delta < 1$, we have*

$$N(n, \delta)A^{2n-2} = \mathbb{F}_q ,$$

where

$$N(n,\delta) = \begin{cases} 10, & \text{if } n = 2\,; \\ 6^{n-3} \max\left\{30\left(3 + \left\lfloor \dfrac{\log \delta^{-1}}{\log 2} \right\rfloor\right), 160\left(1 + \left\lfloor \dfrac{\log n}{\log 2} \right\rfloor\right)\right\}, & \text{if } n \geq 3\,. \end{cases}$$

Furthermore, for several special types of finite fields, such as $\mathbb{F}_p$, $\mathbb{F}_{p^2}$ and $\mathbb{F}_{p^3}$, Glibichuk [107] has also given stronger versions of Theorem 3.11 that hold with the exponent n instead of $2n-2$ (for a slightly different choice of $N(n,\delta)$). On the other hand, it is shown in the comments after [107, Theorem 5] that for infinitely many cases, the exponent $2n-2$ is the best possible regardless of the choice of $N(n,\delta)$.

We also recall a very elegant result of Glibichuk & Rudnev [110, Theorem 6] that extends several previous results of this kind:

Theorem 3.12. *For any two sets $\mathcal{A}, \mathcal{B} \subseteq \mathbb{F}_q$, with $\#A\#\mathcal{B} > 2q$ we have*

$$8(\mathcal{A} \cdot \mathcal{B}) = \mathbb{F}_q .$$

Theorems 3.11 and 3.12, due to their universality and essentially "condition-free" nature, have already found a number of interesting applications, see, for example, Sections 4.2.

In the same direction Bourgain [22] has given the following result:

Theorem 3.13. *For any two sets $\mathcal{A}, \mathcal{B} \subseteq \mathbb{F}_p$,*

$$\#\left(8(\mathcal{A} \cdot \mathcal{B}) - 8(\mathcal{A} \cdot \mathcal{B})\right) > \tfrac{1}{2} \min\left\{\#\mathcal{A}\#\mathcal{B}, p-1\right\} .$$

Several other equations with variables in arbitrary sets have also been considered in the literature. For example, Sárközy [166] has given an asymptotic formula

$$T(\mathcal{A},\mathcal{B},\mathcal{C},\mathcal{D}) = \frac{\#\mathcal{A}\#\mathcal{B}\#\mathcal{C}\#\mathcal{D}}{q-1} + O\left(\left(q\#\mathcal{A}\#\mathcal{B}\#\mathcal{C}\#\mathcal{D}\right)^{1/2}\right) \tag{3.4}$$

for the number $T(\mathcal{A},\mathcal{B},\mathcal{C},\mathcal{D})$ of the solutions of the similar equation

$$a + b = cd\,, \quad a \in \mathcal{A}\,, \quad b \in \mathcal{B}\,, \quad c \in \mathcal{C}\,, \quad d \in \mathcal{D}\,,$$

with arbitrary sets $\mathcal{A},\mathcal{B},\mathcal{C},\mathcal{D} \subseteq \mathbb{F}_q^*$. Hart & Iosevich [116, Theorem 1.1] obtained a similar result for the equation

$$ab + cd = \lambda\,, \quad a \in \mathcal{A}\,, \quad b \in \mathcal{B}\,, \quad c \in \mathcal{C}\,, \quad d \in \mathcal{D}\,. \tag{3.5}$$

The method of [116, 166] is based on bounds of exponential sums.

Clearly (3.4) implies that

$$T(\mathcal{A},\mathcal{B},\mathcal{C},\mathcal{D}) = (1 + o(1))\frac{\#\mathcal{A}\#\mathcal{B}\#\mathcal{C}\#\mathcal{D}}{q-1}\,, \qquad q \to \infty\,, \tag{3.6}$$

if for some fixed $\varepsilon > 0$ we have

$$\#\mathcal{A}\#\mathcal{B}\#\mathcal{C}\#\mathcal{D} \ge q^{3+\varepsilon}\,.$$

It is shown in [180] that using bounds of multiplicative character sums one can obtain a nontrivial asymptotic formula for $T(\mathcal{A},\mathcal{B},\mathcal{C},\mathcal{D})$ for a wider class of possible cardinalities of the sets $\mathcal{A},\mathcal{B},\mathcal{C},\mathcal{D}$. More precisely, by [180, Theorem 1] for any $\varepsilon > 0$ there exists some $\delta > 0$ so that (3.6) holds provided that

$$\#\mathcal{A} \ge q^{\varepsilon}\,, \quad \#\mathcal{B} \ge q^{1/2+\varepsilon} \quad \text{and} \quad \#\mathcal{C}\#\mathcal{D} \ge q^{2-\delta}\,.$$

It also follows from [180, Theorem 1] that under the same condition one has the same asymptotic formula for the number of solutions of the equation (3.5).

Using exponential sums, Hart & Iosevich [116] have shown that for any $2n$ sets $\mathcal{A}_i, \mathcal{B}_i \subseteq \mathbb{F}_q$, $i = 1,\dots,n$, with

$$\prod_{i=1}^{n} \#\mathcal{A}_i\#\mathcal{B}_i > Cq^{n+1}$$

for some absolute constant $C > 0$, the equation

$$\sum_{i=1}^{n} a_i b_i = \lambda\,, \quad a_i \in \mathcal{A}_i\,, \quad b_i \in \mathcal{B}_i\,, \quad i = 1,\dots,n\,, \tag{3.7}$$

has a solution for any $\lambda \in \mathbb{F}_q^*$ (although the proof is given only in the case of $\mathcal{A}_1 = \mathcal{B}_1 = \cdots = \mathcal{A}_n = \mathcal{B}_n$, the method and results immediately extend to the general case, see [116, Remark 1.3]).

Cilleruelo [66] shows that many of the above results can be obtained in elementary fashion via some combinatorial arguments.

Hart, Iosevich & Solymosi [118] have studied the equation

$$(a_1 - b_1) \cdots (a_n - b_n) = \lambda, \quad a_i, b_i \in \mathcal{A}, \; i = 1, \ldots, n, \tag{3.8}$$

where $\mathcal{A} \subseteq \mathbb{F}_q$ and shown that it has a solution for any $\lambda \in \mathbb{F}_q$, provided that $\#\mathcal{A} \geq q^{1/2+1/(2n)}$. Balog [9] has shown that for an odd $n = 2k + 1 \geq 5$, the same result holds for smaller sets $\mathcal{A} \subseteq \mathbb{F}_q$ with $\#\mathcal{A} \geq q^{1/2+1/2^k}$.

A similar result is also given by Hart, Iosevich, Koh & Rudnev [117, Theorem 2.9] for the equation

$$a_1 b_1 + \cdots + a_n b_n = \lambda, \quad a_i, b_i \in \mathcal{A}, \; i = 1, \ldots, n,$$

with $\lambda \in \mathbb{F}_q^*$ for sets $\mathcal{A} \subseteq \mathbb{F}_q^*$ of size $\#\mathcal{A} \geq q^{1/2+1/(2n)}$, see also [64, 201, 205].

Several more interesting results in this direction have been given by Hegyvári [121, 122] and Vinh [198, 201, 205, 206].

Finally, Vinh [200, Theorem 1.2] has considered similar problems for randomly chosen subsets of finite fields.

Theorem 3.14. *For any integer $n \geq 2$ and real $\alpha > 0$ there exists a constant $C > 0$ such that for any integer $t \geq Cq^{1/n}$, a set $\mathcal{A}$ of t elements chosen from $\mathbb{F}_q$ uniformly at random, satisfies $\mathbb{F}_q^* \subseteq d\mathcal{A}^2$ with probability of at least $1 - q\alpha^t$.*

3.5 Incidence Bounds

Let $\mathcal{P}$ be a set of points and let $\mathcal{L}$ be a set of lines in $\mathbb{F}_q^2$. We denote by $I(\mathcal{P}, \mathcal{L})$ the cardinality of the set of incidences between $\mathcal{P}$ and $\mathcal{L}$, that is,

$$I(\mathcal{P}, \mathcal{L}) = \#\{(P, L) \in \mathcal{P} \times \mathcal{L} : P \in L\}.$$

It is easy to prove that

$$I(\mathcal{P}, \mathcal{L}) \ll \#\mathcal{L}\sqrt{\#\mathcal{P}} + \#\mathcal{P}\sqrt{\#\mathcal{L}}. \tag{3.9}$$

Indeed, let $\chi(P, L)$ be the characteristic function of the event $P \in L$. Then, using the Cauchy inequality, we obtain

$$\begin{aligned} I(\mathcal{P}, \mathcal{L})^2 &= \left(\sum_{L \in \mathcal{L}} \sum_{P \in \mathcal{P}} \chi(P, L) \right)^2 \\ &\leq \#\mathcal{L} \sum_{L \in \mathcal{L}} \left(\sum_{P \in \mathcal{P}} \chi(P, L) \right)^2 = \#\mathcal{L} \sum_{P_1, P_2 \in \mathcal{P}} \sum_{L \in \mathcal{L}} \chi(P_1, L)\chi(P_2, L). \end{aligned}$$

For $P_1 = P_2$, we estimate the sum over L trivially as $\#\mathcal{L}$. Otherwise, since two distinct points P_1 and P_2 define a unique line we see that the sum over L is at most 1 in this case. So

$$I(\mathcal{P}, \mathcal{L})^2 \leq \#\mathcal{L} \left(\#\mathcal{P}\#\mathcal{L} + (\#\mathcal{P})^2 \right).$$

The bound (3.9) now follows.

Bourgain, Katz & Tao [36] have obtained a finite field analogue of the celebrated Szemerédi-Trotter theorem and proved that if $N = \max\{\#\mathcal{P}, \#\mathcal{L}\} \le p^{2-\varepsilon}$ then

$$I(\mathcal{P}, \mathcal{L}) \ll N^{3/2-\delta},$$

where $\delta > 0$ depends only on $\varepsilon > 0$, see [86, 193].

Helfgott & Rudnev [127, Theorem 2] have given an explicit version of this result: for prime fields $\mathbb{F}_p$ by proving that for $N < p$ one can take $\delta = 1/10678$. This has been further improved by Jones [138] as follows:

Theorem 3.15. *For any sets of points $\mathcal{P}$ and lines $\mathcal{L}$ in $\mathbb{F}_p^2$ with $\#\mathcal{P} = \#\mathcal{L} = N < p$, we have*

$$I(\mathcal{P}, \mathcal{L}) \ll N^{3/2-1/662+o(1)}.$$

Jones [136] extended the result of Helfgott & Rudnev [127, Theorem 2] to arbitrary finite fields.

As usual, the case of large sets is easier. For instance, Vinh [199] has shown that for any sets $\mathcal{P}$ and $\mathcal{L}$ in $\mathbb{F}_q$, we have

$$I(\mathcal{P}, \mathcal{L}) \le \frac{\#\mathcal{P}\#\mathcal{L}}{q} + \sqrt{q\#\mathcal{P}\#\mathcal{L}}.$$

Cilleruelo [66, Theorem 2.2] has obtained this result via an elementary combinatorial approach.

Finally, we note that Helfgott & Rudnev [127, Theorem 1] give the following elegant result.

Theorem 3.16. *Let $\mathcal{A} \subseteq \mathbb{F}_q$, and let $L(\mathcal{A})$ be the number of lines defined by the points $(a_1, a_2) \in \mathcal{A} \times \mathcal{A}$. If $\#\mathcal{A} < p^{1/2}$ then*

$$L(\mathcal{A}) \gg (\#\mathcal{A})^{1+1/232}.$$

Bourgain [25] has given a remarkable generalization of the Szemerédi-Trotter theorem to so-called *modular hyperbolas*, that is, the set of solutions to a congruence $cxy - ax + dy - b \equiv 0 \pmod p$ which can also be written as

$$\Gamma_g = \{(x, y) \in \mathbb{F}_p^2 : y \equiv \tau_g(x) \pmod p\},$$

where for an nonsingular modulo p matrix

$$g = \begin{pmatrix} a & b \\ c & d \end{pmatrix} \in \mathrm{GL}_2(p)$$

we define

$$\tau_g(x) \equiv \frac{ax + b}{cx + d} \pmod p.$$

It is shown in [25] the for any $\varepsilon > 0$ there exist $\delta > 0$ such that under some natural conditions, for any sets

$$\mathcal{P} \subseteq \mathbb{F}_p \quad \text{and} \quad \mathcal{G} \subseteq \mathrm{GL}_2(p)$$

we have

$$\#\{(x, y, g) \in \mathcal{P} \times \mathcal{P} \times \mathcal{G}\colon (x, y) \in \Gamma_g\} \le (\#\mathcal{P})^{1-\delta}\#\mathcal{G}\,.$$

3.6 Polynomial and Other Nonlinear Functions on Sets

Besides sum, product and inversion, many other operations can be, and have been, applied to a set.

Several such general results are due to Bukh & Tsimerman [49] and Tao [192]. To formulate some of the results from [49], we introduce one more notation. For a polynomial $f \in \mathbb{F}_p[X]$ we define the set

$$\mathcal{A}^{(f)} = \{f(a)\colon a \in \mathcal{A}\}\,.$$

Note that in [49] a different notation $f(\mathcal{A})$ is used, which has a different meaning in this paper. Here we present only two of the results of [49].

For example, by [49, Theorem 1] we have:

Theorem 3.17. *Let $f \in \mathbb{F}_p[X]$ be a polynomial of degree $d \ge 2$. Then for any set $\mathcal{A} \subseteq \mathbb{F}_p$ of size $\#\mathcal{A} \le \sqrt{p}$ we have*

$$\max\{\#(2\mathcal{A}), \#(2\mathcal{A}^{(f)})\} \gg (\#\mathcal{A})^{1+1/16\cdot 6^d}\,.$$

In a similar spirit, Vu [208] has shown that $\max\{\#(2\mathcal{A}), \#\mathcal{A}^{(f)}\}$ is always large, provided that $\#\mathcal{A} > p^{1/2}$, see also [119, 123].

Furthermore, by [49, Theorem 3] we have:

Theorem 3.18. *Let $f \in \mathbb{F}_p[X]$ be a polynomial of degree $d \ge 2$ with k monomials. Then for every fixed integer r and any set $\mathcal{A} \subseteq \mathbb{F}_p$ of size*

$$d^{40r}p^{4/r} \le \#\mathcal{A} \le \sqrt{p}$$

we have

$$\max\{\#(\mathcal{A}^2), \#(2\mathcal{A}^{(f)})\} \gg (\#\mathcal{A})^{1+\eta(d,k,r)}\,,$$

where

$$\eta(d, k, r) = \left(5000(r+k)^2\frac{\log d}{\log 2}\right)^{-k}\,.$$

Open Question 3.19. *Obtain an analogue of Theorems 3.17 and 3.18 for the sets $\mathcal{A}^2$ and $(\mathcal{A}^{(f)})^2$.*

Certainly Question 3.19 may admit a positive answer only under some additional conditions on the polynomial f (for example, for $f(X) = aX^d$ no such bound is possible). As usual, for $\#\mathcal{A} \ge \sqrt{p}$, using the method of Garaev [96], one can obtain a nontrivial bound for the sets of Question 3.19.

Several more results of this type, that apply to large sets and are also based on the method of Garaev [96] have been given by Cilleruelo, Garaev, Ostafe & Shparlinski [68]. The ideas of [68] have found application in the study of the distribution of some families of algebraic curves in isomorphism classes, see [63, 69, 70].

Although clearly one cannot have any nontrivial estimates on the just product set $\mathcal{A}^2$, for the set $\mathcal{A}(\mathcal{A}+1)$, Jones & Roche-Newton [139], improving on a similar result of Garaev & Shen [100], have obtained the following bound:

Theorem 3.20. *For any set $\mathcal{A} \subseteq \mathbb{F}_p$ of size $\#\mathcal{A} \le \sqrt{p}$ we have*

$$\#\left(\mathcal{A}(\mathcal{A}+1)\right) \gg (\#\mathcal{A})^{57/56+o(1)} .$$

We also recall that Bourgain, Glibichuk & Konyagin [35, Theorems 3 and 4] have given the following estimate:

Theorem 3.21. *For any set $\mathcal{A} \subseteq \mathbb{F}_p$ we have*

$$\#\left(2\left(\mathcal{A}\cdot(\mathcal{A}-\mathcal{A})\right)\right) \ge \begin{cases} p/2, & \text{if } \#\mathcal{A} \ge p^{1/2}, \\ 0.1\,(\#\mathcal{A})^{3/2}, & \text{if } \#\mathcal{A} < p^{1/2}. \end{cases}$$

A very general result of this kind is also given by Shen [170].

Bukh & Tsimerman [49, Theorem 6] have obtained a remarkable estimate on the set $F(\mathcal{A},\mathcal{B})$ of some wide class of polynomials $F(X,Y) \in \mathbb{F}_q[X,Y]$, for sets $\mathcal{A},\mathcal{B} \subseteq \mathbb{F}_q$ of sizes $\#\mathcal{A},\#\mathcal{B} \ge q^{7/8}$. One of the remarkable features of this result is that it applies to sets in arbitrary fields. Hegyvári & Hennecart [123] also give some lower bounds on sets involving polynomials.

Tao [192] gives a natural characterization of bivariate polynomials that do not expand large sets up to a dense set in a finite field, that is, polynomials $F(X,Y) \in \mathbb{F}_q[X,Y]$ for which $F(\mathcal{A},\mathcal{B}) = o(q)$ for some large sets $\mathcal{A},\mathcal{B} \subseteq \mathbb{F}_q$.

It is also shown by Bourgain [19, Bound (0.11)] that for any $\varepsilon > 0$, there is some $\delta > 0$ such that for a set $\mathcal{A} \subseteq \mathbb{F}_p^*$ with $p^\varepsilon < \#\mathcal{A} < p^{1-\varepsilon}$ we have

$$\#\left(\mathcal{A}+\mathcal{A}^{-1}\right) \ge \#\mathcal{A}p^{\delta} . \tag{3.10}$$

Open Question 3.22. *Obtain an explicit form of the bound* (3.10).

Clearly for large sets it is given by Corollary 3.9, so in Question 3.22 one has to consider the sets of cardinality $\#\mathcal{A} < p^{1/2}$.

Finally, for some non-polynomial operations, on set elements have been considered in [182]. More precisely, given an element $g \in \mathbb{F}_p^*$ of order T and two sets

$\mathcal{A}, \mathcal{B} \subseteq \mathbb{Z}_T^*$, one can consider the *exponential sum-product problem* of showing that at least one of the sets

$$\mathcal{U} = \{g^a + g^b : a \in \mathcal{A},\ b \in \mathcal{B}\} \quad \text{and} \quad \mathcal{V} = \{g^{ab} : a \in \mathcal{A},\ b \in \mathcal{B}\} \tag{3.11}$$

is "large". Then by [182, Theorem 1], we have

Theorem 3.23. *For arbitrary sets $\mathcal{A}, \mathcal{B} \subseteq \mathbb{Z}_T^*$ and any integer $\nu \geq 1$, for the sets $\mathcal{U}$ and $\mathcal{V}$, given by (3.11), we have*

$$\max\{\#\mathcal{U}, \#\mathcal{V}\} \geq \min\{\sqrt{p\#\mathcal{B}}, (\#\mathcal{A})^{\alpha_\nu} (\#\mathcal{B})^{\beta_\nu} T^{-\tau_\nu} p^{-\rho_n}\} p^{o(1)},$$

as $p \to \infty$, where

$$\alpha_\nu = \frac{\nu^2 + 2\nu}{3\nu^2 + 2\nu - 1}, \qquad \beta_\nu = \frac{2\nu}{3\nu - 1},$$
$$\tau_\nu = \frac{1}{3\nu - 1}, \qquad \rho_\nu = \frac{\nu}{2(3\nu^2 + 2\nu - 1)}.$$

We finish this section with an open question from [182]:

Open Question 3.24. *Given two sets $\mathcal{X}, \mathcal{Y} \subseteq \{1, \dots, p-1\}$ show that at least one of the sets*

$$\{x^y \pmod p : x \in \mathcal{X},\ y \in \mathcal{Y}\} \quad \textit{and} \quad \{y^x \pmod p : x \in \mathcal{X},\ y \in \mathcal{Y}\}$$

is "large".

We remark that one probably needs to impose some additional conditions of the type that all elements $x \in \mathcal{X}$ and $y \in \mathcal{Y}$ are of the same multiplicative order T and also that $\gcd(xy, T) = 1$. Furthermore, it is necessary to assume that $\#\mathcal{X}, \#\mathcal{Y} \leq T^{1-\delta}$ for some fixed $\delta > 0$.

3.7 Structured Sets

We say that $\mathcal{I}$ is an *interval in* $\mathbb{F}_p$ if it is a set of residues of consecutive integers.

For an interval $\mathcal{I}$ in $\mathbb{F}_p$ and a subgroup $\mathcal{G}$ of $\mathbb{F}_p^*$ we trivially have $\#(2\mathcal{I}) \leq 2\#\mathcal{I}$ and $\#(\mathcal{G}^2) = \#\mathcal{G}$. However, investigating the quantities $\#(\mathcal{I}^2)$ and $\#(2\mathcal{G})$ (or their k-fold analogues) is far from trivial.

We start with a recent result of Shkredov & Vyugin [209]:

Theorem 3.25. *For any subgroup $\mathcal{G} \subseteq \mathbb{F}_p^*$ of order $\#\mathcal{G} \ll p^{1/2}$ we have*

$$\#(\mathcal{G} - \mathcal{G}) \gg (\#\mathcal{G})^{5/3} (\log \#\mathcal{G})^{-1/2}.$$

For the set $\mathcal{G} + \mathcal{G}$ only a weaker lower bound is available, see [173]:

Theorem 3.26. *For any subgroup $\mathcal{G} \subseteq \mathbb{F}_p^*$ of order $\#\mathcal{G} \ll p^{1/2}$, we have*

$$\#(\mathcal{G} + \mathcal{G}) \gg (\#\mathcal{G})^{8/5} (\log \#\mathcal{G})^{-3/5} .$$

Hart [115] extended the bound of Theorem 3.26 to subgroups of order $\#\mathcal{G} \ll p^{5/9}$.

For larger subgroups $\mathcal{G} \subseteq \mathbb{F}_p^*$ of order $p^{1/2} \ll \#\mathcal{G} \ll p^{2/3}$, Schoen & Shkredov [167] have given a series of bounds on $\#(\mathcal{G} \pm \mathcal{G})$, see also [173]. Many of this results are based on upper bounds for additive energy, that is

$$E_+(\mathcal{G}) = \#\{a_1 + a_2 = a_3 + a_4 \colon a_1, a_2, a_3, a_4 \in \mathcal{G}\}$$

of $\mathcal{G}$. Heath-Brown & Konyagin [120] have proved that for $\#\mathcal{G} \ll p^{2/3}$, we have

$$E_+(\mathcal{G}) \ll (\#\mathcal{G})^{5/2} , \tag{3.12}$$

see also [147] for a generalization. For small subgroups with $\#\mathcal{G} \ll p^{3/5}(\log p)^{-6/5}$, Shkredov [173] gives a stronger estimate

$$E_+(\mathcal{G}) \ll (\#\mathcal{G})^{22/9} (\log \#\mathcal{G})^{2/3} . \tag{3.13}$$

Bourgain, Garaev, Konyagin & Shparlinski [32] have proved the following estimate, improving and generalizing previous results of Cilleruelo & Garaev [67]:

Theorem 3.27. *Let $\nu \geq 3$ be a fixed integer and let*

$$e_\nu = \max\{\nu^2 - 2\nu - 2, \nu^2 - 3\nu + 4\} .$$

Assume that for some sufficiently large positive integer h and prime p we have

$$h < p^{1/e_\nu} .$$

Then for any interval $\mathcal{I}$ in $\mathbb{F}_p$ of length h we have

$$\#(\mathcal{I}^\nu) > \exp\left(-c(\nu)\frac{\log h}{\log\log h}\right) h^\nu ,$$

where $c(\nu)$ depends only on ν.

Using the results of [33], one obtain similar estimates in a large region of values of h for almost all primes p and also for almost all intervals $\mathcal{I}$.

For a set of rational fractions

$$\mathcal{A} = \left\{\frac{x+s}{x+t} \colon x \in \mathcal{I} \setminus \{-t\}\right\} \subseteq \mathbb{F}_p ,$$

where, $s, t \in \mathbb{F}_p$, $s \neq t$ and, as before, $\mathcal{I}$ is an interval in $\mathbb{F}_p$ of length h, Bourgain, Garaev, Konyagin & Shparlinski [31, Lemma 2.35] have also given an analogue of Theorem 3.27 for smaller values of h, namely for

$$h < p^{c\nu^{-4}} ,$$

where c is a certain absolute constant.

We also recall an observation of Bourgain [19, Bound (0.12)] that Theorem 3.7 implies that for any $\varepsilon > 0$ there is some $\delta > 0$ so that for an arbitrary interval $\mathcal{I}$ in $\mathbb{F}_p$ with $\#\mathcal{I} < p^{1-\varepsilon}$ we have

$$\#(2\mathcal{I}^{-1}) \gg (\#\mathcal{I})^{1+\delta} .$$

Using explicit versions of Theorem 3.7 which, as we have mentioned, seem to be easily obtainable nowadays, one can also have an explicit form of the above inequality.

There are however more direct and powerful methods to investigate the size of $2\mathcal{I}^{-1}$ and more generally of $k\mathcal{I}^{-1}$, which have recently been introduced and used by Bourgain & Garaev [30]. For example, it has been observed in [30] that some results of [67] lead to the bound

$$\#(2\mathcal{I}^{-1}) \ge \min\left\{(\#\mathcal{I})^2, \sqrt{p\#\mathcal{I}}\right\}(\#\mathcal{I})^{o(1)} .$$

Indeed, let $T(\lambda)$ be the number of solutions of the congruence

$$x^{-1} + y^{-1} \equiv \lambda \pmod p , \quad x, y \in \mathcal{I} ,$$

that for $\lambda \not\equiv 0 \pmod p$ can be written in the form

$$(x - \lambda^{-1})(y - \lambda^{-1}) \equiv \lambda^{-2} \pmod p .$$

By [67, Theorem 1] for $\lambda \not\equiv 0 \pmod p$ we have

$$T(\lambda) \le (\#\mathcal{I})^{3/2+o(1)} p^{-1/2} + (\#\mathcal{I})^{o(1)} .$$

Since $T(0) \le \#\mathcal{I}$, the result now follows from

$$(\#\mathcal{I})^2 \le \sum_{\lambda \in 2\mathcal{I}^{-1}} T(\lambda) \ll \#\mathcal{I} + \#(2\mathcal{I}^{-1}) \max_{\gcd(\lambda,p)=1} T(\lambda) .$$

3.8 Elliptic Curve Analogues

We assume that $\mathbb{E}$ is given by an affine Weierstraß equation

$$\mathbb{E} : y^2 + (a_1x + a_3)y = x^3 + a_2x^2 + a_4x + a_6 ,$$

with some $a_1, \dots, a_6 \in \mathbb{F}_q$, see [188].

We recall that the set of all points on $\mathbb{E}$ forms an Abelian group, with the point at infinity $\mathcal{O}$ as the neutral element. We also use $\oplus$ to denote the group operation.

As usual, we write every point $Q \ne \mathcal{O}$ on $\mathbb{E}$ as $Q = (x(Q), y(Q))$.

Let $\mathbb{E}(\mathbb{F}_q)$ denote the set of $\mathbb{F}_q$-rational points on $\mathbb{E}$.

Let $P \in \mathbb{E}(\mathbb{F}_q)$ be a fixed point of order T. Then given two sets $\mathcal{A}, \mathcal{B} \subseteq \mathbb{Z}_T^*$, we consider the following two pair of sets

$$\{x(aP) + x(bP) : a \in \mathcal{A}, b \in \mathcal{B}\} , \quad \{x(abP) : a \in \mathcal{A} , \ b \in \mathcal{B}\} , \qquad (3.14)$$

and

$$\{x(aP)x(bP) \colon a \in \mathcal{A},\ b \in \mathcal{B}\}\,, \quad \{x(abP) \colon a \in \mathcal{A}\,,\ b \in \mathcal{B}\}\,. \tag{3.15}$$

Ahmadi & Shparlinski [2, 3] have shown that the method of Garaev [96] and some bounds of bilinear exponential sums over points of elliptic curves, imply that at least one of the sets (3.14) and at least one of the sets (3.15) is "large".

Similar arguments are used in [179] to show that for two sets $\mathcal{P}, \mathcal{Q} \subseteq \mathbb{E}(\mathbb{F}_q)$ at least one of the sets

$$\{x(P) + x(Q) \colon P \in \mathcal{P},\ Q \in \mathcal{Q}\}, \quad \text{and} \quad \{x(P \oplus Q) \colon P \in \mathcal{P}\,,\ Q \in \mathcal{Q}\}\,,$$

is also "large".

Certainly one can consider several more problems of this type. For example, one can consider the sets

$$\begin{aligned} \mathcal{U} &= \{x(P) + x(Q) \colon P \in \mathcal{P},\ Q \in \mathcal{Q}\}\,, \\ \mathcal{V} &= \{y(P) + y(Q) \colon P \in \mathcal{P}\,,\ Q \in \mathcal{Q}\}. \end{aligned} \tag{3.16}$$

In fact, the last question can be considered in the more general setting of arbitrary affine curves. That is, we assume that $\mathcal{P}$ and $\mathcal{Q}$ are subsets of the set $C(\mathbb{F}_q)$ of $\mathbb{F}_q$-rational points on an affine curve C over $\mathbb{F}_q$:

$$C \colon\ f(X, Y) = 0\,, \tag{3.17}$$

defined by a nonlinear absolutely irreducible polynomial $f \in \mathbb{F}_q[X, Y]$.

In order to give yet another example of application of the method of Garaev [96] and show its versatility, we present the following estimate (for simplicity we consider only the case of prime fields).

Theorem 3.28. *Let $\mathcal{P}, \mathcal{Q} \subseteq C(\mathbb{F}_p)$, where C is an affine curve over $\mathbb{F}_p$ given by the equation*

$$C \colon\ f(X, Y) = 0\,,$$

with a nonlinear absolutely irreducible polynomial $f \in \mathbb{F}_p[X, Y]$. Then for the sets $\mathcal{U}$ and $\mathcal{V}$ given by (3.16) *we have*

$$\#\mathcal{U}\#\mathcal{V} \gg \min\left\{p\#\mathcal{Q},\ \frac{(\#\mathcal{P})^2\,(\#\mathcal{Q})^2}{p}\right\}\,,$$

where the implied constant depends only on $\deg f$.

Proof. Given a subset $\mathcal{S} = \{S = (x(S), y(S)) \in C(\mathbb{F}_p)\}$ of the set $C(\mathbb{F}_p)$, we denote

$$\mathcal{X}_\mathcal{S} = \{x(S)\ :\ S \in \mathcal{S}\} \quad \text{and} \quad \mathcal{Y}_\mathcal{S} = \{y(S)\ :\ S \in \mathcal{S}\}\,.$$

Note that

$$\#\mathcal{X}_\mathcal{S} \gg \#\mathcal{S} \quad \text{and} \quad \#\mathcal{Y}_\mathcal{S} \gg \#\mathcal{S} \tag{3.18}$$

(since each of the maps $S \to x(S)$ and $S \to y(S)$ has at most $\deg f$ preimages).

Thus, $\mathcal{U} = \mathcal{X}_\mathcal{P} + \mathcal{X}_\mathcal{Q}$, $\mathcal{V} = \mathcal{Y}_\mathcal{P} + \mathcal{Y}_\mathcal{Q}$.

Let J be the number of solutions $(u, v, x, y) \in \mathcal{U}\times\mathcal{V}\times\mathcal{X}_\mathcal{Q}\times\mathcal{Y}_\mathcal{Q}$ to the equation

$$f(u - x, v - y) = 0 .$$

Clearly the 4-tuples

$$(x(P) + x(Q), y(P) + y(R), x(Q), y(R)), \quad P \in \mathcal{P}, \ Q, R \in \mathcal{Q},$$

are all solutions to this equation. Therefore, by (3.18),

$$J \gg \#\mathcal{P}(\#\mathcal{Q})^2 . \tag{3.19}$$

On the other hand, as in Section 3.1, we have

$$J = \sum_{(s,t)\in C(\mathbb{F}_p)} \sum_{u\in\mathcal{U}} \sum_{v\in\mathcal{V}} \sum_{x\in\mathcal{X}_\mathcal{Q}} \sum_{y\in\mathcal{Y}_\mathcal{Q}} \frac{1}{p^2} \sum_{\lambda,\mu\in\mathbb{F}_p} \mathbf{e}_p\left(\lambda(s - u + x) + \mu(t - v + y)\right) .$$

Thus, changing the order of summation and separating the term corresponding to $a = b = 0$ we obtain

$$\left| J - \frac{\#C(\mathbb{F}_p)\#\mathcal{U}\#\mathcal{V}\#\mathcal{X}_\mathcal{Q}\#\mathcal{Y}_\mathcal{Q}}{p^2} \right| \le \frac{1}{p^2}\Delta , \tag{3.20}$$

where

$$\Delta = \sum_{\substack{\lambda,\mu\in\mathbb{F}_p \\ (\lambda,\mu)\neq(0,0)}} \left| \sum_{(s,t)\in C(\mathbb{F}_p)} \mathbf{e}_p(\lambda s + \mu t) \right| \left| \sum_{u\in\mathcal{U}} \mathbf{e}_p(\lambda u) \right| \left| \sum_{v\in\mathcal{V}} \mathbf{e}_p(\mu v) \right|$$
$$\times \left| \sum_{x\in\mathcal{X}_\mathcal{Q}} \mathbf{e}_p(\lambda x) \right| \left| \sum_{y\in\mathcal{Y}_\mathcal{Q}} \mathbf{e}_p(\mu y) \right| .$$

Applying the Bombieri bound [18] for character sums along a curve to the first sum and then expanding the summation to all pairs $(\lambda, \mu) \in \mathbb{F}_q^2$, we derive

$$\Delta \ll p^{1/2} \sum_{\lambda\in\mathbb{F}_p} \left| \sum_{u\in\mathcal{U}} \mathbf{e}_p(\lambda u) \right| \left| \sum_{x\in\mathcal{X}_\mathcal{Q}} \mathbf{e}_p(\lambda x) \right|$$
$$\times \sum_{\mu\in\mathbb{F}_p} \left| \sum_{v\in\mathcal{V}} \mathbf{e}_p(\mu v) \right| \left| \sum_{y\in\mathcal{Y}_\mathcal{Q}} \mathbf{e}_p(\mu y) \right| .$$

Again as in Section 3.1, using the Cauchy inequality and using the orthogonality of characters, we derive

$$\Delta \ll p^{5/2}\sqrt{\#\mathcal{U}\#\mathcal{V}\#\mathcal{X}_\mathcal{Q}\#\mathcal{Y}_\mathcal{Q}} .$$

Thus, together with (3.20), we obtain

$$\left| J - \frac{\#C(\mathbb{F}_p)\#\mathcal{U}\#\mathcal{V}\#\mathcal{X}_\mathcal{Q}\#\mathcal{Y}_\mathcal{Q}}{p^2} \right| \ll \sqrt{p\#\mathcal{U}\#\mathcal{V}\#\mathcal{X}_\mathcal{Q}\#\mathcal{Y}_\mathcal{Q}} . \tag{3.21}$$

Comparing (3.19) and (3.21), recalling that by the Weil bound

$$\#C(\mathbb{F}_p) = p + O(p^{1/2}) ,$$

see [159, Section X.5], and using (3.18), we conclude the proof. □

3.9 Matrix Analogues

Let $\mathcal{M}_n(q)$ denote the ring of $n \times n$ matrices over $\mathbb{F}_q$. We also use $\mathrm{GL}_n(q)$ and $\mathrm{SL}_n(q)$ in their standard meaning as the groups of nonsingular matrices and the special linear group over $\mathbb{F}_q$.

Noncommutativity of the matrix multiplication leads to rather unusual phenomena in $\mathrm{GL}_n(q)$ and $\mathrm{SL}_n(q)$, see also [190] for a general point of view on additive combinatorics in noncommutative groups.

It is important to notice that in matrix settings one has results that apply to just one set. A typical example is a result of Gill & Helfgott [104], generalizing the previous results of Helfgott [125, 126].

Theorem 3.29. *For any $\varepsilon > 0$ there exist some C and $\delta > 0$ so that for any prime p, integer $n \geq 2$ and set $\mathcal{R} \subseteq \mathcal{M}_n(q)$ that generates $\mathrm{SL}_n(p)$, either $\mathcal{R} \geq p^{n+1-\varepsilon}$ or $\#(\mathcal{R}^3) \geq (\#\mathcal{R})^{1+\delta}$.*

We also note that Kowalski [152] has given an explicit version of the result of [125].

This direction is now a very active and exciting area of research, several more very significant results in this direction are given in [28, 40–48, 105, 164, 190, 191]. In particular, Tao [190] has introduced a very important notion of an *approximate group*, that has been developed by Breuillard, Green & Tao in [42–44, 46–48].

Chapman & Iosevich [65, Theorem 3] have considered the sets $\mathcal{R}(\mathcal{A}) \subseteq \mathrm{SL}_2(q)$ of 2×2 matrices with entries from some fixed set $\mathcal{A} \subseteq \mathbb{F}_q$.

Theorem 3.30. *For any $\mathcal{A} \subseteq \mathbb{F}_q$ with $\mathcal{A} \gg q^{5/6}$, we have $\#(\mathcal{R}(\mathcal{A})^2) \gg q^3$.*

Ferguson, Hoffman, Luca, Ostafe & Shparlinski [91] have considered matrix analogues of some equations considered in Section 3.4, such as equation (3.8) with $n = 2$. Recently, Hu & Li [129] have improved some of the results of [91].

The following question has been asked by Maze, Monico & Rosenthal [160] in the settings of matrices over semirings, however it is probably natural to start with studying it over finite fields. The question is: given three $n \times n$ matrices A, B, S over $\mathbb{F}_q$, obtain a nontrivial lower bound on the size of the set of the matrix products

$$\mathcal{M}_k(A, B, S) = \{f(A)Sg(B) \colon f, g \in \mathbb{F}_q[X], \deg f, g < k\},$$

where f and g run through all polynomials over $\mathbb{F}_q$ of degree less than k. Certainly some further conditions on A, B and S are necessary. For example, one can assume that minimal polynomials of A and B are of degree n and $k < n$. Clearly, in this case we have the following trivial inequalities

$$\#\mathcal{M}_k(A, B, S) \geq q^k \quad \text{and} \quad \#\mathcal{M}_k(A, A, A) \leq q^n.$$

The question has a flavour of additive combinatorics and can possibly be studied by its methods. For example, let

$$\mathcal{F} = \{f(A)\colon f \in \mathbb{F}_q[X],\ \deg f < k\}\,,$$
$$\mathcal{G} = \{Sg(B)\colon g \in \mathbb{F}_q[X],\ \deg g < k\}\,.$$

Clearly the cardinalities of sum sets

$$\#(\mathcal{F} + \mathcal{F}) = \#\mathcal{F} = q^k \quad \text{and} \quad \#(\mathcal{G} + \mathcal{G}) = \#\mathcal{G} = q^k$$

are "small". Therefore, one can expect that the cardinality of the product set $\#(\mathcal{F} \cdot \mathcal{G}) = \#\mathcal{M}_k(A, B, S)$ is "large".

In this direction, Chang [62] has recently obtained a series of lower bounds on $\#\mathcal{M}_k(A, B, S)$ that depend on spectral properties of A and B and also on the dimension of the kernel of $AB - BA$.

4 Applications

4.1 Exponential and Character Sums

We first recall the following well-known bound that applies to arbitrary subsets $\mathcal{X}, \mathcal{Y} \subseteq \mathbb{F}_p$:

$$\left|\sum_{x\in\mathcal{X}}\sum_{y\in\mathcal{Y}} e_p(xy)\right| \le \sqrt{p\#\mathcal{X}\#\mathcal{Y}}\,, \tag{4.1}$$

see, for example, [29, Bound (1.4)], or [122, Bound (0.1)], or [166, Lemma7]. The proof is very simple, and uses nothing but the Cauchy inequality and the orthogonality relation between exponential functions:

$$\begin{aligned}\left|\sum_{x\in\mathcal{X}}\sum_{y\in\mathcal{Y}} e_p(xy)\right|^2 &\le \left(\sum_{x\in\mathcal{X}}\left|\sum_{y\in\mathcal{Y}} e_p(xy)\right|\right)^2 \le \#\mathcal{X}\sum_{x\in\mathcal{X}}\left|\sum_{y\in\mathcal{Y}} e_p(xy)\right|^2\\ &\le \#\mathcal{X}\sum_{x\in\mathbb{F}_p}\left|\sum_{y\in\mathcal{Y}} e_p(xy)\right|^2 = \#\mathcal{X}\sum_{y,z\in\mathcal{Y}}\sum_{x\in\mathbb{F}_p} e_p(x(y-z))\\ &= p\#\mathcal{X}\#\mathcal{Y}.\end{aligned}$$

Despite the above essentially trivial argument, no direct improvement of the inequality (4.1) is known. In fact, it is easy to see that if, say, $\mathcal{X} = \mathcal{Y} = \{1, \dots, h\}$ with an integer $h = o(p^{1/2})$ then $xy = o(p)$ and, thus, $e_p(xy) = 1 + o(1)$ for $x \in \mathcal{X}$, $y \in \mathcal{Y}$. Thus, no nontrivial bound is possible in this case as,

$$\sum_{x\in\mathcal{X}}\sum_{y\in\mathcal{Y}} e_p(xy) = (1 + o(1))\#\mathcal{X}\#\mathcal{Y}\,. \tag{4.2}$$

However, the methods of additive combinatorics have succeeded in deriving a series of very interesting and useful results for some related sums.

Perhaps the most striking application of additive combinatorics to the bounds of exponential sums has been given by Bourgain, Glibichuk & Konyagin [35]. The result of [35, Theorem 5] has been improved and extended in numerous directions and giving a detailed description of these results requires a separate survey. In fact, such a comprehensive survey has recently been given by Garaev [98] which we highly recommend to the reader.

Here we limit ourselves only to a very few illustrative examples which may hopefully give some taste of this beautiful and challenging direction and also demonstrate the generality of its results.

In particular, we recall that Garaev [98, Theorem 4.1] gives the following estimate (see also [22, Theorems 3 and 5] and [24, Theorem 3]):

Theorem 4.1. *Let $\mathcal{X}_1, \dots, \mathcal{X}_k \subseteq \mathbb{F}_p$ be $k \le 1.44 \log\log p$ arbitrary sets with*

$$\#\mathcal{X}_1 \#\mathcal{X}_2 \left(\#\mathcal{X}_3 \dots \#\mathcal{X}_k\right)^{1/81} > p^{1+\eta},$$

where $\eta > 0$ is an arbitrary fixed real number. Then

$$\sum_{x_1 \in \mathcal{X}_1} \dots \sum_{x_k \in \mathcal{X}_k} \mathbf{e}_p(x_1 \dots x_k) \ll \#\mathcal{X}_1 \dots \#\mathcal{X}_k p^{-0.45\eta/2^k}.$$

Theorem 4.1 immediately implies a nontrivial bound on exponential sums over very small subgroups of $\mathbb{F}_p$, see [98, Corollary 4.1]. For example, here we present the result of Garaev [98, Corollary 4.1] in a slightly more general and explicit form.

Corollary 4.2. *Let $g \in \mathbb{F}_p^*$ of multiplicative order t. Then for*

$$t \ge N \ge \exp\left(57 \frac{\log p}{\log\log p}\right),$$

we have

$$\max_{\gcd(a,p)=1} \left| \sum_{n=1}^{N} \mathbf{e}_p\left(ag^n\right) \right| \ll N \exp\left(-(\log p)^{0.0018}\right).$$

Bourgain & Glibichuk [34] have given analogues of Theorem 4.1 and Corollary 4.2 for arbitrary finite fields (where many additional complications arise).

For exponential sums along short segments of consecutive powers of a primitive root (instead of a small subgroup), Konyagin & Shparlinski [150, Theorem 2], improving some previous estimates of Bourgain & Garaev [29], have derived the following bound:

Theorem 4.3. *For any primitive root $g \in \mathbb{F}_p$ and an integer N, we have*

$$\max_{\gcd(a,p)=1} \left| \sum_{n=1}^{N} \mathbf{e}_p\left(ag^n\right) \right| \le \begin{cases} p^{1/8+o(1)} N^{71/96}, & \text{if } N \le p^{1/2}, \\ p^{23/96+o(1)} N^{49/96}, & \text{if } p^{1/2} < N < p, \end{cases}$$

as $p \to \infty$.

Kerr [143] has extended the bound of Theorem 4.3 to arbitrary $g \in \mathbb{F}_p$.

Finally, we recall the following result of Bourgain & Garaev [29, Theorem 1.2]:

Theorem 4.4. *For arbitrary subsets $\mathcal{X}, \mathcal{Y}, \mathcal{Z} \subseteq \mathbb{F}_p$,*

$$\left| \sum_{x\in\mathcal{X}} \sum_{y\in\mathcal{Y}} \sum_{z\in\mathcal{Z}} \mathbf{e}_p(xyz) \right| \le (\#\mathcal{X}\#\mathcal{Y}\#\mathcal{Z})^{13/16} p^{5/18+o(1)} ,$$

as $p \to \infty$.

Bourgain [19, Propositions 3.6 and 3.7] has also given nontrivial estimates of the sums

$$\sum_{x\in\mathcal{X}} \sum_{y\in\mathcal{Y}} \mathbf{e}_p\left(a\left(xy + x^2y^2\right)\right) \quad \text{and} \quad \sum_{x\in\mathcal{X}} \sum_{y\in\mathcal{Y}} \mathbf{e}_p\left(a\left(xy + g^{x+y}\right)\right) . \tag{4.3}$$

where g is a primitive root of $\mathbb{F}_p$ and $\mathcal{X}, \mathcal{Y} \subseteq \{0, \dots, p-1\}$ are of size about $p^{1/2}$. The key ingredient of the proof is the incidence bound of Bourgain, Katz & Tao [36]. So, using Theorem 3.15, one can now obtain more explicit forms of the bounds of [19, Propositions 3.6 and 3.7].

Open Question 4.5. *Assuming that $\mathcal{X}$ and $\mathcal{Y}$ are of size at least p^δ, how small can δ be taken in order to have nontrivial bounds on the sums?*

Clearly for the first sums we need at least $\delta \ge 1/4$, see the argument that has led to (4.2). However, for the second sum the situation is not so clear.

Finally, Bourgain [19, Theorem A.1] uses the methods and results from additive combinatorics to give the following bound of double Kloosterman-like sums.

Theorem 4.6. *For any $\varepsilon > 0$ there exists some $\delta > 0$ such that for arbitrary intervals $\mathcal{I}$ and $\mathcal{J}$ in $\mathbb{F}_p$, with*

$$\#\mathcal{I}\#\mathcal{J} > p^{1/2+\varepsilon} ,$$

we have

$$\max_{\gcd(a,b,p)=1} \left| \sum_{x\in\mathcal{I}} \sum_{y\in\mathcal{J}} \mathbf{e}_p(axy + bx^{-1}y^{-1}) \right| \ll \#\mathcal{I}\#\mathcal{J}p^{-\delta} .$$

Open Question 4.7. *Obtain an explicit form of Theorem 4.6.*

More recently, Bourgain & Garaev [30], have derive a series of really amazing results on bilinear and single Kloosterman sums. This work is based, in particular, on several methods from additive combinatorics, coupled with a scope of ideas. Some arguments from [31, 32] have been used in [30] as well. In turn these bounds have led to a new version of the Brun–Titchmarsh theorem, improving that of Friedlander & Iwaniec [93].

Shkredov [173–175], using methods and results of additive combinatorics over finite fields, has recently achieved a remarkable progress in estimating *Gauss* and *Heilbronn* exponential sums (individually and on average), improving and extending previous estimates of Heath-Brown & Konyagin [120, 147]. These new bounds immediately lead to improvements of several previous results. For example, the new estimates of [173–175] can be substituted in the arguments of [27, 37, 59, 161, 184, 186] and instantly yield several stronger results (we also note that the results of [187] are based on [173]).

Chang [54, 56, 57] has obtained a series of very interesting estimates for various sums of multiplicative characters, including the generalization of the *Burgess bound* (see [135, Theorems 12.6]) to arbitrary finite fields, which has been an open problem for decades. The results of [54, 56] on the Burgess bound have been further refined by Konyagin [148] who has also used the ideas from additive combinatorics (such as bounds of multiplicative energy, see (1.2)). Let $\omega_1, \ldots, \omega_n$ be a basis of $\mathbb{F}_q = \mathbb{F}_{p^n}$ over $\mathbb{F}_p$. Given some integers K_i and H_i with $0 \le K_i < K_i + H_i < p$, $i = 1, \ldots, n$. Then for any $\varepsilon > 0$ there exists some $\delta > 0$ such that uniformly over all non-principal multiplicative characters of $\mathbb{F}_q$, we have

$$\sum_{x_1=K_1+1}^{K_1+H_1} \ldots \sum_{x_n=K_n+1}^{K_n+H_n} \chi\left(\sum_{i=1}^{n} x_i\omega_i\right) \ll H_1 \ldots H_n p^{-\delta}$$

(where the implied constant depends only on n) provided that either

$$H_1 \ldots H_n \ge p^{(2/5+\varepsilon)n}$$

or

$$H_i \ge p^{1/4+\varepsilon}, \quad i = 1, \ldots, n,$$

by the results of Chang [54] and Konyagin [148], respectively. It still remains to see whether the above condition can be replaced by the just one weaker condition

$$H_1 \ldots H_n \ge p^{(1/4+\varepsilon)n}.$$

Methods of additive combinatorics have also been used to estimate sums of multiplicative characters, see the survey [58]. For example, Bourgain & Chang [26] obtain a version of the Burgess bound for multiple character sums with products of linear forms.

Several more bounds of multiplicative character sums that are based on the methods of additive combinatorics have been given in [31, 32].

4.2 Waring, Erdős–Graham and Other Additive Problems in Finite Fields

Additive combinatorics has played a crucial role in the recent very substantial progress in the classical Waring problem for finite fields, see [71–73].

For example, let $g(n,p)$ denote the smallest k such that for any integer c there is a solution to the congruence:

$$\sum_{i=1}^{k} x_i^n \equiv c \pmod p, \quad 1 \le x_i \le p-1, \quad i = 1,\dots,k.$$

Cipra, Cochrane & Pinner [72] have shown that for some absolute constant C_0, for $k < (p-1)/2$ the bound $g(n,p) \le C_0 n^{1/2}$ holds, which corresponds to one of the *Heilbronn conjectures*. Cochrane & Pinner [73] have used some sum-product estimates (in particular, on weaker versions of Theorems 3.25 and 3.26) to give a form of this inequality with an explicit value of $C_0 = 83$.

Cipra [71] used a version of Theorem 3.12 to extend some of the above results to arbitrary finite fields and, in particular, improved some estimates of Winterhof [210] (see also [146, 211]).

Results of [167, 173] about additive representation of $\mathbb{F}_p*$ but sums of a few elements of a small subgroup $\mathcal{G} \subseteq \mathbb{F}_p^*$ also have a natural interpretation as bounds on $g(n,p)$ for a large n.

Ostafe & Shparlinski [162] have noticed that similar ideas, in particular, Theorems 3.11 and 3.12, can be adjusted to apply to the analogue of the Waring problem with Dickson polynomials in finite fields. In particular, the results of [162] improve some of the previous results of Gomez and Winterhof [111] that are based on the Weil bound of exponential sums, see [135]. One can probably get further improvements by using the bound of Bourgain & Glibichuk [34, Lemma 5] that has been presented in Theorem 3.3.

Erdős & Graham [89] have asked whether for any $\varepsilon > 0$ there exists $k(\varepsilon)$ such that for any prime p and any integer c there exist $k \le k(\varepsilon)$ pairwise distinct integers x_i with

$$1 \le x_i \le p^{\varepsilon}, \quad i = 1,\dots,k, \tag{4.4}$$

and such that

$$\sum_{i=1}^{k} \frac{1}{x_i} \equiv c \pmod p.$$

A positive answer to this question has been given in [177] with $k(\varepsilon) = O(\varepsilon^{-3})$.

Glibichuk [106] has used the methods of additive combinatorics and improved the bound on $k(\varepsilon)$ as $k(\varepsilon) = O(\varepsilon^{-2})$.

Croot [76] has applied the sum-product theorem of Bourgain, Katz & Tao [36] to a more general congruence

$$\sum_{i=1}^{k} \frac{1}{x_i^j} \equiv c \pmod p \tag{4.5}$$

with an arbitrary fixed integer $j \ge 1$. Note that for $j \ge 2$ the approach of [177] does not seem to apply to the congruence (4.5). The undelying idea of [76] is based on the observation that both sums and products of the expressions on the type as on the left-hand side of (4.5) are expressions of the same type. However, unfortunately, in the case of the product, the size of the variables grow, so to make this idea work, one has to be quite inventive with the choice of parameters, see [76] for details.

Finally, Bourgain [20] has shown that the methods of additive combinatorics are powerful enough to deal with a system of equations (4.5) with $j = 1, \dots, J$ and show it has a solution satisfying (4.4), with $k \le K(\varepsilon, J)$, where $K(\varepsilon, J)$ depends only on $\varepsilon > 0$ and $J \ge 1$.

It is quite feasible that the ideas of [20, 76, 106] can be applied to congruences of the type (4.5) with different restrictions on the variables. For instance, instead of (4.4) one can ask for solutions in very smooth integers.

In [101, 102], Theorem 3.12 has been applied to some additive problems in $\mathbb{F}_p$ involving binary recurrence sequences.

Open Question 4.8. *Investigate whether the methods and results of Section* 3.9 *can be used to study the Waring problem with matrices, that is, the solvability of the equation*

$$\sum_{i=1}^{k} X_i^n = C\,, \quad X_i \in \mathcal{M}_n(q)\,, \quad i = 1, \dots, k\,.$$

We also note that the bound of Bourgain, Katz & Tao [36] has been used in [178] to study the analogue of the Waring problem with the *Ramanujan* τ*-function.* Garaev, Garcia & Konyagin [99] have used a result of Glibichuk [106], (which has now been generalized in [110], see Theorem 3.12) and improved some results of [178].

4.3 Intersections of Almost Arithmetic and Geometric Progressions

The questions of Section 3.7 are also closely related to the problem of estimating the intersection of intervals $\mathcal{I}$ and groups $\mathcal{G}$ in $\mathbb{F}_p$. In particular, a variety of estimates on $\#(\mathcal{I} \cap \mathcal{G})$ has been given in [21, 38, 39, 124, 173].

Furthermore, let integers a and g satisfy $\gcd(ag, p) = 1$. Given two intervals $\mathcal{I}$ and $\mathcal{J}$, we denote by $R_{a,g,p}(\mathcal{I}, \mathcal{J})$ the number of solutions of the congruence

$$x \equiv ag^z \pmod p\,, \quad (x, z) \in \mathcal{I} \times \mathcal{J}\,. \tag{4.6}$$

Investigation of $R_{a,g,p}(\mathcal{I}, \mathcal{J})$ is heavily based on the methods of additive combinatorics. For example, let $H_{g,p}(N)$ be the largest H such that $R_{a,g,p}(\mathcal{I}, \mathcal{J}) = 0$ for some a with $\gcd(a, p) = 1$ and intervals $\mathcal{I}$ and $\mathcal{J}$ of lengths H and N, respectively.

In the case where the length of $\mathcal{J}$ is equal to the multiplicative order t modulo p of g, Bourgain, Konyagin & Shparlinski [38, Theorem 7] have given the following estimate of $H_{g,p}(t)$ (improving that of Konyagin & Shparlinski [149, Theorem 7.10]):

Theorem 4.9. *For any element $g \in \mathbb{F}_p^*$ of multiplicative order $t \geq p^{1/2}$ we have*

$$H_{g,p}(t) \leq p^{463/504+o(1)}$$

as $p \to \infty$.

It is easy to see that using the new bound (3.13) of Shkredov [173] in the argument of [38] (where only (3.12) been used), one can improve Theorem 4.9.

Theorem 4.9 and other results of [38] have found several more number theoretic applications. These applications include a series of results on Fermat quotients [27, 59, 161, 184, 186], pseudopowers [37] and the distribution of digits in g-ary expansions of rational fractions [187].

Recent results of Shkredov [173–175] immediately lead to improvements of the estimates from [27, 37, 59, 161, 184, 186] and probably have many other applications (the results of [187] are already based on [173]).

Combining the ideas from [29] and [38], Konyagin & Shparlinski [150, Theorem 2] obtained the following result.

Theorem 4.10. *Let $\nu \geq 1$ be a fixed integer. Then for any primitive root $g \in \mathbb{F}_p$ and $p > N > p^{1/2}$ the following bound holds:*

$$H_{g,p}(N) \leq N^{-47/72+(2\nu+1)/6\nu(\nu+1)+o(1)} p^{95/72-1/6(\nu+1)} + N^{-47/96+1/4\nu+o(1)} p^{119/96-1/4\nu}$$

as $p \to \infty$.

In particular, in the most interesting case of $N = p^{1/2+o(1)}$, the optimal value of ν is $\nu = 72$; thus

$$H_{g,p}(p^{1/2+o(1)}) \leq p^{62635/63072+o(1)} = p^{0.9930714\ldots}.$$

In the symmetric case, when both $\mathcal{I}, \mathcal{J}$ are of the same length h, Chan & Shparlinski [50] have noticed that Theorem 3.2 gives a nontrivial estimate on $R_{a,g,p}(\mathcal{I}, \mathcal{J})$ for any h. Stronger upper bounds on $R_{a,g,p}(\mathcal{I}, \mathcal{J})$, that are given by Bourgain, Garaev, Konyagin & Shparlinski [32] and Cilleruelo & Garaev [67], are also based on the ideas that stem from the methods of additive combinatorics.

Some of the above results can be presented in a more general setting of almost arithmetic and geometric progressions. We say that sets $\mathcal{I} \subseteq \mathbb{F}_p$ and $\mathcal{G} \subseteq \mathbb{F}_p$ are an *almost arithmetic progression* and an *almost geometric progression* if for every fixed positive integer k we have

$$\#(k\mathcal{I}) = \#\mathcal{I} p^{o(1)} \quad \text{and} \quad \#(\mathcal{G}^k) = \#\mathcal{G} p^{o(1)},$$

as $p \to \infty$. It has been shown in [11] that Theorems 3.4 and 3.10 immediately imply the following result:

Theorem 4.11. *For any almost arithmetic progressions $\mathcal{I}, \mathcal{J} \subseteq \mathbb{F}_p^*$ and almost geometric progression $\mathcal{G} \subseteq \mathbb{F}_p^*$ we have,*

$$\#(\mathcal{I} \cap \mathcal{G}) \le \left(\frac{\#\mathcal{I}\#\mathcal{G}}{p} + p^{1/2}\right) p^{o(1)},$$

$$\#(\mathcal{I} \cap \mathcal{J}^{-1}) \le \left(\frac{\#\mathcal{I}\#\mathcal{G}}{p} + p^{1/2}\right) p^{o(1)},$$

as $p \to \infty$.

Besides upper bounds on intersections between small groups and intervals, Bourgain [21] have also found applications of methods of additive combinatorics to studying the distribution of products xu where $x \in \mathcal{I}$ is taken from a small interval $\mathcal{I}$ and $u \in \mathcal{G}$ is taken from a small subgroup $\mathcal{G} \subseteq \mathbb{F}_p^*$, see [21]. Hegyvári & Hennecart [124] have considered more general products $f(x)u$ with a polynomial f.

4.4 Exponential Congruence

For a prime p and an integer $a \in \mathbb{Z}$ with $\gcd(a, p) = 1$ we denote by $N(p; a)$ the number of solutions to the congruence

$$x^x \equiv a \pmod{p}, \quad 1 \le x \le p - 1.$$

Balog, Broughan & Shparlinski [10] have shown that the methods of additive combinatorics can be used to derive various estimates on $N(p; a)$. For instance, by [10, Theorem 2] we have, uniformly for $t \mid p - 1$,

$$\sum_{\substack{a \in \mathbb{Z}_p^* \\ \operatorname{ord} a \mid t}} N(p; a) \le \max\{t, p^{1/2} t^{1/4}\} p^{o(1)}, \tag{4.7}$$

as $p \to \infty$, where $\operatorname{ord} a$ denotes the multiplicative order of $a \in \mathbb{F}_p^*$.

Furthermore, for small values of t by [10, Theorem 4] we also have

$$\sum_{\substack{a \in \mathbb{Z}_p^* \\ \operatorname{ord} a \mid t}} N(p; a) \le p^{1/3 + o(1)} t^{2/3}, \tag{4.8}$$

as $p \to \infty$. In [11] the estimates (4.7) and (4.8) have been improved for $p^{1/4} \le t \le p^{2/3}$ as follows: uniformly over $t \mid p - 1$, we have,

$$\sum_{\substack{a \in \mathbb{Z}_p^* \\ \operatorname{ord} a \mid t}} N(p; a) \le \max\{t, p^{1/2}\} p^{o(1)},$$

as $p \to \infty$.

4.5 Hidden Shifted Power Problem

For a positive integer $e \mid q-1$ and an element $s \in \mathbb{F}_q$, we use $\mathcal{O}_{e,s}$ to denote an oracle that on every input $x \in \mathbb{F}_q$ outputs $\mathcal{O}_{e,s}(x) = (x+s)^e$ for some "hidden" element $s \in \mathbb{F}_q$.

Bourgain, Garaev, Konyagin & Shparlinski [31] have used methods from additive combinatorics, for example, Theorem 3.27 (together with some other results from commutative algebra and analytic number theory) to study the *Hidden Shifted Power Problem*:

Given an oracle $\mathcal{O}_{e,s}$ for some unknown $s \in \mathbb{F}_q$, find s.

Besides, a similar method has also been applied in [31] to the following two versions of the *Shifted Power Identity Testing*:

Given an oracle $\mathcal{O}_{e,s}$ for some unknown $s \in \mathbb{F}_q$ and known $t \in \mathbb{F}_q$, decide whether $s = t$ provided that the call $x = -t$ is forbidden;

and

Given two oracles $\mathcal{O}_{e,s}$ and $\mathcal{O}_{e,t}$ for some unknown $s, t \in \mathbb{F}_q$ decide whether $s = t$.

Certainly these problems are special cases of the more general problems of oracle (also sometimes called "black-box") polynomial interpolation and identity testing for arbitrary polynomials, see [13] and references therein.

4.6 Sum-Product Estimates and Multiplicative Orders of γ and $\gamma + \gamma^{-1}$ in Finite Fields

In [16, Research Problem 3.1], a question has been posed about the possibility of finding $\operatorname{ord}(\gamma + \gamma^{-1})$ from the known value of $\operatorname{ord} \gamma$, (where as before $\operatorname{ord} \gamma$ denotes the multiplicative order of $\gamma \in \mathbb{F}_q^*$), see also [16, Research Problem 5.1].

It has been shown in [176] that no such algorithm can possibly exist and in fact $\operatorname{ord} \gamma$ and $\operatorname{ord}(\gamma + \gamma^{-1})$ are independent. In particular, for any fixed $\varepsilon > 0$ and sufficiently large q, for any positive divisors n and m of $q-1$ with $nm \geq q^{3/2+\varepsilon}$ there exists $\gamma \in \mathbb{F}_q^*$ with

$$\operatorname{ord} \gamma = n \quad \text{and} \quad \operatorname{ord}\left(\gamma + \gamma^{-1}\right) = m .$$

More results about multiplicative orders of $\operatorname{ord} \gamma$ and $\operatorname{ord}(\gamma + \gamma^{-1})$ can be found in [4, 103, 163, 207], in the context of estimating multiplicative orders of Gauss periods. Interestingly enough, many of these results rely on the estimates on the size of the intersections between subgroups and intervals, see [103]. Thus, it is quite possible that the results of Section 4.3 can lead to further progress here. Note that in Section 4.3

we mostly concentrated on upper bounds, but similar techniques also lead to lower bounds, see [27, 37, 38].

A related question has also be studied in [185], however, the argument of the proof of [185, Theorem 1] is unfortunately invalid (the author is grateful to Moubariz Garaev for pointing out this). Here we present and prove a corrected statement.

Let us fix some $n \mid q-1$ and define $\Gamma_q(n)$ as the subgroup of $\mathbb{F}_q^*$ generated by the elements $\gamma + \gamma^{-1}$ for $\gamma \in \mathbb{F}_q$ with $\operatorname{ord} \gamma \mid n$. Clearly $\#\Gamma_q(n) \gg n$. For a prime $q = p$, we now obtain a stronger bound.

Theorem 4.12. *For a prime p and a positive integer $n \le p^{1/2}$, we have*

$$\#\Gamma_p(n) \ge n^{12/11+o(1)} ,$$

as $n \to \infty$.

Proof. We define the sets

$$S = \{\gamma : \operatorname{ord} \gamma \mid n\} \quad \text{and} \quad \mathcal{A} = \{\gamma^2 + \gamma^{-2} : \gamma \in S\} .$$

Clearly if $\gamma \in S$ then we also have $\gamma^2 \in S$. Thus,

$$\#\mathcal{A} \ge \tfrac{1}{8}\#S \gg n \tag{4.9}$$

Note that

$$\mathcal{A}^2 \subseteq \Gamma_p(n) . \tag{4.10}$$

Now let us take $\alpha, \beta \in S$.

Then

$$\alpha^2 + \alpha^{-2} + \beta^2 + \beta^{-2} = \left(\alpha\beta + \alpha^{-1}\beta^{-1}\right)\left(\alpha\beta^{-1} + \alpha^{-1}\beta\right) .$$

Therefore, we also have

$$2\mathcal{A} \subseteq \Gamma_p(n) . \tag{4.11}$$

Combining (4.10) and (4.11), we obtain

$$\Gamma_p(n) \ge \max\left\{\#(2\mathcal{A}), \#(\mathcal{A}^2)\right\} .$$

Now, using Theorem 3.2 and the inequality (4.9), we conclude the proof. □

Certainly the argument of the proof of Theorem 4.12 works also for $n \ge p^{1/2}$. For large values of n, one can also use the result of [176] to get a lower bound on $\Gamma_p(n)$.

4.7 Expansion of Dynamical Systems

Given a polynomial $f \in \mathbb{F}_p[X]$ and an element $u_0 \in \mathbb{F}_p$, we consider the sequence of elements of $\mathbb{F}_p$ generated by iterations $u_n = f(u_{n-1})$, $n = 0, 1, \ldots$. Clearly the sequence u_n is eventually periodic. In particular, let T be the full trajectory length, that is, the smallest integer t such that $u_t = u_s$ for some $s < t$. The study of the diameter

$$D(N) = \max_{0 \le k,m \le N-1} |u_k - u_m|$$

where $N \le T$, has been initiated by Gutierrez & Shparlinski [114] using bounds of exponential sums and also some ideas from additive combinatorics. This direction has been continued in [60, 61, 63, 68] using further tools and ideas from additive combinatorics.

References

[1] O. Ahmadi and I. E. Shparlinski, Geometric progressions in sumsets over finite fields, *Monatsh. Math.* **152** (2007), 177–185.

[2] O. Ahmadi and I. E. Shparlinski, Bilinear character sums and the sum-product problem on elliptic curves, *Proc. Edinb. Math. Soc.* **53** (2010), 1–12.

[3] O. Ahmadi and I. E. Shparlinski, *Exponential sums over points of elliptic curves*, preprint, 2013, http://arxiv.org/abs/1302.4210.

[4] O. Ahmadi, I. E. Shparlinski, and J. F. Voloch, Multiplicative order of Gauss periods, *Intern. J. Number Theory* **6** (2010), 877–882.

[5] N. Alon, Combinatorial Nullstellensatz, *Combin., Probab. and Computing* **8** (1999), 7–29.

[6] N. Alon, Large sets in finite fields are sumsets, *J. Number Theory* **126** (2007), 110–118.

[7] N. Alon, O. Angel, I. Benjamini, and E. Lubetzky, Sums and products along sparse graphs, *Israel J. Math.* **188** (2012), 353–384.

[8] N. Alon, A. Granville, and A. Ubis, The number of sumsets in a finite field., *Bull. Lond. Math. Soc.* **42** (2010), 784–794.

[9] A. Balog, Another sum-product estimate in finite fields, *Sovrem. Probl. Math.*, vol. 16, Seklov Math. Inst., RAS, Moscow, 2012, 31–37.

[10] A. Balog, K. A. Broughan, and I. E. Shparlinski, On the number of solutions of exponential congruences, *Acta Arith.* **148** (2011), 93–103.

[11] A. Balog, K. A. Broughan, and I. E. Shparlinski, Sum-products estimates with several sets and applications, *Integers* **12** (2012), 895–906.

[12] M. Bateman and N. H. Katz, New bounds on cap sets, *J. Amer. Math. Soc.* **25** (2012), 585–613.

[13] M. Beecken, J. Mittmann, and N. Saxena, Algebraic independence and blackbox identity testing, *Inform. and Comput.*, **222** (2013) 2–19.

[14] M. Bennett, A. Iosevich, and J. Pakianathan, *Three-point configurations determined by subsets of $\mathbb{F}_q^2$ via the Elekes–Sharir paradigm*, preprint, 2012, http://arxiv.org/abs/1201.5039.

[15] K. Bibak, Additive combinatorics with a view towards computer science and cryptography: An exposition, *Number Theory and Related Areas*, Springer, 2013, to appear.

[16] I. F. Blake, S. Gao, A. J. Menezes, R. Mullin, S. Vanstone, and T. Yaghoobian, *Applications of finite fields*, Kluwer A. P., Dordrecht, 1993.

[17] T. F. Bloom and T. G. F. Jones, *A sum-product theorem in function fields*, preprint, 2012, http://arxiv.org/abs/1211.5493.

[18] E. Bombieri, On exponential sums in finite fields, *Amer. J. Math.* **88** (1966), 71–105.

[19] J. Bourgain, More on the sum-product phenomenon in prime fields and its applications, *Int. J. Number Theory*, **1** (2005), 1–32.

[20] J. Bourgain, *Some arithmetical applications of the sum-product theorems in finite fields*, Geometric aspects of functional analysis, Lecture Notes in Math. 1910, pp. 99–116, Springer-Verlag, Berlin, 2007.
[21] J. Bourgain, On the distribution of the residues of small multiplicative subgroups of $\mathbb{F}_p$, *Israel J. Math.* **172** (2009), 61–74.
[22] J. Bourgain, Multilinear exponential sums in prime fields under optimal entropy condition on the sources, *Geom. and Func. Anal.* **18** (2009), 1477–1502.
[23] J. Bourgain, *Sum-product theorems and applications*, Additive Number Theory, pp. 9–38, Springer-Verlag, Berlin, 2010.
[24] J. Bourgain, *On exponential sums in finite fields*, An Irregular Mind, Bolyai Soc. Math. Stud. 21, pp. 219–242, János Bolyai Math. Soc., Budapest, 2010.
[25] J. Bourgain, A modular Szemerédi-Trotter theorem for hyperbolas, *Comp. Rend. Acad. Sci. Paris* **350** (2012), 793–796.
[26] J. Bourgain and M.-C. Chang, On a multilinear character sums of Burgess, *Comp. Rend. Acad. Sci. Paris* **348** (2010), 115–120.
[27] J. Bourgain, K. Ford, S. V. Konyagin, and I. E. Shparlinski, On the divisibility of Fermat quotients, *Michigan Math. J.* **59** (2010), 313–328.
[28] J. Bourgain and A. Gamburd, Uniform expansion bounds for Cayley graphs of $\mathrm{SL}_2(\mathbb{F}_p)$, *Annals Math.* **167** (2008), 625–642.
[29] J. Bourgain and M. Z. Garaev, On a variant of sum-product estimates and explicit exponential sum bounds in prime fields, *Math. Proc. Cambr. Phil. Soc.* **146** (2008), 1–21.
[30] J. Bourgain and M. Z. Garaev, *Sumsets of reciprocals in prime fields and multilinear Kloosterman sums*, preprint, 2012, http://arxiv.org/abs/1211.4184.
[31] J. Bourgain, M. Z. Garaev, S. V. Konyagin, and I. E. Shparlinski, On the hidden shifted power problem, *SIAM J. Comp.* **41** (2012), 1524–1557.
[32] J. Bourgain, M. Z. Garaev, S. V. Konyagin, and I. E. Shparlinski, On congruences with products of variables from short intervals and applications, *Proc. Steklov Math. Inst.*, **280** (2013), 67–96.
[33] J. Bourgain, M. Z. Garaev, S. V. Konyagin, and I. E. Shparlinski, *Multiplicative congruences with variables from short intervals*, *J. d'Analyse Math.*, to appear.
[34] J. Bourgain and A. A. Glibichuk, Exponential sum estimate over subgroup in an arbitrary finite field, *J. d'Analyse Math.* **115** (2011), 51–70.
[35] J. Bourgain, A. A. Glibichuk, and S. V. Konyagin, Estimates for the number of sums and products and for exponential sums in fields of prime order, *J. Lond. Math. Soc.* **73** (2006), 380–398.
[36] J. Bourgain, N. Katz, and T. Tao, A sum product estimate in finite fields and applications, *Geom. Funct. Analysis* **14** (2004), 27–57.
[37] J. Bourgain, S. Konyagin, C. Pomerance, and I. E. Shparlinski, On the smallest pseudopower, *Acta Arith.* **140** (2009), 43–55.
[38] J. Bourgain, S. V. Konyagin, and I. E. Shparlinski, *Product sets of rationals, multiplicative translates of subgroups in residue rings and fixed points of the discrete logarithm*, Intern. Math. Res. Notices, 2008, Article rnn090, 1–29. (Corrigenda: *Intern. Math. Res. Notices*, 2009, 3146–3147).
[39] J. Bourgain, S. V. Konyagin, and I. E. Shparlinski, *Distribution of elements of cosets of small subgroups and applications*, Intern. Math. Res. Notices, 2012, Article rnn097, 1968–2009.
[40] E. Breuillard, Y. de Cornulier, A. Lubotzky, and C. Meiri, On conjugacy growth of linear groups, *Math. Proc. Cambr. Phil. Soc.*, to appear.
[41] E. Breuillard and A. Gamburd, Strong uniform expansion in $\mathrm{SL}(2,p)$, *Geom. and Funct. Anal.* **20** (2010), 1201–1209.

[42] E. Breuillard and B. Green, Approximate groups, I: The torsion-free nilpotent case, *J. d'Inst. Math. Jussieu* **10** (2011), 37–57.
[43] E. Breuillard and B. Green, Approximate groups, II: The solvable linear case, *Quart. J. Math.* **62** (2011), 513–521.
[44] E. Breuillard and B. Green, Approximate groups, III: The unitary case, *Turkish J. Math.* **36** (2012), 199–215.
[45] E. Breuillard, B. Green, R. Guralnick, and T. Tao, Strongly dense free subgroups of semisimple algebraic groups, *Israel J. Math.*, to appear.
[46] E. Breuillard, B. Green, and T. Tao, Linear approximate group, *Electron. Res. Announc.* **17** (2010), 57–67.
[47] E. Breuillard, B. Green, and T. Tao, Approximate subgroups of linear groups, *Geom. Funct. Anal.* **4** (2011), 774–819.
[48] E. Breuillard, B. Green, and T. Tao, The structure of approximate groups, *Publ. Math. de l'IHÉSV*, **116** (2012), 115–221.
[49] B. Bukh and J. Tsimerman, Sum-product estimates for rational functions, *Proc. Lond. Math. Soc.* **104** (2012), 1–26.
[50] T. H. Chan and I. E. Shparlinski, On the concentration of points on modular hyperbolas and exponential curves, *Acta Arith.* **142** (2010), 59–66.
[51] M.-C. Chang, *Some problems related to sum-product theorems*, Additive combinatorics CRM Proc. Lecture Notes 43, pp. 235–240, Amer. Math. Soc., Providence, RI, 2007.
[52] M.-C. Chang, Additive and multiplicative structure in matrix spaces, *Combin. Probab. Comput.* **16** (2007), 219–238.
[53] M.-C. Chang, Product theorems in SL_2 and SL_3, *J. Inst. Math. Jussieu* **7** (2008), 1–25.
[54] M.-C. Chang, On a question of Davenport and Lewis and new character sum bounds in finite fields, *Duke Math. J.* **145** (2008), 409–442.
[55] M.-C. Chang, Some problems in combinatorial number theory, *Integers* **8** (2008), Article A1, 1–11.
[56] M.-C. Chang, Burgess inequality in $\mathbb{F}_{p^2}$, *Geom. Funct. Anal.* **19** (2009), 1001–1016.
[57] M.-C. Chang, On character sums of binary quadratic forms, *J. Number Theory* **129** (2009), 2064–2071.
[58] M.-C. Chang, *Character sums in finite fields*, Finite Fields: Theory and applications, Contemp. Math., vol. 518, Amer. Math. Soc., Providence, RI, 2010, 83–98.
[59] M.-C. Chang, Short character sums with Fermat quotients, *Acta Arith.* **152** (2012), 23–38.
[60] M.-C. Chang, Polynomial iteration in characteristic p, *J. Functional Analysis* **263** (2012), 3412–3421.
[61] M.-C. Chang, Expansions of quadratic maps in prime fields, *Proc. Amer. Math. Soc.*, to appear.
[62] M.-C. Chang, *On a matrix product question in cryptography*, preprint, 2012.
[63] M.-C. Chang, J. Cilleruelo, M. Z. Garaev, J. Hernández, I. E. Shparlinski, and A. Zumalacárregui, *Points on curves in small boxes and applications*, preprint, 2011, http://arxiv.org/abs/1111.1543.
[64] J. Chapman, M. B. Erdoğan, A. Iosevich, and D. Koh, Pinned distance sets, k-simplices, Wolff's exponent in finite fields and sum-product estimates, *Math. Zeit.*, to appear.
[65] J. Chapman and A. Iosevich, On a rapid generation of $\mathrm{SL}_2(\mathbb{F}_q)$, *Integers* **9** (2009), 47–52.
[66] J. Cilleruelo, Combinatorial problems in finite fields and Sidon sets, *Combinatorica*, to appear.
[67] J. Cilleruelo and M. Z. Garaev, Concentration of points on two and three dimensional modular hyperbolas and applications, *Geom. and Func. Anal.* **21** (2011), 892–904.
[68] J. Cilleruelo, M. Z. Garaev, A. Ostafe, and I. E. Shparlinski, On the concentration of points of polynomial maps and applications, *Math. Zeit.* **272** (2012) 825–837.

[69] J. Cilleruelo and I. E. Shparlinski, *Concentration of points on curves in finite fields*, preprint, 2012.

[70] J. Cilleruelo, I. E. Shparlinski, and A. Zumalacárregui, Isomorphism classes of elliptic curves over a finite field in some thin families, *Math. Res. Letters* **19** (2012), 335–343.

[71] J. Cipra, Waring's number in a finite field, *Integers* **9** (2009), 435–440.

[72] J. Cipra, T. Cochrane, and C. G. Pinner, Heilbronn's Conjecture on Waring's number $\bmod p$, *J. Number Theory* **125** (2007), 289–297.

[73] T. Cochrane and C. Pinner, Sum-product estimates applied to Waring's problem $\bmod p$, *Integers* **8** (2008), A46, 1–18.

[74] D. Covert, D. Hart, A. Iosevich, D. Koh, and M. Rudnev, Generalized incidence theorems, homogeneous forms and sum-product estimates in finite fields, *European J. Combin.* **31** (2010), 306–319.

[75] D. Covert, D. Hart, A. Iosevich, S. Senger, and I. Uriarte-Tuero, A Furstenberg–Katznelson–Weiss type theorem on $(d+1)$-point configurations in sets of positive density in finite field geometries, *Discrete Math.* **311** (2011), 423–430.

[76] E. S. Croot, Sums of the form $1/x_1^k + \cdots + 1/x_n^k$ modulo a prime, *Integers* **4** (2004), A20, 1–6.

[77] E. S. Croot, The minimal number of three-term arithmetic progressions modulo a prime converges to a limit, *Canad. Math. Bull.* **51** (2008), 47–56.

[78] E. S. Croot and D. Hart, h-fold sums from a set with few products, *SIAM J. Discr. Math.* **24** (2010), 505–519.

[79] E. S. Croot and D. Hart, On sums and products in $\mathbb{C}[x]$, *Ramanujan J.* **22** (2010), 33–54.

[80] E. S. Croot and V. F. Lev, *Open problems in additive combinatorics*, Additive combinatorics CRM Proc. Lecture Notes 43, pp. 207–233, Amer. Math. Soc., Providence, RI, 2007.

[81] J.-M. Deshouillers and V. F. Lev, A refined bound for sum-free sets in groups of prime order, *Bull. Lond. Math. Soc.* **40** (2008), 863–875.

[82] J. A. Dias da Silva and Y. O. Hamidoune, Cyclic spaces for Grassman derivatives and additive theory, *Bull. Lond. Math. Soc.* **26** (1994), 140–146.

[83] R. Dietmann, *On the Erdős–Falconer distance problem for two sets of different size in vector spaces over finite fields*, preprint, 2011, http://arxiv.org/abs/1110.3502.

[84] Z. Dvir, On the size of Kakeya sets in finite fields, *J. Amer. Math. Soc.* **22** (2009), 1093–1097.

[85] Z. Dvir, From randomness extraction to rotating needles, *SIGACT News* **40** (2009), 46–61.

[86] Z. Dvir, Incidence theorems and their applications, *Foundations and Trends in Theor. Comp. Sci. V* **6** (2012), 257–393.

[87] G. Elekes, On the number of sums and products, *Acta Arith.* **81** (1997), 365–367.

[88] J. Ellenberg, R. Oberlin, and T. Tao, The Kakeya set and maximal conjectures for algebraic varieties over finite fields, *Mathematika* **56** (2010), 1–25.

[89] P. Erdős and R. L. Graham, *Old and new problems and results in combinatorial number theory*, Monographies de L'Enseignement Math. 28. de Genéve, Genéve, 1980.

[90] P. Erdős and E. Szemerédi, *On sums and products of integers*, Studies Pure Math., Birkhäuser, Basel, 1983, 213–218.

[91] R. Ferguson, C. Hoffman, F. Luca, A. Ostafe, and I. E. Shparlinski, Some additive combinatorics problems in matrix rings, *Revista Matem. Complutense* **23** (2010), 501–513.

[92] K. Ford, The distribution of integers with a divisor in a given interval, *Annals Math.* **168** (2008), 367–433.

[93] J. B. Friedlander and H. Iwaniec, *The Brun–Titchmarsh theorem*, Analytic Number Theory, Lond. Math. Soc. Lecture Note Series 247, pp. 363–372, 1997.

[94] J. Garibaldi, A. Iosevich, and S. Senger, *The Erdős distance problem*, Amer. Math. Soc., Providence, RI, 2011.

[95] M. Z. Garaev, *An explicit sum-product estimate in* $\mathbb{F}_p$, Intern. Math. Res. Notices, 2007, Article rnm035, 1–11.

[96] M. Z. Garaev, The sum-product estimate for large subsets of prime fields, *Proc. Amer. Math. Soc.* **136** (2008), 2735–2739.

[97] M. Z. Garaev, A quantified version of Bourgain's sum-product estimate in Fp for subsets of incomparable sizes, *Electron. J. Combin.* **15** (2008), Article R58.

[98] M. Z. Garaev, Sums and products of sets and estimates of rational trigonometric sums in fields of prime order, *Russian Math. Surveys* **65** (2010), 599–658, translation from *Uspekhi Mat. Nauk*.

[99] M. Z. Garaev, V. C. Garcia, and S. V. Konyagin, The Waring problem with Ramanujan's τ-function, *Izvestiya Mathem.* **72** (2010), 35–46, translation from *Izv. Ross. Akad. Nauk Ser. Mat.*

[100] M. Z. Garaev and C.-Y. Shen, On the size of the set $A(A+1)$, *Math. Zeit.* **265** (2010), 125–132.

[101] V. C. Garcia, On the distribution of sparse sequences in prime fields and applications, *Integers*, to appear.

[102] V. C. Garcia, F. Luca, and V. J. Mejia, On sums of Fibonacci numbers modulo p, *Bull. Aust. Math. Soc.* **83** (2011), 413–419.

[103] J. von zur Gathen and I. Shparlinski, *Gauss periods in finite fields, Proc. 5th Conference of Finite Fields and their Applications, Augsburg, 1999*, pp. 162–177, Springer-Verlag, Berlin, 2001.

[104] N. Gill and H. A. Helfgott, *Growth of small generating sets in* $\mathrm{SL}_n(\mathbb{Z}/p\mathbb{Z})$, Intern. Math. Res. Notices 2011, pp. 4226–4251.

[105] N. Gill and H. A. Helfgott, *Growth in solvable subgroups of* $\mathrm{GL}_r(\mathbb{Z}/p\mathbb{Z})$, preprint, 2011, http://arxiv.org/abs/1008.5264.

[106] A. Glibichuk, Combinational properties of sets of residues modulo a prime and the Erdős–Graham problem, *Math. Notes*, **79** (2006), 356–365, translation from *Matem. Zametki*.

[107] A. Glibichuk, Sums of powers of subsets or arbitrary finite fields, *Izvestiya Mathem.* **75** (2011), 253–285, translation from *Izv. Ross. Akad. Nauk Ser. Mat.*

[108] A. Glibichuk, Average estimate for additive energy in prime field, *Moscow J. Comb. and Number Theory* **1** (2011), 258–276.

[109] A. Glibichuk and S. V. Konyagin, *Additive properties of product sets in fields of prime order*, Additive combinatorics, CRM Proc. Lecture Notes 43, pp. 279–286, Amer. Math. Soc., Providence, RI, 2007.

[110] A. Glibichuk and M. Rudnev, On additive properties of product sets in an arbitrary finite field, *J. d'Analyse Math.* **108** (2009), 159–170.

[111] D. Gomez and A. Winterhof, *Waring's problem in finite fields with Dickson polynomials*, Finite Fields: Theory and applications, Contemp. Math., vol. 477, Amer. Math. Soc., Providence, RI, 2010, 185–192.

[112] B. Green, *Finite field models in additive combinatorics*, Surveys in Combin., London Math. Soc. Lecture Notes 327, pp. 1–27, 2005.

[113] B. Green and I. Z. Ruzsa, Counting sumsets and sum-free sets modulo a prime, *Studia Sci. Math. Hungar.* **41** (2004), 285–293.

[114] J. Gutierrez and I. E. Shparlinski, Expansion of orbits of some dynamical systems over finite fields, *Bull. Aust. Math. Soc.* **82** (2010), 232–239.

[115] D. Hart, *A note on sumsets of subgroups in* $\mathbb{Z}^*$, preprint, 2013, http://arxiv.org/abs/1303.2729.

[116] D. Hart and A. Iosevich, *Sums and products in finite fields: an integral geometric viewpoint*, Radon Transforms, Geometry, and Wavelets, Contemp. Math. vol. 464, Amer. Math. Soc., Providence, RI, 2008, 129–135.

[117] D. Hart, A. Iosevich, D. Koh, and M. Rudnev, Averages over hyperplanes, sum-product theory in finite fields, and the Erdős–Falconer distance conjecture, *Trans. Amer. Math. Soc.* **363** (2011), 3255–3275.
[118] D. Hart, A. Iosevich, and J. Solymosi, *Sums and products in finite fields via Kloosterman sums*, Intern. Math. Res. Notices, 2007, Article rnm007, 1–14.
[119] D. Hart, L. Li, and C.-Y. Shen, Fourier analysis and expanding phenomena in finite fields, *Proc. Amer. Math. Soc.* **141** (2013), 461–473.
[120] D. R. Heath-Brown and S. V. Konyagin, New bounds for Gauss sums derived from kth powers, and for Heilbronn's exponential sum, *Quart. J. Math.* **51** (2000), 221–235.
[121] N. Hegyvári, On sum-product bases, *Ramanujan J.* **19** (2009), 1–8.
[122] N. Hegyvári, Some remarks on multilinear exponential sums with an application, *J. Number Theory*, **132** (2012), 94–102.
[123] N. Hegyvári and F. Hennecart, Explicit constructions of extractors and expanders., *Acta Arith.* **140** (2009), 233–249.
[124] N. Hegyvári and F. Hennecart, Distribution of residues in approximate subgroups of $\mathbb{F}_p$, *Proc. Amer. Math. Soc.* **140** (2012), 1–6.
[125] H. A. Helfgott, Growth and generation in $\mathrm{SL}_2(\mathbb{Z}/p\mathbb{Z})$, *Ann. of Math.* **167** (2008), 601–623.
[126] H. A. Helfgott, Growth in $\mathrm{SL}_3(\mathbb{Z}/p\mathbb{Z})$, *J. Eur. Math. Soc.* **13** (2011), 761–851.
[127] H. A. Helfgott and M. Rudnev, An explicit incidence theorem in $\mathbb{F}_p$, *Mathematika* **57** (2011), 135–145.
[128] S. Hu and Y. Li, Bilinear character sums over norm groups, *Publ. Math. Debrecen* **78** (2011), 405–412.
[129] S. Hu and Y. Li, *On a uniformly distributed phenomena in matrix groups*, preprint, 2011, http://arxiv.org/abs/1103.3928.
[130] A. Iosevich and D. Koh, Erdős–Falconer distance problem, exponential sums, and Fourier analytic approach to incidence theorems in vector spaces over finite fields, *SIAM J. Disc. Math.* **23** (2008), 123–135.
[131] A. Iosevich, H. Morgan, and J. Pakianathan, On directions determined by subsets of vector spaces over finite fields, *Integers* **11** (2011), 815—825.
[132] A. Iosevich and M. Rudnev, Erdős distance problem in vector spaces over finite fields, *Trans. Amer. Math. Soc.* **359** (2007), 6127–6142.
[133] A. Iosevich and S. Senger, Orthogonal systems in vector spaces over finite fields, *Electronic J. Combin.* **15** (2008), Article R151, 1–10.
[134] A. Iosevich, I. E. Shparlinski, and M. Xiong, Sets with integral distances in finite fields, *Trans. Amer. Math. Soc.* **362** (2010), 2189–2204.
[135] H. Iwaniec and E. Kowalski, *Analytic number theory*, Amer. Math. Soc., Providence RI, 2004.
[136] T. G. F. Jones, Explicit incidence bounds over general finite fields, *Acta Arith.* **150** (2011), 241–262.
[137] T. G. F. Jones, *An improved incidence bound for fields of prime order*, preprint, 2011, http://arxiv.org/abs/1110.4752.
[138] T. G. F. Jones, *Further improvements to incidence and Beck-type bounds over prime finite fields*, preprint, 2012, http://arxiv.org/abs/1206.4517.
[139] T. G. F. Jones and O. Roche-Newton, Improved lower bound on the set $A(A+1)$, *J. Combin. Theory* **120** (2013), 515–526.
[140] N. H. Katz and C.-Y. Shen, Garaev's inequality in finite fields not of prime order, *J. Anal. Combin.* **3** (2008), Article #3, 1–6.
[141] N. H. Katz and C.-Y. Shen, A slight improvement to Garaev's sum product estimate, *Proc. Amer. Math. Soc.* **136** (2008), 2499–2504.

[142] N. M. Katz, I. E. Shparlinski, and M. Xiong, On character sums with distances on the upper half plane over a finite field, *Finite Fields and Their Appl.* **15** (2009), 738–747.
[143] B. Kerr, Incomplete exponential sums over exponential functions, *Bull. Aust. Math. Soc.*, to appear.
[144] D. Koh and C.-Y. Shen, The generalized Erdoős–Falconer distance problems in vector spaces over finite fields, *J. Number Theory* **132** (2012), 2455–2473.
[145] D. Koh and H.-S. Sun, *Cardinalities of distance sets determined by two sets in vector spaces over finite fields*, preprint, 2012, http://arxiv.org/abs/1212.5305.
[146] K. Kononen, More exact solutions to Waring's problem for finite fields, *Acta Arith.* **145** (2010), 209–212.
[147] S. V. Konyagin, *Bounds of exponential sums over subgroups and Gauss sums, Proc. 4th Intern. Conf. Modern Problems of Number Theory and Its Applications*, pp. 86–114, Moscow Lomonosov State Univ., Moscow, 2002, in Russian.
[148] S. V. Konyagin, Bounds of character sums in finite fields, *Mathem. Notes* **88** (2010), 503–515, translation from *Matem. Zametki*.
[149] S. V. Konyagin and I. E. Shparlinski, *Character sums with exponential functions and their applications*, Cambridge Univ. Press, Cambridge, 1999.
[150] S. V. Konyagin and I. E. Shparlinski, On the consecutive powers of a primitive root: Gaps and exponential sums, *Mathematika* **58** (2012), 11–20.
[151] S. Kopparty, V. F. Lev, S. Saraf, and M. Sudan, Kakeya-type sets in finite vector spaces, *J. Algebr. Combin.* **34** (2011), 337–355.
[152] E. Kowalski, Explicit growth and expansion in SL_2 *Intern. Math. Res. Notices*, to appear.
[153] V. F. Lev, Large sum-free sets in $\mathbb{Z}/p\mathbb{Z}$, *Israel J. Math.* **154** (2006), 221–233.
[154] V. F. Lev, Character-free approach to progression-free sets, *Finite Fields Appl.* **18** (2012), 378–383.
[155] L. Li, Slightly improved sum-product estimates in fields of prime order, *Acta Arith.* **147** (2011), 153–160.
[156] L. Li, *Multi-fold sums from a set with few products*, preprint, 2011, http://arxiv.org/abs/1106.6074
[157] L. Li and O. Roche-Newton, An improved sum-product estimate for general finite fields, *SIAM J. Discr. Math.* **25** (2011), 1285–1296.
[158] Y. Lin and J. Wolf, On subsets of $\mathbb{F}_q^n$ containing no k-term progressions, *European J. Combin.* **31** (2010), 1398–1403.
[159] D. Lorenzini, *An invitation to arithmetic geometry*, Amer. Math. Soc., 1996.
[160] G. Maze, C. Monico, and J. Rosenthal, Public key cryptography based on semigroup actions, *Adv. Math. of Commun.* **1** (2007), 489–507.
[161] A. Ostafe and I. E. Shparlinski, Pseudorandomness and dynamics of Fermat quotients, *SIAM J. Discr. Math.* **25** (2011), 50–71.
[162] A. Ostafe and I. E. Shparlinski, On the Waring problem with Dickson polynomials in finite fields, *Proc. Amer. Math. Soc.* **139** (2011), 3815–3820.
[163] R. Popovych, Elements of high order in finite fields of the form $\mathbb{F}_q[x]/\Phi_r(x)$, *Finite Fields Appl.* **18** (2012), 700–710.
[164] L. Pyber and E. Szabó, *Growth in finite simple groups of Lie type of bounded rank*, preprint, 2010, http://arxiv.org/abs/1005.1858.
[165] M. Rudnev, *An improved sum-product inequality in fields of prime order*, Intern. Math. Res. Notices, 2012, Article rnr158, 3693–3705.
[166] A. Sárközy, On sums and products of residues modulo p, *Acta Arith.* **118** (2005), 403–409.
[167] T. Schoen and I. D. Shkredov, Additive properties of multiplicative subgroups of $\mathbb{F}_p$, *Quart. J. Math.*, to appear.

[168] C.-Y. Shen, Quantitative sum product estimates on different sets, *Electronic J. Combin.* **15** (2008), Article N40.
[169] C.-Y. Shen, An extension of Bourgain and Garaev's sum-product estimates, *Acta Arith.* **135** (2008), 351–256.
[170] C.-Y. Shen, On the sum product estimates and two variables expanders, *Publ. Mat.* **54** (2010), 149–157.
[171] C.-Y. Shen, Sum-product phenomenon in finite fields not of prime order, *Rocky Mountain J. Math.* **41** (2011), 941–948.
[172] I. D. Shkredov, Fourier analysis in combinatorial number theory, *Russian Math. Surveys* **65** (2010), 513–567, translation from *Uspekhi Mat. Nauk*.
[173] I. D. Shkredov, *Some new inequalities in additive combinatorics*, preprint, 2012, http://arxiv.org/abs/1208.2344.
[174] I. D. Shkredov, On Heilbronn's exponential sum, *Quart. J. Math.*, to appear.
[175] I. D. Shkredov, *New bounds for Heilbronn's exponential sum*, preprint, 2013, http://arxiv.org/abs/1302.3839.
[176] I. E. Shparlinski, On the multiplicative orders of γ and $\gamma + \gamma^{-1}$ over finite fields, *Finite Fields Appl.* **7** (2001), 327–331.
[177] I. E. Shparlinski, On a question of Erdős and Graham, *Arch. Math. (Basel)* **78** (2002), 445–448.
[178] I. E. Shparlinski, On the value set of the Ramanujan function, *Arch. Math. (Basel)* **85** (2005), 508–513.
[179] I. E. Shparlinski, On the elliptic curve analogue of the sum-product problem, *Finite Fields and Their Appl.* **14** (2008), 721–726.
[180] I. E. Shparlinski, On the solvability of bilinear equations in finite fields, *Glasgow Math. J.* **50** (2008), 523–529.
[181] I. E. Shparlinski, Arithmetic and geometric progressions in product sets over finite fields, *Bull. Aust. Math. Soc.* **78** (2008), 357–364.
[182] I. E. Shparlinski, On the exponential sum-product problem, *Indag. Math.* **19** (2009), 325–331.
[183] I. E. Shparlinski, On point sets in vector spaces over finite fields that determine only acute angle triangles, *Bull. Aust. Math. Soc.* **81** (2010), 114–120.
[184] I. E. Shparlinski, On the value set of Fermat quotients, *Proc. Amer. Math. Soc.* **140** (2012), 1199–1206.
[185] I. E. Shparlinski, Sum-product estimates and multiplicative orders of γ and $\gamma + \gamma^{-1}$ in finite fields, *Bull. Aust. Math. Soc.* **85** (2012), 505–508.
[186] I. E. Shparlinski, On vanishing Fermat quotients and a bound of the Ihara sum, *Kodai Math. J.*, to appear.
[187] I. E. Shparlinski and W. Steiner, On digit patterns in expansions of rational numbers with prime denominator, *Quart. J. Math.*, to appear.
[188] J. H. Silverman, *The arithmetic of elliptic curves*, 2nd ed., Springer, Dordrecht, 2009.
[189] J. Solymosi, Bounding multiplicative energy by the sumset, *Adv. Math.* **222** (2009), 402–408.
[190] T. Tao, Product set estimates in noncommutative groups, *Combinatorica* **28** (2008), 547–594.
[191] T. Tao, The sum-product phenomenon in arbitrary rings, *Contrib. Discrete Math.* **4** (2009), 59–82.
[192] T. Tao, *Expanding polynomials over finite fields of large characteristic, and a regularity lemma for definable sets*, preprint, 2012, http://arxiv.org/abs/1211.2894.
[193] T. Tao and V. Vu, *Additive combinatorics*, Cambridge Univ. Press, Cambridge, 2006.
[194] L. A. Vinh, On the number of orthogonal systems in vector spaces over finite fields, *Electronic J. Combin.* **15** (2008), Article N32.
[195] L. A. Vinh, Explicit Ramsey graphs and Erdős distance problem over finite Euclidean and non-Euclidean spaces, *Electronic J. of Combin.* **15** (2008), Article R5.

[196] L. A. Vinh, Distribution of determinant of matrices with restricted entries over finite fields, *J. Comb. Number Theory* **1** (2009), 203–212.

[197] L. A. Vinh, *On the distribution of permanents of matrices over finite fields*, European Conf. on Combin., Graph Theory and Appl. (EuroComb 2009), Electron. Notes Discrete Math. 34, pp. 519–523, Elsevier Sci. B. V., Amsterdam, 2009.

[198] L. A. Vinh, On the solvability of bilinear equations in finite fields, *Proc. Amer. Math. Soc.* **137** (2009), 2889–2898.

[199] L. A. Vinh, Szemerédi-Trotter type theorem and sum-product estimate in finite fields, *European J. Combin.* **32** (2011), 1177–1181.

[200] L. A. Vinh, On sum of products and the Erdős distance problem over finite fields, *Bull. Aust. Math. Soc.* **84** (2011), 1–9.

[201] L. A. Vinh, On the solvability of systems of sum-product equations in finite fields, *Glasg. Math. J.* **53** (2011), 427–435.

[202] L. A. Vinh, The Erdős–Falconer distance problem on the unit sphere in vector spaces over finite fields, *SIAM J. Discrete Math.* **25** (2011), 681–684.

[203] L. A. Vinh, Explicit Ramsey graphs and Erdős distance problem over finite Euclidean and non-Euclidean spaces, *Electronic J. of Combin.* **18** (2011), Article P213.

[204] L. A. Vinh, On a Furstenberg–Katznelson–Weiss type theorem over finite fields, *Ann. Comb.* **15** (2011), 541–547

[205] L. A. Vinh, The solvability of norm, bilinear and quadratic equations over finite fields via spectra of graphs, *Forum Mathematicum*, to appear.

[206] L. A. Vinh, On some problems of Gyarmati and Sárközy, *Integers*, to appear.

[207] J. F. Voloch, On the order of points on curves over finite fields, *Integers* **7** (2007), Article A49, p. 4.

[208] V. H. Vu, SumÐproduct estimates via directed expanders, *Math. Res. Lett.* **15** (2008), 375–388.

[209] I. V. Vyugin and I. D. Shkredov, On additive shifts of multiplicative subgroups, *Sbornik: Mathematics* **203** (2012), 844–863, translation from *Sbornik: Mathematics*

[210] A. Winterhof, A note on Waring's problem in finite fields, *Acta Arith.* **96** (2001), 365–368.

[211] A. Winterhof and C. van de Woestijne, Exact solutions to Waring's problem in finite fields, *Acta Arith.* **141** (2010), 171–190.

Index

Radon Series on Computational and Applied Mathematics

Volume 10
Thomas Schuster, Barbara Kaltenbacher, Bernd Hofmann, Kamil S. Kazimierski, 2012
Regularization Methods in Banach Spaces
ISBN 978-3-11-025524-9, e-ISBN 978-3-11-025572-0, Set-ISBN 978-3-11-220450-4

Volume 9
Massimo Fornasier (Ed.), 2010
Theoretical Foundations and Numerical Methods for Sparse Recovery
ISBN 978-3-11-022614-0, e-ISBN 978-3-11-022615-7, Set-ISBN 978-3-11-174177-2

Volume 8
Hansjörg Albrecher, Wolfgang J. Runggaldier, Walter Schachermayer (Eds.), 2009
Advanced Financial Modelling
ISBN 978-3-11-021313-3, e-ISBN 978-3-11-021314-0, Set-ISBN 978-3-11-173185-8

Volume 7
Jan H. Maruhn, 2009
Robust Static Super-Replication of Barrier Options
ISBN 978-3-11-020468-1, e-ISBN 978-3-11-020851-1, Set-ISBN 978-3-11-916585-3

Volume 6
Barbara Kaltenbacher, Andreas Neubauer, Otmar Scherzer, 2008
Iterative Regularization Methods for Nonlinear Ill-Posed Problems
ISBN 978-3-11-020420-9, e-ISBN 978-3-11-020827-6, Set-ISBN 978-3-11-916135-0

Volume 5
Johannes Kraus, Svetozar Margenov, 2009
Robust Algebraic Multilevel Methods and Algorithms
ISBN 978-3-11-019365-7, e-ISBN 978-3-11-021483-3, Set-ISBN 978-3-11-173898-7

Volume 4
Sergey Repin, 2008
A Posteriori Estimates for Partial Differential Equations
ISBN 978-3-11-019153-0, e-ISBN 978-3-11-020304-2, Set-ISBN 978-3-11-916169-5

www.degruyter.com